Andrew Mackay

The Theory and Practice of Finding the Longitude at Sea or Land

Andrew Mackay

The Theory and Practice of Finding the Longitude at Sea or Land

ISBN/EAN: 9783741180286

Manufactured in Europe, USA, Canada, Australia, Japa

Cover: Foto ©berggeist007 / pixelio.de

Manufactured and distributed by brebook publishing software
(www.brebook.com)

Andrew Mackay

The Theory and Practice of Finding the Longitude at Sea or Land

THE

THEORY AND PRACTICE

OF FINDING THE

LONGITUDE

AT SEA OR LAND:

TO WHICH ARE ADDED,

VARIOUS METHODS OF DETERMINING

THE

LATITUDE OF A PLACE,

AND

VARIATION OF THE COMPASS;

WITH

NEW TABLES.

BY

ANDREW MACKAY, A. M. F. R. S. E.

IN TWO VOLUMES.

VOLUME II.

CONTENTS
OF THE
SECOND VOLUME.

XXXIII.

CONTENTS.

LI

C O N T E N T S.

E R R A T A.

Page 1. line 5 from bottom, for 5h. 40′ 0″ read 4h. 40′ 0″
last line, for 6h. 7′ 25″ read 5h. 7′ 25″
8 line 8 from bottom, for *increasing or decreasing*, read *decreasing or increasing*.
65, 12° 28′ 50″ for 3731 read 3631
53 40 — .5257 —— 5257
16 53 50 — 9302 —— 9202

THE

THEORY AND PRACTICE

OF FINDING THE

LONGITUDE AT SEA OR LAND.

EXPLANATION OF THE TABLES.

TABLES I. AND II.

For converting Longitude into Time, and conversely.

TABLE I. is intended to facilitate Prob. I. page 72, and is constructed according to the principles there delivered. Col. first contains hours, and the second column the degrees answering thereto; the third contains either minutes, seconds, or thirds, and the fourth column the corresponding degrees and minutes, minutes and seconds, or seconds and thirds—the fifth and sixth columns are a continuation of the two former. Table II. being the converse of table I. requires no farther explanation.

EXAMPLES.

I. Required the degrees, &c. answering to 3 h. 18′ 56″ ?

3 hours in col. first, tab. I.	=	45°	0′	in col. second.
18 minutes in col. third	=	4	30	in col. fourth.
56 seconds in col. fifth	=		14	in col. sixth.
hence 3 h. 18′ 56″	=	49	44	

II. What time answers to 76° 50′ 45″?

70 degrees in col. fifth, table II.	=	4 h.	40′	0″	in col. sixth.
6 degrees in col. first,	=		24	0	in col. second.
50 minutes in col. third,	=		3	20	in col. fourth.
45 seconds in col. third,	=			3	in col. fourth.
hence 76° 50′ 45″	=	5	7	23	

TABLE III.

Depression of the Horizon.

This table is calculated by the rule given for that purpose in page 58. Each number is diminished by one tenth, for the effect of terrestrial refraction, and wrote out to the nearest tenth of a minute, which is sufficiently accurate for any purpose to which it can be applied at sea.

In order to render this table generally useful, the dip is given to the height of the eye above the water, expressed both in English and French feet. The first column contains the height of the eye above the sea, the second the dip answering to the height expressed in English feet, and the third column contains the dip to the height in French feet. Thus if the height is 30 feet, the dip will be 5′.2 or 5′.1, according as the height is expressed in English or French measure.

TABLES IV. and V.

Correction of the Altitude of the Sun or a fixed Star.

These tables are intended to correct the altitude of the Sun or a fixed star, observed at sea, for the purpose of ascertaining the latitude or apparent time, and are therefore wrote out to the nearest tenth of a minute. The corrections are subtractive from the apparent altitude, in order to obtain the true altitude. If the given altitude is not found in the table, a proportional part is to be taken.

TABLE VI.

Mean Astronomical Refraction.

Contains the mean refraction expressed in minutes and seconds. It is intended to be used for correcting the observed distance between any two celestial objects, and for other observations where accuracy is required.

TABLE VII.

Parallax of the Sun in altitude.

In this table, the Sun's parallax in altitude to every third degree is contained. The parallax is additive to the apparent altitude.

TABLE VIII.

Correction of the Mean Refraction.

This table is intended to correct table VI. which is adapted to a mean temperature of the atmosphere. The corrections, contained therein, depend on the heat and weight of the atmosphere, which are pointed out by the thermometer and barometer. Hence the arguments

guments of these corrections, besides the apparent altitude of the object which is expressed in the side column, are the heights of the thermometer and barometer—the first of which is given at the top, and the second at the bottom of the table, within probable limits. The correction depending on the height of the thermometer is additive, while that height is below 50°, and that answering to the barometer is additive, when its height is above 29.6; in other cases, the corrections are subtractive. Proportional parts are to be taken, when the apparent altitude, or the heights of the thermometer and barometer, fall between the tabular arguments.

EXAMPLE.

Let the apparent altitude be 7° 0', the height of the thermometer 72°, and that of the barometer 29.75. Required the true refraction?

Mean refraction, — —	7' 20"
Opposite to 7° and under 72° is	− 23
and above 29.75 is	+ 2
True refraction, — —	6 59

TABLE IX.

Correction of the Apparent Altitude of the Moon.

In this table, the correction of the Moon's apparent altitude is given to every tenth minute of altitude, and to each minute of horizontal parallax. The proportional part for the excess of the given altitude above the next less tabular number is readily found by inspection, and that answering to the seconds of parallax is contained in the right hand part of the table. This correction is additive to the apparent altitude.

EXAMPLE.

Let the Moon's apparent altitude be 32° 52', and horizontal parallax 56' 48". Required the corresponding correction?

Correction to alt. 32° 50' and hor. par. 56' - - - - - -	45' 35"
Proportional part to 2' of altitude, by inspection, is	− 1
to 48" of hor. par. in part 2d. of tab. is	+ 40
Correction required — — — —	46 14

TABLES X. XI. XII.

Reduction of the True to the Apparent Altitude.

These tables are intended to facilitate the reduction of the true to the apparent altitude. They are particularly useful in that method of finding the longitude by the Moon, where the distance only, or the distance and the altitude of one of the objects, are given. Table X. contains the reduction of the Sun's true altitude. It is additive, to obtain the apparent altitude. In table XI. the reduction of the Moon's altitude is contained; and in table XII. that of a fixed star. The first of these is subtractive, and the other additive, to obtain the apparent altitude.

TABLES XIII XIV. XV. XVI.

To reduce the Sun's Declination to a given Time and Meridian.

The declination of the Sun, as given in the Nautical Almanac, is adapted to the meridian of Greenwich; and because the declination is subject to a daily variation, a correction becomes necessary, in order to reduce it to any other meridian, or to any given time of the day. Table XIII. contains the reduction answering to the Sun's declination, when increasing northwards, or while the Sun is between the vernal equinox and summer solstice. Tables XIV. XV. XVI. contain the change of reduction corresponding to the same declination, when the Sun is in the other quarters of the ecliptic. In these tables, the declination is expressed at the top of the page, the time from noon in the first column, and the ship's longitude in the last column.—Their use is shown in prob. v. page 74.

TABLE XVII.

To select a proper Star, from which the Moon's distance is to be observed, in order to find the Longitude.

It sometimes happens, that the star in the Ephemeris with which the Moon ought to be compared, in order to find the ship's longitude, is either obscured, or in an unfavourable position for observation. In this case, it will be proper to select another star, in such a position with respect to the Moon, that as little as possible of the Moon's proper motion may be lost. For this purpose, table XVII. is given, in which the difference of latitude of the Moon and star is contained in the first column, and the difference of longitude in the second. These columns are so adapted to each other, that no more than one eighth of the Moon's proper motion is lost.

TABLE XVIII.

Proportional part of the Daily Variation of the Sun's Right Ascension.

This table is intended to facilitate the reduction of the Sun's right ascension to any given time and place. It is particularly useful in computing the apparent time from the altitude of a star, and in the computation of a star's altitude. Its use has been fully explained in page 75.

TA-

TABLES XIX. and XX.

Proportional part of the Variation of the Moon's Declination in 12 hours, and of the daily Variation of the Moon's passage over the meridian.

The first of these tables contains the proportional part of the variation of the Moon's declination in twelve hours, answering to a given time from noon or midnight, and to a given longitude. The other table serves to reduce the time of the Moon's passage over the meridian of Greenwich, to the time of passage over any other meridian.

TABLE XXI.

Error resulting from an Inclination in the Surfaces of the Central Mirror of the Circular Instrument.

This table contains the error of observation arising from the inclination of the surfaces of the central mirror of the circular instrument, adapted to an inclination of *one minute*, and to an instrument of the Chevalier de Borda's construction, wherein the axis of the telescope makes an angle of 80° with the horizon mirror. The use of this table is shewn in book II. chap. III.

TABLE XXII.

Error resulting from an error in the line of Collimation.

If the line of sight is not parallel to the plane of the instrument, the angle shown by the instrument will exceed the true angle. This table contains the error arising from the above cause, adapted to every tenth minute of an inclination of the line of collimation, and to any angle under 120°.

TABLES XXIII. and XXIV.

To find the Equation of equal Altitudes.

These tables are intended to facilitate the method of computing the equation of equal altitudes. This has been fully explained in the rules given for finding the apparent time by that method, in chap. III. book III.

In taking the numbers from these tables, allowance is to be made for the excess of the given above the next less tabular arguments as usual.

EXAMPLES.

I. Required the number from table XXIII. answering to latitude 57° 9', and interval between the observations 5 h. 17'?

Num-

Number to lat. 57°, and interval 4h. 40′ = 1740

Difference to 1° of lat. = + 68 and $\frac{68 \cdot 9'}{60'}$ = + 10

Difference to 40′ of int. = + 35 and $\frac{35 \times 37'}{40'}$ = + 32

Number required — — — 1782

II. Required the number from tab. XXIV. answering to declination 17° 48′, and interval between the observations 5h. 17′?

Number to decl. 17° 30′ and interval 4h. 40′ = 292

Difference to 30′ of dec. = + 9, and $\frac{9 \times 18'}{30}$ = + 5

Diff. to 40′ of interval = − 14 and $\frac{14 \times 37'}{40}$ = − 13

Number required — — — 284

TABLE XXV.

Altitude most proper to be observed for finding the Apparent Time.

As the change of altitude is quickest, when the object is in the prime vertical, the most proper time for observing an altitude, from which the apparent time is to be inferred, is therefore when the object is due east or west; other errors are also in this case avoided. This table contains the altitude of an object, when in the above position. The declination is marked at the top, and the latitude of the place of observation in the side column.

Altitudes under 3° should not be used in the computation of the apparent time, upon account of the uncertainty of the refraction.

TABLE XXVI.

Reduction of Altitude upon account of the spheroidal figure of the Earth.

This table contains the correction of altitude arising from the spheroidal figure of the Earth. The arguments are, the azimuth of the observed object, and the latitude of the place of observation.— This correction is additive to the observed altitude, when the azimuth reckoned from the meridian is less than 90°, and subtractive when it exceeds that quantity. Its use is explained at the end of chap. III. book III.

TABLES XXVII. XXVIII. XXIX.

For computing the Apparent Time.

These tables are intended to facilitate the computation of the apparent

parent time at sea, from the observed altitude of a known celestial object.— The arguments of tables XXVII. and XXVIII. are, the declination of the observed object at the top, and the latitude of the ship in the side column. Table XXIX. is to be entered with a number deduced from the former, opposite to which in the adjacent column is the minute and part of a minute, and at the top of the table is the hour. Their use for the above purpose is explained in chap. III. book III.

These tables may be also used for computing the true azimuth, the latitude from double altitudes of the Sun, with the intermediate time, and for clearing the apparent distance from the effects of refraction and parallax.

TABLE XXX.

Augmentation of the Moon's Semidiameter.

The apparent magnitude of the Moon is in the inverse ratio of its distance from the observer; and as the Moon's distance diminishes, as its altitude increases, the apparent semidiameter of the Moon therefore increases with its altitude. Table XXX. contains the augmentation of the Moon's semidiameter depending on its altitude.

TABLE XXXI.

Contraction of the Semidiameters of the Sun and Moon.

Since the lower limb of the Sun or Moon is more affected by refraction than the upper limb, and because refraction makes an object appear higher than it really is, it hence follows, that the vertical diameter will be contracted by refraction, altho' the horizontal diameter remains invariable; and the object, especially if it be near the horizon, will assume an elliptical figure. The diameter of the Sun or Moon, measured in any other direction than that parallel to the horizon, will therefore be less, by a certain quantity, than the true diameter. This contraction of semidiameter is contained in table XXXI. the arguments of which are, the apparent altitude of the object's center at the top, and in the side column, the angle contained between the measured diameter and that parallel to the horizon.

TABLES XXXII. XXXIII. XXXIV. XXXV. XXXVI.

Correction of the Moon's Parallax, agreeable to the spheroidal hypothesis.

These tables containing the equations of the Moon's parallax, which arise from the spheroidal figure of the Earth, are adapted to Sir Isaac Newton's hypothesis.

Table XXXII. contains the reduction of the Moon's equatorial horizontal parallax, answering to the latitude of the place of observation.

When

When the method of calculating solar eclipses, and occultations of fixed stars by the Moon, by means of the nonagesimal, is used, the Moon's reduced horizontal parallax is to be increased by a certain quantity depending on the latitude of the place, to which the calculations are adapted. This augmentation of parallax is contained in table XXXIII. The following table contains the correction of the Moon's parallax in latitude, which, in order to obtain the Moon's apparent latitude, is additive or subtractive, according as the latitudes of the Moon and place of observation, are of the same or of a contrary name. In table XXXV. is the greatest effect of the correction of the Moon's parallax in longitude, which takes place, when the Moon's longitude is either 0, III. VI. or IX signs. The correction answering to any other longitude may be found thus:

To the log. secant of the given longitude, and the proportional log. of the greatest correction, the sum will be the proportional log. of the required correction, which is additive, when the Moon is in the first and last quarters of the ecliptic, and subtractive, if the Moon is in the second and third quarters. The contrary rule is to be applied, when the place of observation is south of the equator.

The reduction of the latitude of any given place is contained in table XXXVI.

TABLE XXXVII.

Equation of Second Difference.

In computing the Moon's place from the Nautical Almanac for a given instant, a correction, resulting from the Moon's unequal motion, must be applied to the proportional part of the Moon's motion in longitude or latitude, answering to the given time after noon or midnight. This correction is contained in table XXXVII. the arguments of which are, the mean of the two second differences of the Moon's motion at the top, and the apparent time after noon or midnight in the respective side column. This equation must be added to, or subtracted from the proportional part of the first difference of the Moon's motion in twelve hours, according as that difference is increasing or decreasing. Hence the corrected proportional part, answering to the given interval, will be obtained.

If the given second difference is not found in the table, the sum of the equations, answering to the several terms which make up the second difference, is to be taken.

EXAMPLE.

Let the apparent time after noon be 3h. 10′ and the mean second difference 7′ 46″. Required the corresponding equation?

Equa-

Equation to 3h. 10′ and 7′ 0″ = 40″.8
and 0 40 = 3.9
0 6 = 0.6

Equation to 3h. 10′ and 7′ 46″ = 45.3

This table may alſo be applied in the computation of the place of a planet.

TABLE XXXVIII.

Correction of Apparent Time, ariſing from the Equation of Second Difference.

The computation of the apparent time, anſwering to a given longitude of the Moon, is by this table greatly facilitated. The equation of ſecond difference is the argument at the top, and the Moon's motion in longitude in twelve hours is the argument in the ſide column. This equation is to be applied to the proportional part, anſwering to the exceſs of the given, above the next leſs longitude, by addition or ſubtraction, according as the Moon's motion in longitude in twelve hours, is increaſing or decreaſing.

TABLES XXXIX. and XL.

Moon's Equatorial Semidiameter in Time, and Increaſe of the ſame.

By theſe tables, the interval of time between the paſſage of the Moon's center and either limb over the meridian, is eaſily found. The firſt table contains the equatorial ſemidiameter of the Moon in time, the arguments of which are, the daily variation of the Moon's paſſing the meridian, at the top, and the apparent ſemidiameter in the ſide column. The other table contains the increaſe of the ſemidiameter, anſwering to the Moon's declination and equatorial ſemidiameter in time.

Example.

Required the time the Moon's ſemidiameter takes to paſs the meridian, its ſemidiameter being 16′ 10″, diurnal retardation 52′, and declination 21°?

To ſemidiam. 16′ 10″, and diur. retard. 52′, equat. ſemidia. is 67″.0
To equat. ſemidia. 67″.0, and declinat. 21°, the increaſe is 4.7

Time of Moon's ſemidiameter paſſing the meridian, 71.7

TABLE XLI.

Natural Verſed Sines.

As the ſolution of the problem, for correcting the obſerved diſtance between the Moon and the Sun or a fixed ſtar, is rendered very ſhort, when the operation is performed by means of a table of natural verſed ſines; upon this account the above table is inſerted; and in order to make the operation as ſimple as poſſible, this table is extended to every tenth ſecond; the proportional part to each intermediate ſecond is given at the bottom of the page. Hence the natural verſed ſine of any given arch may be taken out at ſight.

The natural co-verſed ſine of a given degree and minute is found on the ſame line with the next leſs minute, in the column marked 60″ at the bottom of the page; and that anſwering to an arch, expreſſed in degrees, minutes, and ſeconds, is obtained, by deducting the proportional part, anſwering to the intermediate ſeconds, from the co-verſed ſine of the given degree, minute, and next leſs tenth ſecond.

Altho' this table does not extend beyond 90°, yet the natural verſed ſine of an obtuſe angle may be found, by ſubtracting the natural verſed ſine of its ſupplement from 2000000.

Beſide the application of this table to the ſolution of the above problem, it may be uſed very ſucceſsfully in many other calculations.

TABLES XLII. XLIII. XLIV.

Logarithmic Difference.

Theſe tables are alſo intended to facilitate the method of reducing the apparent to the true diſtance. The firſt is to be entered with the Moon's apparent altitude at the top, and its horizontal parallax in the ſide column. The other two, which ſerve to correct the former, according as the Moon is compared with the Sun or a fixed ſtar, are to be entered with the apparent altitude of the Sun or obſerved ſtar reſpectively. Proportional parts are to be taken as uſual for the exceſs of the given altitude, or parallax, above the next leſs tabular arguments.

EXAMPLE.

Let the Sun's apparent altitude be 27° 43′, that of the Moon 46° 18′, and horizontal parallax 59′ 33″. Required the logarithmic difference?

Log. diff. to ap. alt. 46° 0′ and hor. par. 59′ 30″ - - -	9.994770
Proportional part to 18′ of altitude - - - - - - - - - —	27
— — to 3′ of parallax - - - - - - - - —	4
Number from table XLIII. anſwering to the Sun's alt. —	9
Logarithmic difference required, - - - - - - - - - -	9.994730

TABLE XLV.

Proportional Logarithms.

This table was calculated by Dr Maſkelyne, in order to render the

the operation more eafy, for finding the apparent time at Greenwich, anfwering to a given diftance between the Moon and the Sun or a fixed ftar. It is extended to three hours, on account of the diftances being fet down in the Ephemeris, at fuch intervals. As degrees and hours are fimilarly divided, each hour may therefore be efteemed a degree.

The purpofes to which this table may be applied are many. It is particularly ufeful in calculations where fexagefimals are concerned, and where the arch may be fubftituted for its fine or tangent.

The multiplication of a fexagefimal by a common number is performed, by fubtracting the proportional log. of one minute from the fum of the proportional log's of the factors.

A fexagefimal is divided, by fubtracting the proportional log. of the divifor from the fum of the proportional log's. of the dividend and one minute. The proportional log. of a geometrical mean between two numbers is equal to half the fum of the proportional log's of thefe numbers. That of the fquare root is found, by taking half the fum of the proportional logs. of the given quantity and one minute; and the P. Log. of the fquare of a given quantity, by fubtracting the P. Log. of one minute from twice the P. Log. of the propofed quantity. If this quantity is greater than $13\frac{28}{60}$, its fquare will exceed the limits of the table. In order, therefore, to extend this problem, fubtract the proportional log. of 2.3.4 minutes, &c. from twice the proportional log. of the given quantity; the remainder will be the prop. log. of a half, third, fourth, &c. of the required quantity.

Proportion is performed, by fubtracting the P. Log. of the firft term from the fum of the P. Logs. of the other two. If the firft term is 3° or 3 hours, the fum of the P. Logs of the two laft is that of the anfwer. If the firft term is not 180°, it will be found convenient to reduce it, if poffible, to that quantity, and one of the other terms muft be increafed or diminifhed in the fame ratio. If either the fecond or third term is 180°, the P. Log. of the fourth term is obtained, by fubtracting the P. Log. of the firft term from that of the other.

A fexagefimal is multiplied by the fine, tangent, or fecant of an arch, by adding to its P. Log. the log. co-fecant, co-tangent, or co-fine of that arch; and a fexagefimal is divided by a fine, tangent, or fecant, when the log. of the given fine, tangent, or fecant, is added to its P. Log.

It may be remarked, that in the ufe of this table, one minute may alfo be efteemed either one degree or one fecond; by which means, its ufe will be rendered more extenfive.

TABLES XLVI. XLVII. XLVIII. XLIX. L. LI. LII. LIII. LIV. LV. LVI.

Of the Right Ascension and Declination of Fixed Stars.

In the first of these tables, are contained, the mean right ascension and declination of 180 fixed stars, adapted to the beginning of the year 1793, with their annual variations in right ascension and declination, by which, their mean place may be found with tolerable accuracy, for a few years preceding or following the epoch of the tables. The proper annual motions of a few of those stars are allowed for.

The mean place of a star may be accurately ascertained for a future period, by applying its proper motion and the equations from tables XLVII. and XLVIII. answerable to the given interval. Table XLVII. contains part first of the precession in right ascension for complete years, common to all the stars. Part second of the precession in right ascension is found by multiplying the equation from table XLVIII. answering to the given interval, and the right ascension of the star at the mean interval, by the natural tangent of its declination, and changing the sign, if the star is south of the equator. The precession in declination is found, by entering the same table with the right ascension of the star increased by 90°, and changing the sign, if the star is south. If the given interval is not found in the table, the sum of the equations answering to the years that make up the interval is to be taken. The precession in right ascension, answering to months and days, is found, by multiplying the annual precession by the decimal from table XLIX. answering to the given time, wherein the effect of the semiannual equation of precession is allowed. The precession in declination, answering to months and days, is found, by taking a part of the annual precession proportional to the time. The semiannual solar equation of northern stars in declination is contained in table L. and the sign is to be changed, if the declination of the star is south.— These two are taken from Dr. Maskelyne's tables.

The Nutation of a star in right ascension and declination is found, by means of tables LI. LII. and LIII. as follows:

From the right ascension of the star, subtract the longitude of the Moon's node; the remainder will be the argument of table LI. and their sum will be the argument of table LII. Multiply the sum or difference of the corresponding equations, according as they have the same or contrary signs, by the natural tangent of the star's declination, and change the sign of the product, if the declination is south; to which apply the equation from table LIII. answering to the longitude

tude of the Moon's node; and the sum or difference will be the nutation in right ascension.

To each of the two first arguments, add three signs, and the sum or difference of the resulting equations, from tables LI. and LII. will be the nutation in declination.

EXAMPLE.

Required the nutation of α Aquilæ in right ascension and declination, January 1, 1793?

R. A. α Aquilæ =	9s. 25° 10′					
Lon. moon's node =	5 18 34					
Remainder	4 6 36	tab. LI. =	+ 4″.96	7s. 6° 36′	tab. LI. =	+ 6″.69
Sum	3 13 44	tab. LII. =	+ 0.28	6 13 44	tab. LII. =	+ 1.18
			+ 5.24 log. 0.719		nut. in dec. =	+ 7.87
Declination of α Aquilæ 8° 20′ N.		log. tangent	9.166			
			+ 0.8 = 9.885			
Equat. from tab. LIII. to long. moon's node			− 3.2			
Nutation in right ascension,			− 2.4			

Tables LIV. LV. LVI. for computing the aberration of a fixed star in right ascension and declination, were constructed by M. de Lambre. The aberration may be found as follows:

Subtract the Sun's longitude from the right ascension of the star; the remainder will be the argument of table LIV. and their sum will be that of table LV.— Now the sum or difference of the equations answering thereto, according as they have the same or contrary signs, being multiplied by the natural secant of the star's declination, will be its aberration in right ascension.

Increase the arguments of tables LIV. and LV. each by three signs; take out the corresponding equations, and the product of their sum or difference by the natural sine of the star's declination, will be part first of the aberration in declination. The equation from table LVI. answering to the sum of the Sun's longitude, and the declination of the star, is part second, and that from the same table answering to the Sun's longitude, diminished by the star's declination, is part third of the aberration in declination, which parts being connected according to their signs will give the absolute aberration in declination. If the declination of the star is south, the sign of the final result is to be changed.

EXAMPLE.

Required the aberration of α Aquilæ in right ascension and declination, January 1, 1793?

Right

R. A. of α Aquilæ	= 9s. 25° 10′				
Sun's longitude	= 9 11 39				
Remainder, —	0 13 31	tab. LIV. =	— 18″.64		
Sum — —	7 6 49	tab. LV. =	— 0.66		
			— 19.30	log. =	1.2855
Declination of α Aquilæ = 8° 20′ N.			log. secant,		0.0046
Aberr. in R. A.			— 19.30	log.	1.2901
Rem. + III. signs, =	3 13 31	tab. LIV. =	+ 4.48		
Sum + III. signs, =	10 6 49	tab. LV. =	+ 0.50		
			+ 4.98	log.	0.6972
Declination of α Aquilæ = 8 20 N.			log. sine,		9.1612
Sun's longitude, =	9 11 39		+ 0.72	log.	9.8584
Sum — —	9 19 59	tab. LVI.	— 1.36		
Remainder, —	9 3 19	table LVI.	— 0.23		
Aberration in declination			— 0.87		

In computing the place of a ſtar, its right aſcenſion and declination at the mean interval is to be uſed. Theſe are found, by applying to the tabular place, the multiples of the annual variations anſwering to half the interval.

Application of the preceding rules to the computation of the apparent right aſcenſion and declination of a fixed ſtar.

Example.

Required the apparent right aſcenſion and declination of Arcturus, 12th March, 1853?

Mean R. A. Arcturus 1793, =	211° 33′ 17″	Mean dec.	20° 15′ 58″ N.
Variation in half interval,	+ 20 24		— 9 33
R. A. at mean interval,	211 53 41	dec. at m. int.	20° 6 25 N.

R. A. of Arcturus at beginning of 1793,		211 33 17
Part firſt of preceſſion, tab. XLVII.	+	46 1.2
Part ſecond of preceſſion, tab. XLVIII.	—	3 52.6
Preceſſion for days, including the ſemi-annual equat.	+	8.4
Proper motion of Arcturus in R. A. for 60¼ years,		1 4.3
		212 14 29.7
Nutation per tables LI. LII. LIII. —	—	14.6
Aberration per tables LIV. LV. LVI. —	+	14.5
Apparent R. A. of Arcturus, 12th March 1853,		212 14 29.6

Mean

Mean declination of Arcturus, at beginning of 1793,		20°	15′	58″ N.
Preceſſion in declination for 60 years	−		17	1.4
Preceſſion in declination for days,	−			3.3
Proper motion of Arcturus in decl. for 60¼ years,	−		2	1.0
Semi-annual equation,	−			.4
		19	56	51.9
Nutation	+			5.5
Aberration	−			11.6
Apparent declination of Arcturus, 12th March 1853,		19	56	45.8

TABLES LVII. LVIII. LIX. LX. LXI. LXII.

Of the Longitude and Latitude of the Fixed Stars.

The firſt of theſe tables contains the mean longitude and latitude of 120 fixed ſtars, adapted to the beginning of the year 1793. The longitude and latitude of 34 of theſe ſtars were computed from a table containing their right aſcenſion, communicated by Dr Maſkelyne; and the places of the other ſtars were inferred from the catalogues of Meſſrs Bradley and Mayer. The ſtars in this table are moſtly zodiacal, being intended principally for aſcertaining the longitude, from occultations of fixed ſtars by the Moon.

The apparent longitude and latitude of a ſtar may be found at any given time from its mean place, by means of the four following tables. The firſt of them, or table LVIII. contains the preceſſion in longitude for complete years, and table LIX. contains the preceſſion for months and days. The ſecular equation in longitude is found, by multiplying the equation from table LX. anſwering to the longitude of the ſtar, by the natural tangent of the ſtar's latitude, and to the product, applying the conſtant quantity − 15″.4. The equation from the ſame table, anſwering to the longitude of the ſtar, increaſed by 3 ſigns, will be the ſecular equation in latitude. The equation anſwering to any number of years is found by taking the proportional part of the ſecular equation. Table LXI. contains the equation of the equinoxes in longitude, common to fixed ſtars and planets; and in table LXII. is the aberration of points of the ecliptic in longitude: from which the aberration of a ſtar in longitude and latitude may be found thus:

From the ſun's longitude, ſubtract that of the ſtar; find the correſponding equation, which multiplied by the natural ſecant of the ſtar's latitude, will be the aberration in longitude. Increaſe the former argument by 3 ſigns, and the equation anſwering thereto, being multiplied by the natural ſine of the ſtar's latitude, will be the aberration of the ſtar in latitude.

In

In computing the apparent place of a ſtar, for a propoſed time, its mean longitude, as ſettled at ſome certain epoch, muſt be corrected by the preceſſion of the equinoxes in longitude anſwering to the interval from tables LVIII. LIX; the proportional part of the ſecular variation, table LX. the equation of the equinoxes in longitude, table LXI. and the aberration table LXII. The apparent latitude is found by applying the proportional part of the ſecular equation, and the aberration. The proper motion of the ſtar in longitude and latitude ſhould alſo be applied.

EXAMPLE.

What is the apparent longitude and latitude of α Andromedæ, 1ſt June 1807?

Mean longitude of α Andromedæ, beginning of 1793 ——— ———	=	0^s	11°	$25'$	$37''$
Preceſſion in longitude for 14 years, table LVIII. ——— ———	= +			11	44.9
for 1ſt June, t. LIX.	= +				20.2
Proportional part of the ſecular equat. t. LX.	= −				5.2
Equation of the equinoxes in long. tab. LXI.	= −				16.9
		0	11	37	20.0
Aberration, table LXII. ———	−				11.5
Apparent longitude α Androm. 1ſt June 1807	=	0	11	37	8.5
Mean latitude of α Andromedæ, begin. of 1793	=		25	41	3 N.
Prop. part of ſecular equation, table LX.	=		+		2.2
Aberration, table LXII. ———			+		7.4
App. latitude of α Andromedæ, 1ſt June 1807			25	41	12.6 N.

The maxima of the equations of the above tables for correcting the mean place of a ſtar, are agreeable to the determinations of Drs. Bradley and Maſkelyne. The reſults, as given by the French aſtronomers differ a little therefrom; for, according to M. de la Lande, the mean annual preceſſion of the equinoxes in longitude is 50".25 in the preſent century, and, by the late calculations of M. de la Place, the combined action of the planets makes the annual advance of the equinoctial points to be 0".2016 along the equator, and 0".1849 along the ecliptic. Agreeable to theſe elements, M. de Lambre has given very full tables of the preceſſion in right aſcenſion and declination.

TABLE I.

To convert Time into Longitude.

Hours.	Degrees.	M / S / T	D M / M S / S T	M / S / T	D M / M S / S T
1	15	1	0 15	31	7 45
2	30	2	0 30	32	8 0
3	45	3	0 45	33	8 15
4	60	4	1 0	34	8 30
5	75	5	1 15	35	8 45
6	90	6	1 30	36	9 0
7	105	7	1 45	37	9 15
8	120	8	2 0	38	9 30
9	135	9	2 15	39	9 45
10	150	10	2 30	40	10 0
11	165	11	2 45	41	10 15
12	180	12	3 0	42	10 30
13	195	13	3 15	43	10 45
14	210	14	3 30	44	11 0
15	225	15	3 45	45	11 15
16	240	16	4 0	46	11 30
17	255	17	4 15	47	11 45
18	270	18	4 30	48	12 0
19	285	19	4 45	49	12 15
20	300	20	5 0	50	12 30
21	315	21	5 15	51	12 45
22	330	22	5 30	52	13 0
23	345	23	5 45	53	13 15
24	360	24	6 0	54	13 30
		25	6 15	55	13 45
		26	6 30	56	14 0
		27	6 45	57	14 15
		28	7 0	58	14 30
		29	7 15	59	14 45
		30	7 30	60	15 0

TABLE II.

To reduce Longitude to Time.

D / M / S	H M / M S / S T	D / M / S	H M / M S / S T	Degrees.	Hours.	Minutes.
1	0 4	31	2 4	70	4	40
2	0 8	32	2 8	80	5	20
3	0 12	33	2 12	90	6	0
4	0 16	34	2 16	100	6	40
5	0 20	35	2 20	110	7	20
6	0 24	36	2 24	120	8	0
7	0 28	37	2 28	130	8	40
8	0 32	38	2 32	140	9	20
9	0 36	39	2 36	150	10	0
10	0 40	40	2 40	160	10	40
11	0 44	41	2 44	170	11	20
12	0 48	42	2 48	180	12	0
13	0 52	43	2 52	190	12	40
14	0 56	44	2 56	200	13	20
15	1 0	45	3 0	210	14	0
16	1 4	46	3 4	220	14	40
17	1 8	47	3 8	230	15	20
18	1 12	48	3 12	240	16	0
19	1 16	49	3 16	250	16	40
20	1 20	50	3 20	260	17	20
21	1 24	51	3 24	270	18	0
22	1 28	52	3 28	280	18	40
23	1 32	53	3 32	290	19	20
24	1 36	54	3 36	300	20	0
25	1 40	55	3 40	310	20	40
26	1 44	56	3 44	320	21	20
27	1 48	57	3 48	330	22	0
28	1 52	58	3 52	340	22	40
29	1 56	59	3 56	350	23	20
30	2 0	60	4 0	360	24	0

TABLE III.

Depression of the Horizon.

Height.	Dip to height in Eng. ft. ′	Dip to height in Fr. ft. ′	Height.	Dip to height in Eng. ft. ′	Dip to height in Fr. ft. ′	Height.	Dip to height in Eng. ft. ′	Dip to height in Fr. ft. ′
0 4	0.6	0.6	18	4.1	3.7	110	10.0	9.7
0 8	0.8	0.8	19	4.2	4.0	120	10.5	10.1
1 0	1.0	0.9	20	4.3	4.1	130	10.9	10.5
1 6	1.2	1.1	22	4.5	4.3	140	11.3	10.9
2 0	1.4	1.3	24	4.7	4.5	150	11.7	11.3
3 0	1.7	1.6	26	4.9	4.7	160	12.0	11.6
4 0	1.9	1.8	28	5.0	4.9	170	12.4	12.0
5 0	2.1	2.1	30	5.2	5.1	180	12.8	12.4
6 0	2.3	2.3	35	5.6	5.4	190	13.1	12.7
7 0	2.5	2.4	40	6.0	5.8	200	13.5	13.0
8 0	2.7	2.6	45	6.4	6.2	210	13.8	13.4
9 0	2.9	2.8	50	6.7	6.5	220	14.1	13.7
10 0	3.0	2.9	55	7.1	6.8	230	14.4	14.0
11 0	3.2	3.1	60	7.4	7.2	240	14.7	14.3
12 0	3.3	3.2	65	7.7	7.4	250	15.1	14.6
13 0	3.4	3.3	70	8.0	7.7	260	15.4	14.9
14 0	3.6	3.5	75	8.2	8.0	270	15.7	15.2
15 0	3.7	3.6	80	8.5	8.2	280	16.0	15.5
16 0	3.8	3.7	90	9.0	8.8	290	16.2	15.7
17 0	4.0	3.8	100	9.5	9.2	300	16.5	16.0

TABLE IV.

Correction of the Sun in Altitude.

Alt. ° ′	Cor. ′	Alt. ° ′	Cor. ′	Alt. °	Cor. ′
0 0	32.9	2 40	15.6	18	2.8
0 5	32.0	3 0	14.4	20	2.4
0 10	31.2	3 20	13.4	22	2.2
0 15	30.4	3 40	12.5	24	2.0
0 20	29.7	4 0	11.7	26	1.8
0 25	28.9	4 30	10.6	28	1.6
0 30	28.2	5 0	9.7	30	1.5
0 35	27.5	5 30	9.0	32	1.4
0 40	26.8	6 0	8.3	34	1.3
0 45	26.2	6 30	7.7	36	1.2
0 50	25.5	7 0	7.2	38	1.1
0 55	24.[illegible]	8 0	6.3	40	1.0
1 0	24.3	9 0	5.6	45	0.8
1 10	23.2	10 0	5.1	50	0.7
1 20	22.1	11 0	4.6	55	0.6
1 30	21.1	12 0	4.2	60	0.5
1 40	20.2	13 0	3.9	65	0.4
1 50	19.8	14 0	3.6	70	0.3
2 0	18.4	15 0	3.3	80	0.1
2 10	16.9	16 0	3.1	90	0.0

[18]

TABLE V. Correction of a fixed star in Altitude.

TABLE VI. MEAN REFRACTION.

Alt.	Cor.	Alt.	Refr.	Alt.	Refr.	Altit.	Refr.	Altit.	Refr.	Altit.	Refr.	Alti.	Refr.
° ′	″	° ′	′ ″	° ′	′ ″	° ′	′ ″	° ′	′ ″	° ′	′ ″	°	″
0 0	33.0	0 0	33 0	5 0	9 54	10 0	5 15	20 0	2 35	32 20	1 30	67	24
0 5	32.2	0 5	32 10	5 5	9 46	10 10	5 10	20 10	2 34	32 40	1 29	68	23
0 10	31.4	0 10	31 22	5 10	9 38	10 20	5 5	20 20	2 32	33 0	1 28	69	22
0 15	30.6	0 15	30 35	5 15	9 30	10 30	5 0	20 30	2 31	33 20	1 26	70	21
0 20	29.8	0 20	29 50	5 20	9 23	10 40	4 56	20 40	2 29	33 40	1 25	71	19
0 25	29.1	0 25	29 6	5 25	9 15	10 50	4 51	20 50	2 28	34 0	1 24	72	18
0 30	28.4	0 30	28 23	5 30	9 8	11 0	4 47	21 0	2 27	34 20	1 23	73	17
0 35	27.7	0 35	27 41	5 35	9 1	11 10	4 43	21 10	2 26	34 40	1 22	74	16
0 40	27.0	0 40	27 0	5 40	8 54	11 20	4 39	21 20	2 25	35 0	1 21	75	15
0 45	26.3	0 45	26 20	5 45	8 47	11 30	4 35	21 30	2 24	35 20	1 20	76	14
0 50	25.7	0 50	25 42	5 50	8 41	11 40	4 31	21 40	2 23	35 40	1 19	77	13
0 55	25.1	0 55	25 5	5 55	8 34	11 50	4 27	21 50	2 21	36 0	1 18	78	12
1 0	24.5	1 0	24 29	6 0	8 28	12 0	4 23	22 0	2 20	36 30	1 17	79	11
1 10	23.3	1 5	23 54	6 5	8 21	12 10	4 20	22 10	2 19	37 0	1 16	80	10
1 20	22.3	1 10	23 20	6 10	8 15	12 20	4 16	22 20	2 18	37 30	1 14	81	9
1 30	21.3	1 15	22 47	6 15	8 9	12 30	4 13	22 30	2 17	38 0	1 13	82	8
1 40	20.3	1 20	22 15	6 20	8 3	12 40	4 9	22 40	2 16	38 30	1 11	83	7
1 50	19.4	1 25	21 44	6 25	7 57	12 50	4 6	22 50	2 15	39 0	1 10	84	6
2 0	18.6	1 30	21 15	6 30	7 51	13 0	4 3	23 0	2 14	39 30	1 9	85	5
2 20	17.1	1 35	20 46	6 35	7 46	13 10	4 0	23 10	2 13	40 0	1 8	86	4
2 40	15.7	1 40	20 18	6 40	7 40	13 20	3 57	23 20	2 12	40 30	1 7	87	3
3 0	14.6	1 45	19 51	6 45	7 35	13 30	3 54	23 30	2 11	41 0	1 5	88	2
3 20	13.6	1 50	19 25	6 50	7 30	13 40	3 51	23 40	2 10	41 30	1 4	89	1
3 40	12.7	1 55	19 0	6 55	7 25	13 50	3 48	23 50	2 9	42 0	1 3	90	0
4 0	11.8	2 0	18 35	7 0	7 20	14 0	3 45	24 0	2 8	42 30	1 2		
4 30	10.8	2 5	18 11	7 5	7 15	14 10	3 43	24 10	2 7	43 0	1 1	TAB. VII.	
5 0	9.9	2 10	17 48	7 10	7 11	14 20	3 40	24 20	2 6	43 30	1 0	Parallax of	
5 30	9.1	2 15	17 26	7 15	7 6	14 30	3 38	24 30	2 5	44 0	0 59	the Sun in	
6 0	8.5	2 20	17 4	7 20	7 2	14 40	3 35	24 40	2 4	44 30	0 58	Altitude.	
6 30	7.8	2 25	16 44	7 25	6 57	14 50	3 33	24 50	2 3	45 0	0 57		
7°	7.3	2 30	16 24	7 30	6 53	15 0	3 30	25 0	2 2	45 30	0 56	0°	9″
8	6.5	2 35	16 4	7 35	6 49	15 10	3 28	25 10	2 1	46 0	0 55	3	9
9	5.8	2 40	15 45	7 40	6 45	15 20	3 26	25 20	2 0	46 30	0 54	6	9
10	5.2	2 45	15 27	7 45	6 41	15 30	3 24	25 30	1 59	47 0	0 53	9	9
11	4.8	2 50	15 9	7 50	6 37	15 40	3 21	25 40	1 58	47 30	0 52	12	9
12	4.4	2 55	14 52	7 55	6 33	15 50	3 19	25 50	1 57	48 0	0 51	15	8
13	4.0	3 0	14 36	8 0	6 29	16 0	3 17	26 0	1 56	48 30	0 50	18	8
14	3.7	3 5	14 20	8 5	6 25	16 10	3 15	26 10	1 55	49 0	0 49	21	8
15	3.5	3 10	14 4	8 10	6 22	16 20	3 12	26 20	1 55	49 30	0 49	24	8
16	3.3	3 15	13 49	8 15	6 18	16 30	3 10	26 30	1 54	50 0	0 48	27	8
18	2.9	3 20	13 34	8 20	6 15	16 40	3 8	26 40	1 53	50 30	0 47	30	8
20	2.6	3 25	13 20	8 25	6 11	16 50	3 6	26 50	1 52	51 0	0 46	33	7
22	2.3	3 30	13 6	8 30	6 8	17 0	3 4	27 0	1 51	51 30	0 45	36	7
24	2.1	3 35	12 51	8 35	6 5	17 10	3 3	27 15	1 50	52 0	0 44	39	7
26	1.9	3 40	12 40	8 40	6 1	17 20	3 1	27 30	1 49	52 30	0 44	42	6
28	1.7	3 45	12 27	8 45	5 58	17 30	2 59	27 45	1 48	53 0	0 43	45	6
30	1.6	3 50	12 15	8 50	5 55	17 40	2 57	28 0	1 47	53 30	0 42	48	6
32	1.5	3 55	12 3	8 55	5 52	17 50	2 55	28 15	1 46	54 0	0 41	51	5
34	1.4	4 0	11 51	9 0	5 48	18 0	2 54	28 30	1 45	55 0	0 40	54	5
36	1.3	4 5	11 40	9 5	5 45	18 10	2 52	28 45	1 44	56 0	0 38	57	5
38	1.2	4 10	11 [illegible]	9 10	5 42	18 20	2 51	29 0	1 42	57	0 37	60	4
40	1.1	4 15	11 18	9 15	5 39	18 30	2 49	29 20	1 41	58	0 35	63	4
45	0.9	4 20	11 8	9 20	5 36	18 40	2 47	29 40	1 40	59	0 34	66	4
50	0.8	4 25	10 58	9 25	5 34	18 50	2 46	30 0	1 38	60	0 33	69	3
55	0.7	4 30	10 48	9 30	5 31	19 0	2 44	30 20	1 37	61	0 32	72	3
60	0.5	4 35	10 39	9 35	5 28	19 10	2 43	30 40	1 36	62	0 30	75	2
65	0.4	4 40	10 29	9 40	5 25	19 20	2 41	31 0	1 35	63	0 29	78	2
70	0.3	4 45	10 20	9 45	5 23	19 30	2 40	31 20	1 33	64	0 28	81	1
80	0.2	4 50	10 11	9 50	5 20	19 40	2 38	31 40	1 32	65	0 26	84	1
90	0.0	4 55	10 2	9 55	5 18	19 50	2 37	32 0	1 31	66	0 25	90	0

TABLE VIII. Correction of the Mean Refraction.

Argument. Height of the Thermometer.

Ap. Alt.	20°	24°	28°	32°	36°	40°	44°	48°	52°	56°	60°	64°	68°	72°	76°	80°
° ′	+ ′ ″	+ ′ ″	+ ′ ″	+ ′ ″	+ ′ ″	+ ″	+ ″	+ ″	− ″	− ″	− ″	− ′ ″	− ′ ″	− ′ ″	− ′ ″	− ′ ″
0 0	2 40	2 18	1 55	1 33	1 11	51	31	10	10	29	48	1 7	1 25	1 43	2 1	2 19
0 10	2 32	2 11	1 49	1 28	1 8	48	29	10	9	27	45	1 4	1 21	1 38	1 54	2 12
0 20	2 25	2 5	1 44	1 24	1 4	46	28	9	9	26	44	1 1	1 17	1 33	1 49	2 5
0 30	2 18	1 59	1 39	1 20	1 1	44	26	9	8	25	41	58	1 13	1 28	1 43	1 59
0 40	2 11	1 53	1 34	1 16	58	42	25	8	8	24	39	55	1 10	1 24	1 38	1 53
0 50	2 5	1 48	1 29	1 12	55	40	24	8	8	23	37	52	1 6	1 20	1 34	1 48
1 0	1 59	1 43	1 25	1 9	53	38	23	8	7	21	36	50	1 3	1 17	1 30	1 43
1 10	1 53	1 38	1 21	1 6	50	36	22	7	7	20	34	48	1 0	1 13	1 26	1 38
1 20	1 48	1 33	1 17	1 3	48	34	21	7	6	19	32	45	57	1 9	1 22	1 33
1 30	1 43	1 29	1 14	1 0	46	32	20	7	6	18	31	43	54	1 6	1 18	1 29
1 40	1 39	1 25	1 11	57	44	31	18	6	6	18	30	41	52	1 4	1 15	1 25
1 50	1 35	1 21	1 8	55	42	30	17	6	6	17	28	39	50	1 1	1 11	1 22
2 0	1 31	1 18	1 5	53	41	29	17	6	5	16	27	37	48	58	1 8	1 18
2 20	1 23	1 11	1 0	48	37	26	16	5	5	15	25	35	44	54	1 3	1 11
2 40	1 17	1 6	55	44	34	24	14	5	5	14	23	32	41	50	58	1 6
3 0	1 11	1 1	51	41	32	22	13	4	4	13	21	30	38	46	54	1 1
3 20	1 6	57	47	38	29	21	13	4	4	12	20	28	35	43	50	57
3 40	1 2	53	44	36	28	20	12	4	4	11	18	26	33	40	47	53
4 0	58	49	41	33	26	18	11	4	4	10	17	24	31	37	44	50
4 30	53	45	38	31	24	17	10	3	3	9	16	22	28	34	40	45
5 0	48	41	35	28	22	16	9	3	3	9	14	20	26	31	36	40
5 30	45	38	32	26	20	14	9	3	3	8	13	19	24	29	34	38
6 0	41	35	30	24	19	13	8	3	2	7	12	17	22	26	31	35
6 30	38	33	28	22	17	12	7	2	2	7	11	15	20	24	29	33
7 0	36	31	26	21	17	12	7	2	2	6	10	14	19	23	27	31
8	32	27	23	19	15	10	6	2	2	5	9	13	16	20	24	27
9	28	24	20	16	13	9	5	2	2	5	8	11	14	18	21	24
10	26	22	18	15	12	8	5	2	1	4	7	10	13	16	19	22
11	23	20	17	14	11	8	5	2	1	4	7	9	12	15	18	20
12	21	18	15	13	10	7	4	1	1	4	6	9	11	13	16	18
13	20	17	15	12	9	7	4	1	1	3	6	8	10	12	15	17
14	18	16	13	11	8	6	4	1	1	3	5	7	9	11	14	16
15	17	15	12	[illegible]	8	6	3	1	1	3	5	7	9	11	13	15
16	16	14	12	9	7	5	3	1	1	3	5	6	8	10	12	14
17	15	13	11	9	7	5	3	1	1	3	4	6	8	9	11	13
18	14	12	10	8	6	5	3	1	1	2	4	6	7	9	10	12
19	13	11	10	8	6	4	3	1	1	2	4	5	7	8	10	11
20	13	11	9	7	6	4	2	1	1	2	4	5	6	8	9	11
21	12	10	9	7	5	4	2	1	1	2	3	5	6	7	9	10
22	11	10	8	7	5	4	2	1	1	2	3	5	6	7	8	10
23	11	9	8	6	5	4	2	1	1	2	3	4	6	7	8	9
24	10	9	7	6	5	3	2	1	1	2	3	4	5	6	8	9
25	10	8	7	6	5	3	2	1	1	2	3	4	5	6	7	8
26	9	8	7	6	4	3	2	1	1	2	3	4	5	6	7	8
27	9	8	6	5	4	3	2	1	1	2	3	4	5	6	7	8
28	9	7	6	5	4	3	2	1	0	1	2	3	5	5	6	7
30	8	7	6	5	4	3	2	1	0	1	2	3	4	5	6	7
32	7	6	5	4	3	2	1	0	0	1	2	3	4	5	5	6
34	7	6	5	4	3	2	1	0	0	1	2	3	3	4	5	6
36	6	5	5	4	3	2	1	0	0	1	2	3	3	4	5	5
38	6	5	4	4	3	2	1	0	0	1	2	2	3	4	4	5
40	6	5	4	3	3	2	1	0	0	1	2	2	3	3	4	5
45	5	4	3	3	2	2	1	0	0	1	1	2	2	3	3	4
50	4	3	3	2	2	1	1	0	0	1	1	2	2	2	3	3
55	3	3	2	2	2	1	1	0	0	1	1	1	2	2	2	3
60	3	2	2	2	1	1	1	0	0	0	1	1	1	2	2	2
65	2	2	1	1	1	1	0	0	0	0	1	1	1	1	2	2
70	2	1	1	1	1	1	0	0	0	0	0	1	1	1	1	1
80	1	1	1	0	0	0	0	0	0	0	0	0	0	1	1	1
90	0	0	0	0	0	0	0	0	0	0	0	0	0	0	0	0
				−	−	−	−	−	+	+	+	+	+			
				28.26	28.56	28.85	29.15	29.45	29.75	30.05	30.35	30.64	30.93			

Argument. Height of the Barometer.

TABLE IX. Correction of the Moon's Apparent Altitude.

Arg. app. altit.		Arg. Horizontal Parallax.																Prop. Part, for Sec. of Par.									
		54′		55′		56′		57′		58′		59′		60′		61′		0″	1″	2″	3″	4″	5″	6″	7″	8″	9″
°	′	′	″	′	″	′	″	′	″	′	″	′	″	′	″	′	″										
0	0	21	0	22	0	23	0	24	0	25	0	26	0	27	0	28	0	0	1	2	3	4	5	6	7	8	9
	10	22	38	23	38	24	38	25	38	26	38	27	38	28	38	29	38	10	11	12	13	14	15	16	17	18	19
	20	24	10	25	10	26	10	27	10	28	10	29	10	30	10	31	10	20	21	22	23	24	25	26	27	28	29
	30	25	37	26	37	27	37	28	37	29	37	30	37	31	37	32	37	30	31	32	33	34	35	36	37	38	39
	40	27	0	28	0	29	0	30	0	31	0	32	0	33	0	34	0	40	41	42	43	44	45	46	47	48	49
	50	28	18	29	18	30	18	31	18	32	18	33	18	34	18	35	18	50	51	52	53	54	55	56	57	58	59
1	0	29	31	30	31	31	31	32	31	33	31	34	31	35	31	36	31	0	1	2	3	4	5	6	7	8	9
	10	30	40	31	40	32	40	33	40	34	40	35	40	36	40	37	40	10	11	12	13	14	15	16	17	18	19
	20	31	44	32	44	33	44	34	44	35	44	36	44	37	44	38	44	20	21	22	23	24	25	26	27	28	29
	30	32	44	33	44	34	44	35	44	36	44	37	44	38	44	39	44	30	31	32	33	34	35	36	37	38	39
	40	33	41	34	41	35	41	36	41	37	41	38	41	39	41	40	41	40	41	42	43	44	45	46	47	48	49
	50	34	34	35	34	36	34	37	34	38	34	39	34	40	34	41	34	50	51	52	53	54	55	56	57	58	59
2	0	35	23	36	23	37	23	38	23	39	23	40	23	41	23	42	23	0	1	2	3	4	5	6	7	8	9
	10	36	10	37	10	38	10	39	10	40	9	41	9	42	9	43	9	10	11	12	13	14	15	16	17	18	19
	20	36	53	37	53	38	53	39	53	40	53	41	53	42	53	43	53	20	21	22	23	24	25	26	27	28	29
	30	37	34	38	34	39	33	40	33	41	33	42	33	43	33	44	33	30	31	32	33	34	35	36	37	38	39
	40	38	12	39	12	40	12	41	11	42	11	43	11	44	11	45	11	40	41	42	43	44	45	46	47	48	49
	50	38	47	39	47	40	47	41	47	42	47	43	47	44	47	45	47	50	51	52	53	54	55	56	57	58	59
3	0	39	21	40	20	41	20	42	20	43	20	44	20	45	20	46	20	0	1	2	3	4	5	6	7	8	9
	10	39	52	40	52	41	52	42	52	43	51	44	51	45	51	46	51	10	11	12	13	14	15	16	17	18	19
	20	40	21	41	21	42	21	43	21	44	21	45	21	46	21	47	20	20	21	22	23	24	25	26	27	28	29
	30	40	49	41	49	42	48	43	48	44	48	45	48	46	48	47	48	30	31	32	33	34	35	36	37	38	39
	40	41	15	42	14	43	14	44	14	45	14	46	14	47	14	48	14	40	41	42	43	44	45	46	47	48	49
	50	41	39	42	39	43	39	44	39	45	39	46	38	47	38	48	38	50	51	52	53	54	55	56	57	58	59
4	0	42	2	43	2	44	2	45	2	46	1	47	1	48	1	49	1	0	1	2	3	4	5	6	7	8	9
	10	42	24	43	23	44	23	45	23	46	23	47	23	48	23	49	22	10	11	12	13	14	15	16	17	18	19
	20	42	44	43	44	44	44	45	43	46	43	47	43	48	43	49	43	20	21	22	23	24	25	26	27	28	29
	30	43	3	44	3	45	3	46	3	47	3	48	2	49	2	50	2	30	31	32	33	34	35	36	37	38	39
	40	43	21	44	21	45	21	46	21	47	21	48	20	49	20	50	20	40	41	42	43	44	45	46	47	48	49
	50	43	39	44	38	45	38	46	38	47	38	48	37	49	37	50	37	50	51	52	53	54	55	56	57	58	59
5	0	43	55	44	55	45	54	46	54	47	54	48	54	49	53	50	53	0	1	2	3	4	5	6	7	8	9
	10	44	10	45	10	46	10	47	9	48	9	49	9	50	9	51	8	10	11	12	13	14	15	16	17	18	19
	20	44	25	45	24	46	24	47	24	48	24	49	23	50	23	51	23	20	21	22	23	24	25	26	27	28	29
	30	44	38	45	38	46	38	47	38	48	37	49	37	50	37	51	36	30	31	32	33	34	35	36	37	38	39
	40	44	51	45	51	46	51	47	51	48	50	49	50	50	50	51	49	40	41	42	43	44	45	46	47	48	49
	50	45	4	46	4	47	3	48	3	49	3	50	2	51	2	52	2	50	51	52	53	54	55	56	57	58	59
6	0	45	16	46	15	47	15	48	15	49	14	50	14	51	14	52	13	0	1	2	3	4	5	6	7	8	9
	10	45	27	46	26	47	26	48	26	49	25	50	25	51	25	52	24	10	11	12	13	14	15	16	17	18	19
	20	45	37	46	37	47	37	48	36	49	36	50	36	51	35	52	35	20	21	22	23	24	25	26	27	28	29
	30	45	47	46	47	47	47	48	46	49	46	50	46	51	45	52	45	30	31	32	33	34	35	36	37	38	39
	40	45	57	46	57	47	56	48	56	49	55	50	55	51	55	52	54	40	41	42	43	44	45	46	47	48	49
	50	46	6	47	6	48	5	49	5	50	4	51	4	52	4	53	3	50	51	52	53	54	55	56	57	58	59
7	0	46	15	47	14	48	14	49	13	50	13	51	13	52	12	53	12	0	1	2	3	4	5	6	7	8	9
	10	46	23	47	23	48	22	49	22	50	21	51	21	52	20	53	20	10	11	12	13	14	15	16	17	18	19
	20	46	31	47	31	48	30	49	30	50	29	51	29	52	28	53	28	20	21	22	23	24	25	26	27	28	29
	30	46	38	47	38	48	37	49	37	50	36	51	36	52	35	53	35	30	31	32	33	34	35	36	37	38	39
	40	46	45	47	45	48	44	49	44	50	43	51	43	52	42	53	42	40	41	42	43	44	45	46	47	48	49
	50	46	52	47	52	48	51	49	51	50	50	51	49	52	49	53	48	50	51	52	53	54	55	56	57	58	58
8	0	46	59	47	58	48	58	49	57	50	56	51	56	52	55	53	55	0	1	2	3	4	5	6	7	8	9
	10	47	5	48	4	49	4	50	3	51	2	52	2	53	2	54	1	10	11	12	13	14	15	16	17	18	19
	20	47	11	48	10	49	9	50	9	51	8	52	8	53	7	54	6	20	21	22	23	24	25	26	27	28	29
	30	47	16	48	16	49	15	50	14	51	14	52	13	53	12	54	12	30	31	32	33	34	35	36	37	38	39
	40	47	22	48	21	49	20	50	20	51	19	52	18	53	17	54	17	40	41	42	43	44	45	45	46	47	48
	50	47	27	48	26	49	25	50	25	51	24	52	23	53	22	54	22	49	50	51	52	53	54	55	56	57	58
9	0	47	31	48	31	49	30	50	29	51	29	52	28	53	27	54	26	0	1	2	3	4	5	6	7	8	9
	10	47	36	48	35	49	35	50	34	51	33	52	32	53	31	54	31	10	11	12	13	14	15	16	17	18	19
	20	47	40	48	40	49	39	50	38	51	37	52	36	53	36	54	35	20	21	22	23	24	25	26	27	28	29
	30	47	45	48	44	49	43	50	42	51	41	52	40	53	40	54	39	30	31	32	33	34	35	36	36	37	38
	40	47	[illegible]	48	46	49	47	50	46	51	45	52	44	53	43	54	43	39	40	41	42	43	44	45	46	47	48
	50	47	52	48	51	49	50	50	50	51	49	52	48	53	47	54	46	49	50	51	[illegible]	[illegible]	[illegible]	[illegible]	[illegible]	[illegible]	[illegible]

TABLE IX. Correction of the Moon's apparent Altitude.

Arg. app. altit.		Arg. Horizontal Parallax.																Prop. part for seconds of par.									
		54'		55'		56'		57'		58'		59'		60'		61'		0"	1"	2"	3"	4"	5"	6"	7"	8"	9"
°	'	'	"	'	"	'	"	'	"	'	"	'	"	'	"	'	"										
10	0	47	56	48	55	49	54	50	53	51	52	52	51	53	50	54	49	0	1	2	3	4	5	6	7	8	9
	10	47	59	48	58	49	57	50	56	51	55	52	54	53	53	54	53	10	11	12	13	14	15	16	17	18	19
	20	48	2	49	1	50	0	50	59	51	58	52	57	53	57	54	56	20	21	22	23	24	25	26	27	28	29
	30	48	5	49	4	50	3	51	2	52	1	53	0	53	59	54	58	29	30	31	32	33	34	35	36	37	38
	40	48	8	49	7	50	6	51	5	52	4	53	3	54	2	55	1	39	40	41	42	43	44	45	46	47	48
	50	48	11	49	10	50	9	51	8	52	7	53	6	54	4	55	3	49	50	51	52	53	54	55	56	57	58
11	0	48	13	49	12	50	11	51	10	52	9	53	8	54	7	55	6	0	1	2	3	4	5	6	7	8	9
	10	48	16	49	15	50	14	51	12	52	11	53	10	54	9	55	8	10	11	12	13	14	15	16	17	18	19
	20	48	18	49	17	50	16	51	15	52	13	53	12	54	11	55	10	20	21	22	23	24	24	25	26	27	28
	30	48	20	49	19	50	18	51	17	52	15	53	14	54	13	55	12	29	30	31	32	33	34	35	36	37	38
	40	48	22	49	21	50	20	51	19	52	17	53	16	54	15	55	14	39	40	41	42	43	44	45	46	47	48
	50	48	24	49	23	50	22	51	20	52	19	53	18	54	16	55	15	49	50	51	52	53	54	55	56	57	58
12	0	48	26	49	25	50	23	51	22	52	21	53	19	54	18	55	17	0	1	2	3	4	5	6	7	8	9
	10	48	27	49	26	50	25	51	23	52	22	53	21	54	19	55	18	10	11	12	13	14	15	16	17	18	19
	20	48	29	49	28	50	26	51	25	52	23	53	22	54	21	55	19	20	21	21	22	23	24	25	26	27	28
	30	48	30	49	29	50	28	51	26	52	25	53	23	54	22	55	20	29	30	31	32	33	34	35	36	37	38
	40	48	32	49	30	50	29	51	27	52	26	53	24	54	23	55	21	39	40	41	42	43	44	45	46	47	48
	50	48	33	49	31	50	30	51	28	52	27	53	25	54	24	55	22	49	50	51	52	53	54	55	56	57	58
13	0	48	34	49	32	50	31	51	29	52	28	53	26	54	25	55	23	0	1	2	3	4	5	6	7	8	9
	10	48	35	49	33	50	32	51	30	52	28	53	27	54	25	55	24	10	11	12	13	14	15	16	17	18	19
	20	48	36	49	34	50	32	51	31	52	29	53	28	54	26	55	24	19	20	21	22	23	24	25	26	27	28
	30	48	36	49	35	50	33	51	31	52	30	53	28	54	26	55	25	29	30	31	32	33	34	35	36	37	38
	40	48	37	49	35	50	34	51	32	52	30	53	29	54	27	55	25	39	40	41	42	43	44	45	46	47	48
	50	48	38	49	36	50	34	51	32	52	31	53	29	54	27	55	25	49	50	51	52	53	53	54	55	56	57
14	0	48	38	49	36	50	35	51	33	52	31	53	29	54	27	55	26	0	1	2	3	4	5	6	7	8	9
	10	48	38	49	37	50	35	51	33	52	31	53	29	54	28	55	26	10	11	12	13	14	15	15	16	17	18
	20	48	39	49	37	50	35	51	33	52	31	53	30	54	28	55	26	19	20	21	22	23	24	25	26	27	28
	30	48	39	49	37	50	35	51	33	52	31	53	30	54	28	55	26	29	30	31	32	33	34	35	36	37	38
	40	48	39	49	37	50	35	51	33	52	31	53	30	54	28	55	26	39	40	41	42	43	44	45	46	46	47
	50	48	39	49	37	50	35	51	33	52	31	53	29	54	27	55	25	48	49	50	51	52	53	54	55	56	57
15	0	48	39	49	37	50	35	51	33	52	31	53	29	54	27	55	25	0	1	2	3	4	5	6	7	8	9
	10	48	39	49	37	50	35	51	33	52	31	53	29	54	27	55	24	10	11	12	13	14	14	15	16	17	18
	20	48	39	49	37	50	35	51	33	52	31	53	28	54	26	55	24	19	20	21	22	23	24	25	26	27	28
	30	48	39	49	37	50	34	51	32	52	30	53	28	54	26	55	24	29	30	31	32	33	34	35	36	37	38
	40	48	38	49	36	50	34	51	32	52	30	53	27	54	25	55	23	39	39	40	41	42	43	44	45	46	47
	50	48	38	49	36	50	34	51	31	52	29	53	27	54	24	55	22	48	49	50	51	52	53	54	55	56	57
16	0	48	38	49	35	50	33	51	31	52	28	53	26	54	24	55	21	0	1	2	3	4	5	6	7	8	9
	10	48	37	49	35	50	32	51	30	52	28	53	25	54	23	55	21	10	11	12	13	14	15	15	16	17	18
	20	48	36	49	34	50	32	51	29	52	27	53	24	54	22	55	20	19	20	21	22	23	24	25	26	27	28
	30	48	36	49	33	50	31	51	29	52	26	53	24	54	21	55	19	29	30	31	32	33	34	35	36	37	38
	40	48	35	49	33	50	30	51	28	52	25	53	23	54	20	55	18	38	39	40	41	42	43	44	45	46	47
	50	48	34	49	32	50	29	51	27	52	24	53	22	54	19	55	16	48	49	50	51	52	53	54	55	56	57
17	0	48	34	49	31	50	28	51	26	52	23	53	21	54	18	55	15	0	1	2	3	4	5	6	7	8	9
	10	48	33	49	30	50	27	51	25	52	22	53	19	54	17	55	14	10	10	11	12	13	14	15	16	17	18
	20	48	32	49	29	50	26	51	24	52	21	53	18	54	15	55	13	19	20	21	22	23	24	25	26	27	28
	30	48	31	49	28	50	25	51	22	52	20	53	17	54	14	55	11	29	30	30	31	32	33	34	35	36	37
	40	48	30	49	27	50	24	51	21	52	18	53	16	54	13	55	10	38	39	40	41	42	43	44	45	46	47
	50	48	29	49	26	50	23	51	20	52	17	53	14	54	12	55	8	48	49	50	51	52	53	53	54	55	56
18	0	48	27	49	24	50	22	51	19	52	16	53	13	54	10	55	7	0	1	2	3	4	5	6	7	8	9
	10	48	26	49	23	50	20	51	17	52	14	53	11	54	8	55	5	9	10	11	12	13	14	15	16	17	18
	20	48	25	49	22	50	19	51	16	52	13	53	10	54	7	55	4	19	20	21	22	23	24	25	26	27	27
	30	48	24	49	20	50	17	51	14	52	11	53	8	54	5	55	2	28	29	30	31	32	33	34	35	36	37
	40	48	22	49	19	50	16	51	13	52	9	53	6	54	3	55	0	38	39	40	41	42	43	44	45	46	46
	50	48	21	49	17	50	14	51	11	52	8	53	5	54	1	54	58	47	48	49	50	51	52	53	54	55	56
19	0	48	19	49	16	50	13	51	9	52	6	53	3	54	0	54	56	0	1	2	3	4	5	6	7	8	9
	10	48	18	49	14	50	11	51	8	52	4	53	1	53	58	54	54	9	10	11	12	13	14	15	16	17	18
	20	48	16	49	13	50	9	51	6	52	2	52	59	53	56	54	52	19	20	21	22	23	24	24	25	26	27
	30	48	14	49	11	50	8	51	4	52	1	52	57	53	54	54	50	28	29	30	31	32	33	34	35	36	37
	40	48	13	49	9	50	6	51	2	51	59	52	55	53	52	54	48	38	39	40	41	41	42	43	44	45	46
	50	48	11	49	7	50	4	51	0	51	57	52	53	53	49	54	46	47	48	49	50	51	52	53	54	55	56

TABLE IX. Correction of the Moon's apparent Altitude.

Arg. app. alt.	Arg. Horizontal Parallax.								Prop. part for seconds of par.									
	54′	55′	56′	57′	58′	59′	60′	61′	0″	1″	2″	3″	4″	5″	6″	7″	8″	9″
° ′	′ ″	′ ″	′ ″	′ ″	′ ″	′ ″	′ ″	′ ″										
20 0	48 9	49 5	50 2	50 58	51 55	52 51	53 47	54 44	0	1	2	3	4	5	6	7	8	8
10	48 7	49 3	50 0	50 56	51 52	52 49	53 45	54 41	9	10	11	12	13	14	15	16	17	18
20	48 5	49 2	49 58	50 54	51 50	52 47	53 43	54 39	19	20	21	22	23	23	24	25	26	27
30	48 3	49 0	49 56	50 52	51 48	52 44	53 41	54 37	28	29	30	31	32	33	34	35	36	37
40	48 1	48 57	49 54	50 50	51 46	52 42	53 38	54 34	37	38	39	40	41	42	43	44	45	46
50	47 59	48 55	49 51	50 48	51 44	52 40	53 36	54 32	47	48	49	50	51	51	52	53	54	55
21 0	47 57	48 53	49 49	50 45	51 41	52 37	53 33	54 29	0	1	2	3	4	5	6	7	7	8
10	47 55	48 51	49 47	50 43	51 39	52 35	53 31	54 27	9	10	11	12	13	14	15	16	17	18
20	47 53	48 49	49 45	50 41	51 36	52 32	53 28	54 24	19	20	20	21	22	23	24	25	26	27
30	47 51	48 46	49 42	50 38	51 34	52 30	53 26	54 21	28	29	30	31	32	33	33	34	35	36
40	47 48	48 44	49 40	50 36	51 31	52 27	53 23	54 19	37	38	39	40	41	42	43	44	45	46
50	47 46	48 42	49 37	50 33	51 29	52 25	53 20	54 16	47	47	48	49	50	51	52	53	54	55
22 0	47 44	48 39	49 35	50 31	51 26	52 22	53 18	54 13	0	1	2	3	4	5	6	6	7	8
10	47 41	48 37	49 32	50 28	51 24	52 19	53 15	54 10	9	10	11	12	13	14	15	16	17	18
20	47 39	48 34	49 30	50 25	51 21	52 16	53 12	54 7	18	19	20	21	22	23	24	25	26	27
30	47 36	48 32	49 27	50 23	51 18	52 14	53 9	54 4	28	29	30	30	31	32	33	34	35	36
40	47 34	48 29	49 25	50 20	51 15	52 11	53 6	54 1	37	38	39	40	41	42	43	43	44	45
50	47 31	48 27	49 22	50 17	51 13	52 8	53 3	53 58	46	47	48	49	50	51	52	53	54	55
23 0	47 29	48 24	49 19	50 14	51 10	52 5	53 0	53 55	0	1	2	3	4	5	6	6	7	8
10	47 26	48 21	49 17	50 12	51 7	52 2	52 57	53 52	9	10	11	12	13	14	15	16	17	17
20	47 23	48 19	49 14	50 9	51 4	51 59	52 54	53 49	18	19	20	21	22	23	24	25	26	27
30	47 21	48 16	49 11	50 6	51 1	51 56	52 51	53 46	28	28	29	30	31	32	33	34	35	36
40	47 18	48 13	49 8	50 3	50 58	51 53	52 48	53 43	37	38	38	39	40	41	42	43	44	45
50	47 15	48 10	49 5	50 0	50 55	51 50	52 45	53 39	46	47	48	49	50	50	51	52	53	54
24 0	47 12	48 7	49 2	49 57	50 52	51 46	52 41	53 36	0	1	2	3	4	5	5	6	7	8
10	47 9	48 4	48 59	49 54	50 48	51 43	52 38	53 33	9	10	11	12	13	14	15	15	16	17
20	47 7	48 1	48 56	49 51	50 45	51 40	52 35	53 28	18	19	20	21	22	23	24	25	25	26
30	47 4	47 58	48 53	49 48	50 42	51 37	52 31	53 26	27	28	29	30	31	32	33	34	35	35
40	47 1	47 55	48 50	49 44	50 39	51 33	52 28	53 22	36	37	38	39	40	41	42	43	44	45
50	46 57	47 52	48 47	49 41	50 36	51 30	52 24	53 19	46	46	47	48	49	50	51	52	53	54
25 0	46 55	47 49	48 43	49 38	50 32	51 27	52 21	53 15	0	1	2	3	4	5	5	6	7	8
10	46 52	47 46	48 40	49 35	50 29	51 23	52 17	53 12	9	10	11	12	13	14	14	15	16	17
20	46 48	47 43	48 37	49 31	50 25	51 20	52 14	53 7	18	19	20	21	22	23	23	24	25	26
30	46 45	47 40	48 34	49 28	50 22	51 16	52 10	53 4	27	28	29	30	31	32	32	33	34	35
40	46 42	47 36	48 30	49 24	50 18	51 13	52 7	53 1	36	37	38	39	40	41	41	42	43	44
50	46 39	47 33	48 27	49 21	50 15	51 9	52 3	52 57	45	46	47	48	49	50	51	51	52	53
26 0	46 36	47 30	48 24	49 17	50 11	51 5	51 59	52 53	0	1	2	3	4	4	5	6	7	8
10	46 32	47 26	48 20	49 14	50 8	51 2	51 55	52 49	9	10	11	12	13	13	14	15	16	17
20	46 29	47 23	48 17	49 10	50 4	50 58	51 52	52 45	18	19	20	21	22	22	23	24	25	26
30	46 26	47 19	48 13	49 7	50 0	50 54	51 48	52 42	27	28	29	30	30	31	32	33	34	35
40	46 22	47 16	48 9	49 3	49 57	50 50	51 44	52 38	36	37	38	38	39	40	41	42	43	44
50	46 19	47 12	48 6	48 59	49 53	50 47	51 40	52 34	45	46	47	47	48	49	50	51	52	53
27 0	46 15	47 9	48 2	48 56	49 49	50 43	51 36	52 30	0	1	2	3	4	4	5	6	7	8
10	46 11	47 5	47 59	48 52	49 45	50 39	51 32	52 26	9	10	11	12	12	13	14	15	16	17
20	46 8	47 2	47 55	48 48	49 42	50 35	51 28	52 21	18	19	20	20	21	22	23	24	25	26
30	46 5	46 58	47 51	48 44	49 38	50 31	51 24	52 17	27	27	28	29	30	31	32	33	34	35
40	46 1	46 54	47 47	48 41	49 34	50 27	51 20	52 13	35	36	37	38	39	40	41	42	43	43
50	45 58	46 51	47 44	48 37	49 30	50 23	51 16	52 9	44	45	46	47	48	49	50	51	51	52
28 0	45 54	46 47	47 40	48 33	49 26	50 19	51 12	52 5	0	1	2	3	4	4	5	6	7	8
10	45 50	46 43	47 36	48 29	49 22	50 15	51 8	52 0	9	10	11	11	12	13	14	15	16	17
20	45 46	46 39	47 32	48 25	49 18	50 11	51 3	51 56	18	18	19	20	21	22	23	24	25	25
30	45 43	46 35	47 28	48 21	49 14	50 6	50 59	51 52	26	27	28	29	30	31	32	33	33	34
40	45 39	46 32	47 24	48 17	49 9	50 2	50 55	51 47	35	36	37	38	39	40	40	41	42	43
50	45 35	46 28	47 20	48 13	49 5	49 58	50 50	51 43	44	45	46	47	47	48	49	50	51	52
29 0	45 31	46 24	47 16	48 9	49 1	49 54	50 46	51 39	0	1	2	3	3	4	5	6	7	8
10	45 27	46 20	47 12	48 5	48 57	49 49	50 42	51 34	9	10	10	11	12	13	14	15	16	17
20	45 23	46 16	47 8	48 0	48 53	49 45	50 37	51 30	17	18	19	20	21	22	23	23	24	25
30	45 20	46 12	47 4	47 56	48 48	49 41	50 33	51 24	26	27	28	29	30	30	31	32	33	34
40	45 16	46 8	47 0	47 52	48 44	49 36	50 28	51 20	35	36	37	37	38	39	40	41	42	43
50	45 12	46 4	46 56	47 47	48 40	49 32	50 24	51 16	44	44	45	46	47	48	49	50	50	51

TABLE IX. Correction of the Moon's apparent Altitude.

Arg. app. altit.	Arg. Horizontal Parallax.								Prop. part for seconds of par.									
	54'	55'	56'	57'	58'	59'	60'	61'	0"	1"	2"	3"	4"	5"	6"	7"	8"	9"
° '	' "	' "	' "	' "	' "	' "	' "	' "										
30 0	45 7	45 59	46 51	47 43	48 35	49 27	50 19	51 11	0	1	2	3	3	4	5	6	7	8
10	45 3	45 55	46 47	47 39	48 31	49 23	50 15	51 6	9	9	10	11	12	13	14	15	16	16
20	44 59	45 51	46 43	47 35	48 26	49 18	50 10	51 2	17	18	19	20	21	22	22	23	24	25
30	44 55	45 47	46 39	47 30	48 22	49 14	50 5	50 57	26	27	28	28	29	30	31	32	33	34
40	44 51	45 43	46 34	47 26	48 17	49 9	50 1	50 52	34	35	36	37	38	39	40	41	41	42
50	44 47	45 38	46 30	47 21	48 13	49 4	49 56	50 47	43	44	45	46	47	47	48	49	50	51
31 0	44 43	45 34	46 25	47 17	48 8	49 0	49 51	50 43	0	1	2	3	3	4	5	6	7	8
10	44 38	45 30	46 21	47 12	48 4	48 55	49 46	50 38	9	9	10	11	12	13	14	15	15	16
20	44 34	45 25	46 17	47 8	47 59	48 50	49 42	50 33	17	18	19	20	20	21	22	23	24	25
30	44 30	45 21	46 12	47 3	47 54	48 46	49 37	50 28	26	26	27	28	29	30	31	32	32	33
40	44 25	45 16	46 8	46 59	47 50	48 41	49 32	50 23	34	35	36	37	38	38	39	40	41	42
50	44 21	45 12	46 3	46 54	47 45	48 36	49 27	50 18	43	44	44	45	46	47	48	49	49	50
32 0	44 17	45 8	45 58	46 49	47 40	48 31	49 22	50 13	0	1	2	3	3	4	5	6	7	8
10	44 12	45 3	45 54	46 45	47 35	48 26	49 17	50 8	8	9	10	11	12	13	13	14	15	16
20	44 8	44 58	45 49	46 40	47 31	48 21	49 12	50 3	17	18	19	19	20	21	22	23	24	24
30	44 3	44 54	45 45	46 35	47 26	48 16	49 7	49 58	25	26	27	28	28	29	30	31	32	33
40	43 59	44 49	45 40	46 30	47 21	48 11	49 2	49 52	33	34	35	36	37	38	38	39	40	41
50	43 55	44 45	45 35	46 26	47 16	48 6	48 57	49 47	42	43	44	45	46	47	47	48	49	50
33 0	43 50	44 41	45 30	46 21	47 11	48 1	48 52	49 42	0	1	2	3	3	4	5	6	7	8
10	43 45	44 35	45 26	46 16	47 6	47 56	48 46	49 37	8	9	10	11	12	13	13	14	15	16
20	43 40	44 31	45 21	46 11	47 1	47 51	48 41	49 31	17	18	18	19	20	21	22	23	23	24
30	43 36	44 26	45 16	46 6	46 56	47 46	48 36	49 26	25	26	27	28	28	29	30	31	32	33
40	43 31	44 21	45 11	46 1	46 51	47 41	48 31	49 21	33	34	35	36	37	38	38	39	40	41
50	43 26	44 16	45 6	45 56	46 46	47 36	48 26	49 15	42	43	43	44	45	46	47	48	48	49
34 0	43 22	44 12	45 1	45 51	46 41	47 30	48 20	49 10	0	1	2	2	3	4	5	6	7	7
10	43 17	44 7	44 56	45 46	46 36	47 25	48 15	49 4	8	9	10	11	12	12	13	14	15	16
20	43 12	44 2	44 51	45 41	46 30	47 20	48 9	48 59	16	17	18	19	20	21	21	22	23	24
30	43 7	43 57	44 46	45 36	46 25	47 15	48 4	48 54	25	26	26	27	28	29	30	31	31	32
40	43 3	43 52	44 41	45 31	46 20	47 9	47 59	48 48	33	34	35	35	36	37	38	39	40	40
50	42 58	43 47	44 36	45 25	46 15	47 4	47 53	48 42	41	42	43	44	44	45	46	47	48	49
35 0	42 53	43 42	44 31	45 20	46 9	46 59	47 48	48 37	0	1	2	2	3	4	5	6	7	7
10	42 48	43 37	44 26	45 15	46 4	46 53	47 42	48 31	8	9	10	11	11	12	13	14	15	15
20	42 43	43 32	44 21	45 10	45 59	46 48	47 37	48 26	16	17	18	19	20	20	21	22	23	24
30	42 38	43 27	44 16	45 5	45 53	46 42	47 31	48 20	24	25	26	27	28	28	29	30	31	32
40	42 33	43 22	44 10	44 59	45 48	46 37	47 25	48 14	33	33	34	35	36	37	37	38	39	40
50	42 28	43 17	44 5	44 54	45 43	46 31	47 20	48 8	41	42	42	43	44	45	46	46	47	48
36 0	42 23	43 11	44 0	44 49	45 37	46 26	47 14	48 3	0	1	2	2	3	4	5	6	6	7
10	42 18	43 6	43 55	44 43	45 32	46 20	47 8	47 57	8	9	10	10	11	12	13	14	14	15
20	42 13	43 1	43 49	44 38	45 26	46 14	47 3	47 51	16	17	18	18	19	20	21	22	23	23
30	42 8	42 56	43 44	44 32	45 21	46 9	46 57	47 45	24	25	26	27	27	28	29	30	31	31
40	42 2	42 51	43 39	44 27	45 15	46 3	46 51	47 39	32	33	34	35	35	36	37	38	39	39
50	41 57	42 46	43 33	44 21	45 9	45 57	46 45	47 33	40	41	42	43	43	44	45	46	47	47
37 0	41 52	42 40	43 28	44 16	45 4	45 52	46 40	47 28	0	1	2	2	3	4	5	6	6	7
10	41 47	42 35	43 22	44 10	44 58	45 46	46 34	47 22	8	9	10	10	11	12	13	13	14	15
20	41 42	42 29	43 17	44 5	44 52	45 40	46 28	47 16	16	17	17	18	19	20	21	21	22	23
30	41 36	42 24	43 12	43 59	44 47	45 34	46 22	47 10	24	25	25	26	27	28	29	29	30	31
40	41 31	42 18	43 6	43 54	44 41	45 28	46 16	47 3	32	33	33	34	35	36	36	37	38	39
50	41 26	42 13	43 0	43 48	44 35	45 23	46 10	46 57	40	40	41	42	43	44	44	45	46	47
38 0	41 20	42 8	42 55	43 42	44 29	45 17	46 4	46 51	0	1	2	2	3	4	5	5	6	7
10	41 15	42 2	42 49	43 36	44 24	45 11	45 58	46 45	8	9	9	10	11	12	13	13	14	15
20	41 10	41 57	42 44	43 31	44 18	45 5	45 52	46 39	16	16	17	18	19	20	20	21	22	23
30	41 4	41 51	42 38	43 25	44 12	44 59	45 46	46 33	23	24	25	26	27	27	28	29	30	31
40	40 59	41 45	42 32	43 19	44 6	44 53	45 40	46 27	31	32	33	34	34	35	36	37	38	39
50	40 53	41 40	42 27	43 13	44 0	44 47	45 34	46 21	39	40	41	41	42	43	44	45	45	46
39 0	40 48	41 34	42 21	43 8	43 54	44 41	45 27	46 14	0	1	2	2	3	4	5	5	6	7
10	40 42	41 29	42 15	43 2	43 48	44 35	45 21	46 8	8	9	9	10	11	12	13	13	14	15
20	40 37	41 23	42 9	42 56	43 42	44 29	45 15	46 2	15	16	17	18	19	19	20	21	22	23
30	40 31	41 17	42 4	42 50	43 36	44 23	45 9	45 55	23	24	25	25	26	27	28	29	29	30
40	40 25	41 12	41 58	42 44	43 30	44 16	45 3	45 49	31	32	32	33	34	35	36	36	37	38
50	40 20	41 6	41 52	42 38	43 24	44 10	44 56	45 42	39	39	40	41	42	42	43	44	45	46

TABLE IX. Correction of the Moon's apparent Altitude.

Arg. app. alt.		Arg. Horizontal Parallax. 54'		55'		56'		57'		58'		59'		60'		61'		Prop. part for seconds of par. 0"	1"	2"	3"	4"	5"	6"	7"	8"	9"
°	'	'	"	'	"	'	"	'	"	'	"	'	"	'	"	'	"										
40	0	40	14	41	0	41	46	42	32	43	18	44	4	44	50	45	36	0	1	1	2	3	4	5	5	6	7
	10	40	8	40	54	41	40	42	26	43	12	43	58	44	44	45	29	8	8	9	10	11	11	12	13	14	14
	20	40	3	40	48	41	34	42	20	43	6	43	51	44	37	45	23	15	16	17	17	18	19	20	21	21	22
	30	39	57	40	43	41	28	42	14	43	0	43	45	44	31	45	17	23	24	24	25	26	27	27	28	29	30
	40	39	51	40	37	41	22	42	8	42	53	43	39	44	24	45	10	30	31	32	33	33	34	35	36	36	37
	50	39	46	40	31	41	16	42	2	42	47	43	33	44	18	45	3	38	39	40	40	41	42	43	44	44	45
41	0	39	40	40	25	41	10	41	56	42	41	43	26	44	11	44	57	0	1	1	2	3	4	4	5	6	7
	10	39	34	40	19	41	4	41	50	42	35	43	20	44	5	44	50	7	8	9	10	10	11	12	13	13	14
	20	39	28	40	13	40	58	41	43	42	28	43	13	43	58	44	43	15	16	16	17	18	19	19	20	21	22
	30	39	22	40	7	40	52	41	37	42	22	43	7	43	52	44	37	22	23	24	25	25	26	27	28	28	29
	40	39	16	40	1	40	46	41	31	42	16	43	0	43	45	44	30	30	31	31	32	33	34	34	35	36	37
	50	39	11	39	55	40	40	41	25	42	9	42	54	43	39	44	23	37	38	39	40	40	41	42	43	43	44
42	0	39	5	39	49	40	34	41	18	42	3	42	48	43	32	44	17	0	1	1	2	3	4	4	5	6	7
	10	38	59	39	43	40	28	41	12	41	56	42	41	43	25	44	10	7	8	9	10	10	11	12	13	13	14
	20	38	53	39	37	40	21	41	6	41	50	42	34	43	19	44	3	15	15	16	17	18	18	19	20	21	21
	30	38	47	39	31	40	15	40	59	41	44	42	28	43	12	43	56	22	23	24	24	25	26	27	27	28	29
	40	38	41	39	25	40	9	40	53	41	37	42	21	43	5	43	49	29	30	31	32	32	33	34	35	35	36
	50	38	35	39	19	40	3	40	47	41	31	42	15	42	59	43	43	37	38	38	39	40	41	41	42	43	43
43	0	38	29	39	12	39	56	40	40	41	24	42	8	42	52	43	36	0	1	1	2	3	4	4	5	6	7
	10	38	22	39	6	39	50	40	34	41	17	42	1	42	45	43	29	7	8	9	9	10	11	12	12	13	14
	20	38	16	39	0	39	44	40	27	41	11	41	55	42	38	43	22	15	15	16	17	17	18	19	20	20	21
	30	38	10	38	54	39	37	40	21	41	4	41	48	42	31	43	15	22	22	23	24	25	25	26	27	28	28
	40	38	4	38	47	39	31	40	14	40	58	41	41	42	24	43	8	29	30	30	31	32	33	33	34	35	36
	50	37	58	38	41	39	24	40	8	40	51	41	34	42	18	43	1	36	37	38	38	39	40	41	41	42	43
44	0	37	52	38	35	39	18	40	1	40	44	41	27	42	11	42	54	0	1	1	2	3	4	4	5	6	6
	10	37	46	38	28	39	11	39	54	40	37	41	20	42	4	42	47	7	8	9	9	10	11	11	12	13	14
	20	37	39	38	22	39	5	39	48	40	31	41	14	41	57	42	40	14	15	16	16	17	18	19	19	20	21
	30	37	33	38	16	38	59	39	41	40	24	41	7	41	50	42	33	21	22	23	24	24	25	26	26	27	28
	40	37	27	38	9	38	52	39	35	40	17	41	0	41	43	42	25	28	29	30	31	31	32	33	34	34	35
	50	37	20	38	3	38	46	39	28	40	11	40	53	41	36	42	18	36	36	37	38	39	39	40	41	42	42
45	0	37	14	37	56	38	39	39	21	40	4	40	46	41	29	42	11	0	1	1	2	3	4	4	5	6	6
	10	37	8	37	50	38	32	39	15	39	57	40	39	41	22	42	4	7	8	8	9	10	11	11	12	13	13
	20	37	1	37	44	38	26	39	8	39	50	40	32	41	14	41	57	14	15	15	16	17	18	18	19	20	20
	30	36	55	37	37	38	19	39	1	39	43	40	25	41	7	41	49	21	22	22	23	24	25	25	26	27	27
	40	36	49	37	31	38	13	38	54	39	36	40	18	41	0	41	42	28	29	29	30	31	32	32	33	34	34
	50	36	42	37	24	38	6	38	48	39	29	40	11	40	53	41	35	35	36	36	37	38	39	39	40	41	41
46	0	36	36	37	17	37	59	38	41	39	22	40	4	40	46	41	27	0	1	1	2	3	4	4	5	6	6
	10	36	29	37	11	37	52	38	34	39	15	39	57	40	39	41	20	7	8	8	9	10	10	11	12	12	13
	20	36	23	37	4	37	46	38	27	39	8	39	50	40	31	41	13	14	14	15	16	17	17	18	19	19	20
	30	36	16	36	58	37	39	38	20	39	1	39	43	40	24	41	5	21	21	22	23	23	24	25	25	26	27
	40	36	9	36	51	37	32	38	13	38	54	39	36	40	17	40	58	28	28	29	30	30	31	32	32	33	34
	50	36	3	36	44	37	26	38	6	38	47	39	28	40	9	40	50	34	35	36	36	37	38	39	39	40	41
47	0	35	57	36	38	37	19	37	59	38	40	39	21	40	2	40	43	0	1	1	2	3	3	4	5	5	6
	10	35	50	36	31	37	12	37	52	38	33	39	14	39	55	40	36	7	7	8	9	9	10	11	11	12	13
	20	35	43	36	24	37	5	37	45	38	26	39	7	39	47	40	28	14	14	15	16	16	17	18	18	19	20
	30	35	37	36	17	36	58	37	38	38	19	38	59	39	40	40	20	20	21	22	22	23	24	24	25	26	26
	40	35	30	36	10	36	51	37	31	38	12	38	52	39	33	40	13	27	28	28	29	30	30	31	32	32	33
	50	35	23	36	4	36	44	37	24	38	5	38	45	39	25	40	5	34	34	35	36	37	37	38	39	39	40
48	0	35	17	35	57	36	37	37	17	37	57	38	37	39	18	39	58	0	1	1	2	3	3	4	5	5	6
	10	35	10	35	50	36	30	37	10	37	50	38	30	39	10	39	50	7	7	8	9	9	10	11	11	12	13
	20	35	3	35	43	36	23	37	3	37	43	38	23	39	3	39	42	13	14	15	15	16	17	17	18	19	19
	30	34	56	35	36	36	16	36	56	37	36	38	15	38	55	39	35	20	21	21	22	23	23	24	25	25	26
	40	34	50	35	29	36	9	36	49	37	28	38	8	38	47	39	27	27	27	28	29	29	30	30	31	32	32
	50	34	43	35	22	36	2	36	41	37	21	38	0	38	40	39	19	33	34	34	35	36	36	37	38	38	39
49	0	34	36	35	16	35	55	36	34	37	14	37	53	38	32	39	12	0	1	1	2	3	3	4	5	5	6
	10	34	29	35	9	35	48	36	27	37	6	37	45	38	25	39	4	6	7	8	8	9	10	10	11	12	12
	20	34	22	35	2	35	41	36	20	36	59	37	38	38	17	38	56	13	14	14	15	16	16	17	18	18	19
	30	34	16	34	55	35	34	36	12	36	51	37	30	38	9	38	48	19	20	21	21	22	23	23	24	25	25
	40	34	9	34	48	35	26	36	5	36	44	37	23	38	2	38	41	26	27	27	28	29	29	30	31	31	32
	50	34	2	34	40	35	19	35	58	36	37	37	15	37	54	38	33	32	33	34	34	35	36	36	37	38	38

TABLE IX. Correction of the Moon's Apparent Altitude.

Arg. app. altit.	Arg. Horizontal Parallax.								Prop. Part, for Sec. of Par.									
	54′	55′	56′	57′	58′	59′	60′	61′	0″	1″	2″	3″	4″	5″	6″	7″	8″	9″
° ′	′ ″	′ ″	′ ″	′ ″	′ ″	′ ″	′ ″	′ ″										
50 0	33 55	34 32	35 12	35 51	36 29	37 8	37 46	38 25	0	1	1	2	3	3	4	4	5	6
10	33 48	34 26	35 5	35 43	36 22	37 0	37 38	38 17	6	7	8	8	9	10	10	11	11	12
20	33 41	34 19	34 58	35 36	36 14	36 52	37 31	38 9	13	13	14	15	15	16	17	17	18	18
30	33 34	34 12	34 50	35 28	36 7	36 45	37 23	38 1	19	20	20	21	22	22	23	24	24	25
40	33 27	34 5	34 43	35 21	35 59	36 37	37 15	37 53	25	26	27	27	28	29	29	30	31	31
50	33 20	33 58	34 36	35 14	35 51	36 29	37 7	37 45	32	32	33	34	34	35	36	36	37	38
51 0	33 13	33 51	34 28	35 6	35 44	36 22	36 59	37 37	0	1	1	2	2	3	4	4	5	6
10	33 6	33 43	34 21	34 59	35 36	36 14	36 52	37 29	6	7	7	8	9	9	10	11	11	12
20	32 59	33 36	34 14	34 51	35 29	36 6	36 44	37 21	12	13	14	14	15	16	16	17	17	18
30	32 52	33 29	34 6	34 44	35 21	35 58	36 36	37 13	19	19	20	21	21	22	22	23	24	24
40	32 45	33 22	33 59	34 36	35 13	35 51	36 28	37 5	25	26	26	27	28	29	29	30	30	31
50	32 37	33 14	33 52	34 29	35 6	35 43	36 20	36 57	31	32	32	33	34	34	35	35	36	37
52 0	32 30	33 7	33 44	34 21	34 58	35 35	36 12	36 49	0	1	1	2	2	3	4	4	5	5
10	32 23	33 0	33 37	34 13	34 50	35 27	36 4	36 41	6	7	7	8	9	9	10	10	11	12
20	32 16	32 53	33 29	34 6	34 43	35 19	35 56	36 33	12	13	13	14	15	15	16	16	17	18
30	32 9	32 45	33 22	33 58	34 35	35 11	35 48	36 24	18	19	19	20	21	21	22	23	23	24
40	32 1	32 38	33 14	33 51	34 27	35 3	35 40	36 16	24	25	26	26	27	27	28	29	29	30
50	31 54	32 30	33 7	33 43	34 19	34 55	35 32	36 8	30	31	32	32	33	33	34	35	35	36
53 0	31 47	32 23	32 59	33 35	34 11	34 48	35 24	36 0	0	1	1	2	2	3	4	4	5	5
10	31 40	32 16	32 52	33 28	34 4	34 40	35 15	35 51	6	6	7	8	8	9	10	10	11	11
20	31 32	32 8	32 44	33 20	33 56	34 32	35 7	35 43	12	12	13	14	14	15	15	16	17	17
30	31 25	32 1	32 36	33 12	33 48	34 24	34 59	35 35	18	18	19	20	20	21	21	22	23	23
40	31 18	31 53	32 29	33 4	33 40	34 15	34 51	35 27	24	24	25	26	26	27	27	28	29	29
50	31 10	31 45	32 21	32 57	33 32	34 7	34 43	35 18	30	30	31	32	32	33	33	34	35	35
54 0	31 3	31 38	32 13	32 49	33 24	33 59	34 35	35 10	0	1	1	2	2	3	3	4	5	5
10	30 56	31 31	32 6	32 41	33 16	33 51	34 26	35 2	6	6	7	8	8	9	9	10	10	11
20	30 48	31 23	31 58	32 33	33 8	33 43	34 18	34 53	12	12	13	13	14	15	15	16	16	17
30	30 41	31 16	31 51	32 25	33 0	33 35	34 10	34 45	17	18	19	19	20	20	21	21	22	23
40	30 33	31 8	31 43	32 18	32 52	33 27	34 2	34 36	23	24	24	25	26	26	27	27	28	28
50	30 26	31 1	31 35	32 10	32 44	33 19	33 53	34 28	29	30	30	31	31	32	32	33	34	34
55 0	30 19	30 53	31 27	32 2	32 36	33 11	33 45	34 19	0	1	1	2	2	3	3	4	5	5
10	30 11	30 45	31 20	31 54	32 28	33 2	33 37	34 11	6	6	7	7	8	8	9	10	11	11
20	30 4	30 38	31 12	31 46	32 20	32 54	33 28	34 2	11	12	12	13	14	14	15	15	16	16
30	29 56	30 30	31 4	31 38	32 12	32 46	33 20	33 54	17	18	18	19	19	20	20	21	22	22
40	29 48	30 21	30 55	31 29	32 3	32 37	33 11	33 45	23	23	24	24	25	25	26	27	27	28
50	29 41	30 14	30 47	31 21	31 55	32 29	33 2	33 36	28	29	29	30	31	31	32	32	33	33
56 0	29 33	30 7	30 40	31 14	31 48	32 21	32 54	33 28	0	1	1	2	2	3	3	4	4	5
10	29 26	29 59	30 33	31 6	31 39	32 13	32 46	33 20	6	6	7	7	8	8	9	9	10	10
20	29 18	29 51	30 25	30 58	31 31	32 4	32 38	33 11	11	12	12	13	13	14	14	15	15	16
30	29 11	29 44	30 17	30 50	31 23	31 56	32 29	33 2	17	17	18	18	19	19	20	20	21	22
40	29 3	29 36	30 9	30 42	31 15	31 48	32 21	32 54	22	23	23	24	24	25	25	26	26	27
50	28 55	29 28	30 1	30 34	31 7	31 39	32 12	32 45	28	28	29	29	30	30	31	31	32	33
57 0	28 48	29 20	29 53	30 26	30 58	31 31	32 4	32 36	0	1	1	2	2	3	3	4	4	5
10	28 40	29 12	29 45	30 18	30 50	31 22	31 55	32 28	5	6	6	7	8	8	9	9	10	10
20	28 32	29 5	29 37	30 9	30 42	31 14	31 47	32 19	11	11	12	12	13	13	14	14	15	16
30	28 24	28 57	29 29	30 1	30 33	31 6	31 38	32 10	16	17	17	18	18	19	19	20	20	21
40	28 17	28 49	29 21	29 53	30 25	30 57	31 29	32 1	21	22	23	23	24	24	25	25	26	26
50	28 9	28 41	29 13	29 45	30 17	30 49	31 21	31 53	27	27	28	28	29	30	30	31	31	32
58 0	28 1	28 33	29 5	29 37	30 9	30 40	31 12	31 44	0	1	1	2	2	3	3	4	4	5
10	27 53	28 25	28 57	29 29	30 0	30 32	31 3	31 35	5	6	6	7	7	8	8	9	9	10
20	27 46	28 17	28 49	29 20	29 52	30 23	30 55	31 26	10	11	11	12	13	13	14	14	15	15
30	27 38	28 9	28 41	29 12	29 43	30 15	30 46	31 17	16	16	17	17	18	18	19	19	20	20
40	27 30	28 1	28 33	29 4	29 35	30 6	30 37	31 9	21	21	22	22	23	23	24	25	25	26
50	27 22	27 53	28 24	28 56	29 27	29 58	30 29	31 0	26	27	27	28	28	29	29	30	30	31
59 0	27 15	27 45	28 16	28 47	29 18	29 49	30 20	30 51	0	1	1	2	2	3	3	4	4	5
10	27 7	27 37	28 8	28 39	29 10	29 41	30 11	30 42	5	6	6	7	7	8	8	9	9	10
20	26 59	27 30	28 0	28 31	29 1	29 32	30 2	30 33	10	11	11	12	12	13	13	14	14	15
30	26 51	27 21	27 52	28 22	28 53	29 23	29 54	30 24	15	16	16	17	17	18	18	19	19	20
40	26 43	27 13	27 44	28 14	28 44	29 14	29 45	30 15	20	21	21	22	22	23	23	24	24	25
50	26 35	27 5	27 35	28 6	28 36	29 6	29 36	30 6	25	26	26	26	27	28	28	29	29	30

TABLE IX. Correction of the Moon's apparent Altitude.

Arg. app. altit. ° ′	Arg. Horizontal Parallax. 54′ ′ ″	55′ ′ ″	56′ ′ ″	57′ ′ ″	58′ ′ ″	59′ ′ ″	60′ ′ ″	61′ ′ ″	Prop. part for seconds of par. 0″	1″	2″	3″	4″	5″	6″	7″	8″	9″
60 0	26 27	26 57	27 27	27 57	28 27	28 57	29 27	29 57	0	0	1	1	2	2	3	3	4	4
10	26 19	26 49	27 19	27 49	28 19	28 48	29 18	29 48	5	5	6	6	7	7	8	8	9	9
20	26 11	26 41	27 11	27 40	28 20	28 40	29 9	29 39	10	10	11	11	12	12	13	13	14	14
30	26 3	26 33	27 2	27 32	28 1	28 31	29 1	29 30	15	15	16	16	17	17	18	18	19	19
40	25 55	26 25	26 54	27 23	27 53	28 22	28 52	29 21	20	20	21	21	22	22	23	23	24	24
50	25 47	26 16	26 46	27 15	27 44	28 13	28 43	29 12	25	25	26	26	27	27	28	28	29	29
61 0	25 39	26 8	26 37	27 6	27 36	28 5	28 34	29 3	0	0	1	1	2	2	3	3	4	4
10	25 31	26 0	26 29	26 58	27 27	27 56	28 25	28 54	5	5	6	6	7	7	8	8	9	9
20	25 23	25 52	26 21	26 49	27 18	27 47	28 16	28 45	10	10	10	11	11	12	12	13	13	14
30	25 15	25 44	26 12	26 41	27 10	27 38	28 7	28 33	14	15	15	16	16	17	17	18	18	19
40	25 7	25 35	26 4	26 32	27 1	27 29	27 58	28 26	19	20	20	21	21	21	22	22	23	23
50	24 59	25 27	25 56	26 24	26 52	27 21	27 49	28 17	24	24	25	25	26	26	27	27	28	28
62 0	24 51	25 19	25 47	26 15	26 43	27 12	27 40	28 8	0	0	1	1	2	2	3	3	4	4
10	24 43	25 11	25 39	26 7	26 35	27 3	27 31	27 59	5	5	6	6	6	7	7	8	8	9
20	24 35	25 2	25 30	25 58	26 26	26 54	27 22	27 50	9	10	10	11	11	12	12	12	13	13
30	24 26	24 54	25 22	25 50	26 17	26 45	27 13	27 40	14	14	15	15	16	16	17	17	18	18
40	24 18	24 46	25 13	25 41	26 8	26 36	27 4	27 31	18	19	19	20	20	21	21	22	22	23
50	24 10	24 38	25 5	25 32	26 0	26 27	26 54	27 22	23	24	24	24	25	25	26	26	27	27
63 0	24 2	24 29	24 56	25 24	25 51	26 18	26 45	27 13	0	0	1	1	2	2	3	3	4	4
10	23 54	24 21	24 48	25 15	25 42	26 9	26 36	27 3	4	5	5	6	6	7	7	8	8	8
20	23 45	24 12	24 39	25 6	25 33	26 0	26 27	26 54	9	9	10	10	11	11	12	12	13	13
30	23 37	24 4	24 31	24 58	25 24	25 51	26 18	26 45	13	14	14	15	15	16	16	17	17	17
40	23 29	23 56	24 21	24 49	25 15	25 42	26 9	26 35	18	18	19	19	20	20	21	21	21	22
50	23 21	23 47	24 14	24 40	25 7	25 33	26 0	26 26	22	23	23	24	24	25	25	25	26	26
64 0	23 12	23 39	24 5	24 31	24 58	25 24	25 50	26 17	0	0	1	1	2	2	3	3	3	4
10	23 4	23 30	23 57	24 23	24 49	25 15	25 41	26 7	4	5	5	6	6	6	7	7	8	8
20	22 56	23 22	23 48	24 14	24 40	25 6	25 32	25 58	9	9	9	10	10	11	11	12	12	12
30	22 48	23 14	23 39	24 5	24 31	24 57	25 23	25 48	13	13	14	14	15	15	15	16	16	17
40	22 39	23 5	23 31	23 56	24 22	24 48	25 13	25 39	17	18	18	18	19	19	20	20	21	21
50	22 31	22 57	23 22	23 48	24 13	24 39	25 4	25 30	22	22	22	23	23	24	24	25	25	25
65 0	22 23	22 48	23 13	23 39	24 4	24 30	24 55	25 20	0	0	1	1	2	2	2	3	3	4
10	22 14	22 40	23 5	23 30	23 55	24 20	24 46	25 11	4	5	5	5	6	6	7	7	7	8
20	22 6	22 31	22 56	23 21	23 46	24 11	24 36	25 1	8	9	9	10	10	10	11	11	12	12
30	21 58	22 23	22 47	23 12	23 37	24 2	24 27	24 52	12	13	13	14	14	14	15	15	16	16
40	21 49	22 14	22 39	23 3	23 28	23 53	24 18	24 42	17	17	17	18	18	19	19	19	20	20
50	21 41	22 5	22 30	22 55	23 19	23 44	24 8	24 33	21	21	22	22	22	23	23	24	24	24
66 0	21 32	21 57	22 21	22 46	23 10	23 34	23 59	24 23	0	0	1	1	2	2	2	3	3	4
10	21 24	21 48	22 13	22 37	23 1	23 25	23 50	24 14	4	4	5	5	6	6	6	7	7	8
20	21 16	21 40	22 4	22 28	22 52	23 16	23 40	24 4	8	8	9	9	10	10	10	11	11	12
30	21 7	21 31	21 55	22 19	22 43	23 7	23 31	23 55	12	12	13	13	14	14	14	15	15	16
40	20 59	21 23	21 46	22 10	22 34	22 58	23 21	23 45	16	16	17	16	18	18	18	19	19	20
50	20 50	21 14	21 37	22 1	22 25	22 48	23 12	23 35	20	20	21	21	21	22	22	23	23	23
67 0	20 42	21 5	21 29	21 52	22 16	22 39	23 2	23 26	0	0	1	1	2	2	2	3	3	3
10	20 33	20 57	21 20	21 43	22 6	22 30	22 53	23 16	4	4	5	5	5	6	6	6	7	7
20	20 25	20 48	21 11	21 34	21 57	22 20	22 44	23 7	8	8	8	9	9	10	10	10	11	11
30	20 16	20 39	21 2	21 25	21 48	22 11	22 34	22 57	11	12	12	13	13	13	14	14	14	15
40	20 8	20 31	20 53	21 16	21 39	22 2	22 25	22 47	15	16	16	16	17	17	18	18	18	19
50	19 59	20 22	20 45	21 7	21 30	21 52	22 15	22 38	19	19	20	20	21	21	21	22	22	22
68 0	19 51	20 13	20 36	20 58	21 21	21 43	22 6	22 28	0	0	1	1	1	2	2	3	3	3
10	19 42	20 4	20 27	20 49	21 11	21 34	21 56	22 18	4	4	4	5	5	5	6	6	7	7
20	19 34	19 56	20 18	20 40	21 2	21 24	21 46	22 9	7	8	8	8	9	9	10	10	10	11
30	19 25	19 47	20 9	20 31	20 53	21 15	21 37	21 59	11	11	12	12	12	13	13	14	14	14
40	19 16	19 38	20 0	20 22	20 44	21 6	21 27	21 49	15	15	15	16	16	16	17	17	18	18
50	19 8	19 29	19 51	20 13	20 34	20 56	21 18	21 40	18	19	19	19	20	20	20	21	21	22
69 0	18 59	19 21	19 42	20 4	20 25	20 47	21 8	21 30	0	0	1	1	1	2	2	2	3	3
10	18 51	19 12	19 33	19 55	20 16	20 37	20 59	21 20	4	4	4	5	5	5	6	6	6	7
20	18 42	19 3	19 24	19 46	20 7	20 28	20 49	21 10	7	7	8	8	8	9	9	9	10	10
30	18 33	18 54	19 15	19 36	19 57	20 18	20 39	21 0	11	11	11	12	12	12	13	13	13	14
40	18 25	18 46	19 6	19 27	19 48	20 9	20 30	20 51	14	14	15	15	15	16	16	16	17	17
50	18 16	18 37	18 57	19 18	19 39	20 0	20 20	20 41	18	18	18	19	19	19	20	20	20	21

TABLE IX. Correction of the Moon's apparent Altitude.

Arg. app. alt.	Arg. Horizontal Parallax.								Prop. part for seconds of par.									
	54′	55′	56′	57′	58′	59′	60′	61′	0″	1″	2″	3″	4″	5″	6″	7″	8″	9″
° ′	′ ″	′ ″	′ ″	′ ″	′ ″	′ ″	′ ″	′ ″										
70 0	18 7	18 28	18 49	19 9	19 30	19 50	20 11	20 31	0	0	1	1	1	2	2	2	3	3
10	17 57	18 19	18 39	19 0	19 20	19 41	20 1	20 21	3	4	4	4	5	5	5	6	6	6
20	17 50	18 10	18 30	18 51	19 11	19 31	19 51	20 11	7	7	7	8	8	8	9	9	9	10
30	17 41	18 1	18 21	18 41	19 1	19 22	19 42	20 2	10	10	11	11	11	12	12	12	13	13
40	17 31	17 53	18 12	18 32	18 52	19 12	19 32	19 52	13	14	14	14	15	[illegible]	15	16	16	16
50	17 24	17 43	18 3	18 23	18 43	19 2	19 22	19 42	17	17	17	18	18	18	19	19	19	20
71 0	17 15	17 33	17 54	18 14	18 33	18 53	19 12	19 32	0	0	1	1	1	2	2	2	3	3
10	17 6	17 26	17 45	18 5	18 24	18 43	19 3	19 22	3	3	4	4	4	5	5	5	6	6
20	16 58	17 17	17 36	17 55	18 15	18 34	18 53	19 12	6	7	7	7	8	8	8	9	9	9
30	16 49	17 8	17 27	17 46	18 5	18 24	18 43	19 2	10	10	10	10	11	11	11	12	12	12
40	16 40	16 59	17 18	17 37	17 56	18 15	18 33	18 52	13	13	13	14	14	14	15	15	15	16
50	16 32	16 50	17 9	17 28	17 46	18 5	18 24	18 42	16	16	16	17	17	17	18	18	18	19
72 0	16 23	16 41	17 0	17 18	17 37	17 55	18 14	18 32	0	0	1	1	1	2	2	2	2	3
10	16 14	16 32	16 51	17 9	17 27	17 46	18 4	18 23	3	3	4	4	4	5	5	5	5	6
20	16 5	16 23	16 42	17 0	17 18	17 36	17 54	18 13	6	6	7	7	7	8	8	8	8	9
30	15 56	16 14	16 32	16 50	17 9	17 27	17 45	18 3	9	9	10	10	10	11	11	11	11	12
40	15 47	16 5	16 23	16 41	16 59	17 17	17 35	17 53	12	12	13	13	13	14	14	14	14	15
50	15 39	15 56	16 14	16 32	16 50	17 7	17 25	17 43	15	15	16	16	16	17	17	17	17	18
73 0	15 30	15 47	16 5	16 22	16 40	16 58	17 15	17 33	0	0	1	1	1	1	2	2	2	3
10	15 21	15 38	15 56	16 13	16 31	16 48	17 5	17 23	3	3	3	4	4	4	5	5	5	5
20	15 12	15 29	15 47	16 4	16 21	16 38	16 55	17 13	6	6	6	7	7	7	7	8	8	8
30	15 3	15 20	15 37	15 54	16 11	16 29	16 46	17 3	9	9	9	9	10	10	10	11	11	11
40	14 55	15 11	15 28	15 45	16 2	16 19	16 36	16 53	11	12	12	12	12	13	13	13	14	14
50	14 46	15 2	15 19	15 36	15 52	16 9	16 26	16 43	14	14	15	15	15	16	16	16	16	17
74 0	14 37	14 53	15 10	15 26	15 43	15 59	16 16	16 32	0	0	1	1	1	1	2	2	2	2
10	14 28	14 44	15 0	15 17	15 33	15 50	16 6	16 22	3	3	3	3	4	4	4	5	5	5
20	14 19	14 35	14 51	15 8	15 24	15 40	15 56	16 12	5	6	6	6	6	7	7	7	7	8
30	14 10	14 26	14 42	14 58	15 14	15 31	15 46	16 2	8	8	9	9	9	9	10	10	10	10
40	14 1	14 17	14 33	14 49	15 5	15 21	15 36	15 52	11	11	11	11	12	12	12	13	13	13
50	13 52	14 8	14 24	14 39	14 55	15 11	15 26	15 42	13	14	14	14	14	15	15	15	15	16
75 0	13 43	13 59	14 14	14 30	14 45	15 1	15 16	15 32	0	0	1	1	1	1	2	2	2	2
10	13 34	13 50	14 5	14 20	14 36	14 51	15 7	15 22	3	3	3	3	4	4	4	4	5	5
20	13 25	13 41	13 56	14 11	14 26	14 41	14 57	15 12	5	5	6	6	6	6	7	7	7	7
30	13 16	13 32	13 47	14 2	14 17	14 32	14 47	15 2	8	8	8	8	9	9	9	9	10	10
40	13 8	13 22	13 37	13 52	14 7	14 22	14 37	14 52	10	10	11	11	11	11	12	12	12	12
50	12 59	13 13	13 28	13 43	13 57	14 12	14 27	14 41	13	13	13	13	14	14	14	14	15	15
76 0	12 49	13 4	13 19	13 33	13 48	14 2	14 17	14 31	0	0	0	1	1	1	1	2	2	2
10	12 41	12 55	13 9	13 24	13 38	13 52	14 7	14 21	2	3	3	3	3	3	4	4	4	4
20	12 32	12 46	13 0	13 14	13 28	13 43	13 57	14 11	5	5	5	5	6	6	6	6	7	7
30	12 23	12 37	12 51	13 5	13 19	13 33	13 47	14 1	7	7	7	8	8	8	8	9	9	9
40	12 13	12 27	12 41	12 55	13 9	13 23	13 36	13 51	9	10	10	10	10	10	11	11	11	11
50	12 4	12 18	12 32	12 45	12 59	13 13	13 26	13 41	12	12	12	12	13	13	13	13	14	14
77 0	11 56	12 9	12 23	12 36	12 50	13 3	13 16	13 30	0	0	0	1	1	1	1	2	2	2
10	11 47	11 59	12 13	12 26	12 40	12 53	13 6	13 20	2	2	3	3	3	3	4	4	4	4
20	11 38	11 50	12 4	12 17	12 30	12 43	12 56	13 10	4	5	5	5	5	6	6	6	6	6
30	11 29	11 41	11 54	12 7	12 21	12 34	12 46	13 0	7	7	7	7	7	8	8	8	8	9
40	11 20	11 32	11 45	11 58	12 11	12 24	12 36	12 50	9	9	9	9	10	10	10	10	10	11
50	11 11	11 23	11 36	11 48	12 1	12 14	12 26	12 39	11	11	11	12	12	12	12	12	13	13
78 0	11 1	11 14	11 26	11 39	11 51	12 4	12 16	12 29	0	0	0	1	1	1	1	1	1	[illegible]
10	10 52	11 5	11 17	11 29	11 42	11 54	12 6	12 19	2	2	2	3	3	3	3	3	4	4
20	10 43	10 55	11 8	11 20	11 32	11 44	11 56	12 [illegible]	4	4	4	5	5	5	5	5	6	6
30	10 34	10 46	10 58	11 10	11 22	11 34	11 46	11 5[illegible]	6	6	6	7	7	7	7	7	8	8
40	10 25	10 37	10 49	11 1	11 12	11 24	11 36	11 48	8	8	8	9	9	9	9	9	10	10
50	10 16	10 28	10 39	10 51	11 3	11 14	11 26	11 38	10	10	10	11	11	11	11	11	12	12
79 0	10 7	10 19	10 30	10 41	10 53	11 4	11 16	11 27	0	0	0	1	1	1	1	1	1	2
10	9 58	10 9	10 21	10 32	10 43	10 54	11 6	11 17	2	2	2	2	3	3	3	3	3	3
20	9 49	10 0	10 11	10 22	10 33	10 44	10 5[illegible]	11 7	4	4	4	4	4	5	5	5	5	5
30	9 40	9 51	10 2	10 13	10 24	10 35	10 5[illegible]	10 5[illegible]	5	6	6	6	6	6	7	7	7	7
40	9 31	9 42	9 52	10 [illegible]	10 14	10 25	10 3	10 4[illegible]	7	7	8	8	8	8	8	9	9	9
50	9 27	9 32	9 43	9 53	10 4	10 15	10 2	10 3[illegible]	9	9	9	10	10	10	10	10	11	11

TABLE IX. Correction of the Moon's apparent Altitude.

Arg. app. altit.	Arg. Horizontal Parallax.								Prop. part for seconds of par.									
	54′	55′	56′	57′	58′	59′	60′	61′	0″	1″	2″	3″	4″	5″	6″	7″	8″	9″
° ′	′ ″	′ ″	′ ″	′ ″	′ ″	′ ″	′ ″	′ ″										
80 0	9 13	9 23	9 33	9 44	9 54	10 5	10 15	10 26	0	0	0	0	1	1	1	1	1	1
10	9 3	9 14	9 24	9 34	9 44	9 55	10 5	10 15	2	2	2	2	2	2	3	3	3	3
20	8 54	9 4	9 14	9 25	9 35	9 45	9 55	10 5	3	3	4	4	4	4	4	4	5	5
30	8 45	8 55	9 5	9 15	9 25	9 35	9 45	9 55	5	5	5	5	6	6	6	6	6	6
40	8 36	8 46	8 56	9 5	9 15	9 25	9 34	9 44	7	7	7	7	7	7	8	8	8	8
50	8 27	8 37	8 46	8 56	9 5	9 15	9 24	9 34	8	8	9	9	9	9	9	9	10	10
81 0	8 18	8 27	8 37	8 46	8 55	9 5	9 14	9 24	0	0	0	0	1	1	1	1	1	1
10	8 9	8 18	8 27	8 36	8 46	8 55	9 4	9 13	1	2	2	2	2	2	2	3	3	3
20	8 0	8 9	8 18	8 27	8 36	8 45	8 54	9 3	3	3	3	3	4	4	4	4	4	4
30	7 50	7 59	8 8	8 17	8 26	8 35	8 44	8 52	4	5	5	5	5	5	5	5	6	6
40	7 41	7 50	7 59	8 7	8 16	8 25	8 33	8 42	6	6	6	6	7	7	7	7	7	7
50	7 32	7 41	7 49	7 58	8 6	8 15	8 23	8 32	7	8	8	8	8	8	8	8	9	9
82 0	7 23	7 31	7 40	7 48	7 56	8 5	8 13	8 21	0	0	0	0	1	1	1	1	1	1
10	7 14	7 22	7 30	7 38	7 46	7 55	8 3	8 11	1	1	2	2	2	2	2	2	2	2
20	7 5	7 13	7 21	7 29	7 37	7 45	7 53	8 1	3	3	3	3	3	3	3	4	4	4
30	6 55	7 3	7 11	7 19	7 27	7 35	7 42	7 50	4	4	4	4	4	5	5	5	5	5
40	6 46	6 54	7 2	7 9	7 17	7 25	7 32	7 40	5	5	5	6	6	6	6	6	6	6
50	6 37	6 45	6 52	6 59	7 7	7 14	7 22	7 29	7	7	7	7	7	7	7	7	8	8
83 0	6 28	6 35	6 42	6 50	6 57	7 4	7 12	7 19	0	0	0	0	0	1	1	1	1	1
10	6 19	6 26	6 33	6 40	6 40	6 54	7 2	7 9	1	1	1	1	2	2	2	2	2	2
20	6 10	6 16	6 23	6 30	6 37	6 44	6 51	6 58	2	2	2	3	3	3	3	3	3	3
30	6 0	6 7	6 14	6 21	6 27	6 34	6 41	6 48	3	4	4	4	4	4	4	4	4	4
40	5 51	5 58	6 4	6 11	6 18	6 24	6 31	6 37	5	5	5	5	5	5	5	5	5	6
50	5 42	5 49	5 55	6 1	6 8	6 14	6 21	6 27	6	6	6	6	6	6	6	6	7	7
84 0	5 32	5 39	5 45	5 52	5 58	6 4	6 10	6 17	0	0	0	0	0	0	1	1	1	1
10	5 23	5 30	5 36	5 42	5 48	5 54	6 0	6 6	1	1	1	1	1	1	2	2	2	2
20	5 14	5 20	5 26	5 32	5 38	5 44	5 50	5 56	2	2	2	2	2	2	2	3	3	3
30	5 5	5 11	5 17	5 22	5 28	5 34	5 40	5 45	3	3	3	3	3	3	3	4	4	4
40	4 56	5 1	5 7	5 13	5 18	5 24	5 29	5 35	4	4	4	4	4	4	4	5	5	5
50	4 47	4 52	4 57	5 3	5 8	5 14	5 19	5 24	5	5	5	5	5	5	5	5	6	6
85 0	4 37	4 43	4 48	4 53	4 58	5 4	5 9	5 14	0	0	0	0	0	0	0	1	1	1
10	4 28	4 33	4 38	4 43	4 48	4 53	4 59	5 4	1	1	1	1	1	1	1	1	1	1
20	4 19	4 24	4 29	4 34	4 38	4 43	4 48	4 53	2	2	2	2	2	2	2	2	2	2
30	4 10	4 14	4 19	4 24	4 29	4 33	4 38	4 43	2	2	2	3	3	3	3	3	3	3
40	4 1	4 5	4 10	4 14	4 19	4 23	4 28	4 32	3	3	3	3	3	4	4	4	4	4
50	3 51	3 56	4 0	4 4	4 9	4 13	4 17	4 22	4	4	4	4	4	4	4	4	5	5
86 0	3 42	3 46	3 50	3 55	3 59	4 3	4 7	4 11	0	0	0	0	0	0	0	0	0	1
10	3 33	3 37	3 41	3 45	3 49	3 53	3 57	4 1	1	1	1	1	1	1	1	1	1	1
20	3 24	3 27	3 31	3 35	3 39	3 43	3 47	3 50	1	1	1	1	1	2	2	2	2	2
30	3 14	3 18	3 22	3 25	3 29	3 33	3 36	3 40	2	2	2	2	2	2	2	2	2	2
40	3 5	3 9	3 12	3 16	3 19	3 23	3 26	3 29	2	3	3	3	3	3	3	3	3	3
50	2 56	2 59	3 2	3 6	3 9	3 12	3 16	3 19	3	3	3	3	3	3	3	3	4	4
87 0	2 47	2 50	2 53	2 56	2 59	3 2	3 5	3 9	0	0	0	0	0	0	0	0	0	0
10	2 37	2 40	2 43	2 46	2 49	2 52	2 55	2 58	0	0	1	1	1	1	1	1	1	1
20	2 28	2 31	2 34	2 36	2 39	2 42	2 45	2 48	1	1	1	1	1	1	1	1	1	1
30	2 19	2 21	2 24	2 27	2 29	2 32	2 35	2 37	1	1	1	1	1	2	2	2	2	2
40	2 10	2 12	2 14	2 17	2 19	2 22	2 24	2 27	2	2	2	2	2	2	2	2	2	2
50	2 0	2 3	2 5	2 7	2 9	2 12	2 14	2 16	2	2	2	2	2	2	2	2	3	3
88 0	1 51	1 53	1 55	1 57	1 59	2 2	2 4	2 6	0	0	0	0	0	0	0	0	0	0
10	1 42	1 44	1 46	1 48	1 50	1 51	1 53	1 55	0	0	0	0	0	0	0	0	0	0
20	1 33	1 34	1 36	1 38	1 40	1 41	1 43	1 45	1	1	1	1	1	1	1	1	1	1
30	1 23	1 25	1 26	1 28	1 30	1 31	1 33	1 34	1	1	1	1	1	1	1	1	1	1
40	1 14	1 15	1 17	1 18	1 20	1 21	1 22	1 24	1	1	1	1	1	1	1	1	1	1
50	1 5	1 6	1 7	1 8	1 10	1 11	1 12	1 13	1	1	1	1	1	1	1	1	2	2
89 0	0 56	0 57	0 58	0 59	1 [illegible]	1 1	1 2	1 3	0	0	0	0	0	0	0	0	0	0
10	0 46	0 47	0 48	0 49	0 50	0 51	0 52	0 52	0	0	0	0	0	0	0	0	0	0
20	0 37	0 38	0 38	0 39	0 40	0 41	0 41	0 42	0	0	0	0	0	0	0	0	0	0
30	0 28	0 28	0 29	0 29	0 30	0 30	0 31	0 32	0	0	0	0	0	0	0	0	0	0
40	0 19	0 19	0 19	0 20	0 20	0 20	0 21	0 21	0	0	0	0	0	0	0	0	0	0
50	0 9	0 9	0 10	0 10	0 10	0 10	0 10	0 10	1	1	1	1	1	1	1	1	1	1

Tables for reducing the true Altitudes of the Sun, Moon, and Stars, to their apparent Altitudes.

TAB. XI. Reduction of the Moon's Altitude.

Arg. Moon's Horizontal Parallax.

True altitude.		TAB. X. Reduction of the Sun's alt.		53′		54′		55′		56′		57′		58′		59′		60′		61′		62′		TAB. XII. Reduction of a Star's altitude.	
°	′	′	″	′	″	′	″	′	″	′	″	′	″	′	″	′	″	′	″	′	″	′	″	′	″
1	0	21	53	25	15	26	8	27	1	27	54	28	47	29	40	30	32	31	25	32	18	33	10	22	2
1	10	20	59	26	28	27	21	28	13	29	6	29	59	30	52	31	45	32	38	33	31	34	24	21	8
1	20	20	7	27	36	28	29	29	23	30	16	31	9	32	3	32	56	33	49	34	42	35	35	20	16
1	30	19	19	28	40	29	34	30	27	31	21	32	15	33	9	34	2	34	56	35	50	36	43	19	28
1	40	18	33	29	42	30	36	31	29	32	23	33	17	34	11	35	5	35	59	36	53	37	46	18	42
1	50	17	49	30	40	31	34	32	28	33	22	34	16	35	11	36	5	36	59	37	53	38	47	17	58
2	0	17	7	31	35	32	29	33	24	34	19	35	13	36	8	37	3	37	57	38	52	39	46	17	16
2	10	16	28	32	27	33	22	34	17	35	12	36	7	37	1	37	56	38	51	39	46	40	41	16	37
2	20	15	51	33	17	34	12	35	7	36	2	36	57	37	53	38	48	39	43	40	38	41	33	16	0
2	30	15	16	34	3	34	59	35	54	36	49	37	45	38	40	39	36	40	31	41	26	42	22	15	25
2	40	14	43	34	48	35	43	36	39	37	35	38	30	39	25	40	20	41	16	42	12	43	7	14	52
2	50	14	12	35	29	36	25	37	21	38	17	39	13	40	8	41	4	42	0	42	56	43	52	14	21
3	0	13	43	36	9	37	5	38	1	38	58	39	54	40	50	41	46	42	42	43	38	44	34	13	52
3	20	12	42	37	22	38	19	39	15	40	12	41	8	42	4	42	59	43	56	44	53	45	49	12	58
3	40	12	1	38	26	39	23	40	19	41	16	42	13	43	10	44	7	45	4	46	1	46	58	12	10
4	0	11	17	39	23	40	20	41	17	42	14	43	11	44	9	45	6	46	3	47	0	47	57	11	26
4	20	10	38	40	13	41	11	42	8	43	6	44	3	45	1	45	58	46	55	47	53	48	50	10	47
4	40	10	2	40	58	41	56	42	54	43	52	44	50	45	47	46	44	47	42	48	40	49	38	10	11
5	0	9	29	41	40	42	38	43	36	44	34	45	32	46	30	47	28	48	26	49	23	50	21	9	38
5	30	8	46	42	33	43	31	44	29	45	27	46	25	47	23	48	22	49	20	50	18	51	16	8	55
6	0	8	8	43	18	44	17	45	15	46	14	47	12	48	11	49	9	50	7	51	5	52	4	8	17
6	30	7	33	43	57	44	56	45	54	46	53	47	51	48	50	49	49	50	46	51	45	52	44	7	42
7	0	7	4	44	32	45	30	46	29	47	28	48	26	49	25	50	24	51	22	52	20	53	19	7	13
7	30	6	38	45	2	46	1	47	0	47	59	48	57	49	56	50	54	51	53	52	42	53	51	6	47
8	0	6	15	45	28	46	27	47	26	48	24	49	23	50	22	51	21	52	20	53	19	54	17	6	24
8	30	5	55	45	49	46	48	47	47	48	46	49	45	50	43	51	42	52	41	53	40	54	39	6	4
9	0	5	36	46	8	47	7	48	6	49	5	50	4	51	2	52	1	53	0	53	59	54	58	5	45
10		5	3	46	38	47	37	48	36	49	35	50	34	51	32	52	31	53	30	54	29	55	28	5	12
11		4	36	47	1	48	0	48	59	49	58	50	57	51	55	52	54	53	53	54	51	55	50	4	45
12		4	12	47	18	48	16	49	15	50	14	51	13	52	11	53	10	54	9	55	8	56	7	4	21
13		3	53	47	29	48	28	49	26	50	25	51	23	52	22	53	21	54	19	55	18	56	16	4	2
14		3	35	47	36	48	35	49	33	50	32	51	30	52	29	53	27	54	25	55	23	56	22	3	44
15		3	20	47	42	48	40	49	38	50	36	51	34	52	38	53	31	54	29	55	27	56	25	3	29
16		3	7	47	42	48	40	49	38	50	36	51	34	52	32	53	30	54	28	55	26	56	24	3	16
17		2	55	47	39	48	37	49	34	50	32	51	30	52	27	53	25	54	23	55	21	56	19	3	4
18		2	45	47	36	48	33	49	31	50	28	51	26	52	23	53	20	54	18	55	15	56	12	2	54
19		2	35	47	29	48	26	49	23	50	21	51	18	52	15	53	12	54	9	55	6	56	3	2	44
20		2	27	47	21	48	18	49	14	50	11	51	8	52	5	53	2	53	59	54	56	55	52	2	35
21		2	19	47	11	48	8	49	4	50	0	50	57	51	53	52	50	53	47	54	43	55	40	2	27
22		2	12	46	59	47	55	48	51	49	47	50	44	51	40	52	36	53	32	54	28	55	25	2	20
23		2	6	46	45	47	41	48	37	49	32	50	28	51	24	52	20	53	16	54	11	55	7	2	14
24		2	0	46	30	47	26	48	21	49	16	50	11	51	7	52	3	52	58	53	53	54	49	2	7
25		1	54	46	14	47	9	48	4	48	59	49	54	50	49	51	44	52	39	53	35	54	29	2	1
26		1	48	45	56	46	51	47	45	48	40	49	35	50	29	51	24	52	18	53	13	54	8	1	56
27		1	43	45	37	46	33	47	26	48	20	49	14	50	8	51	3	51	57	52	51	53	45	1	51
28		1	39	45	17	46	11	47	4	47	58	48	52	49	46	50	39	51	33	52	27	53	21	1	47
29		1	34	44	55	45	49	46	42	47	35	48	28	49	22	50	15	51	8	52	1	52	55	1	42
30		1	30	44	33	45	26	46	19	47	12	48	4	48	57	49	50	50	43	51	35	52	28	1	38
31		1	27	44	9	45	2	45	54	46	46	47	38	48	31	49	53	50	15	51	7	51	59	1	35
32		1	23	43	44	44	36	45	28	46	19	47	11	48	2	48	55	49	46	50	38	51	30	1	31
33		1	20	43	18	44	10	45	1	45	51	46	43	47	34	48	25	49	17	50	8	50	59	1	28
34		1	17	42	51	43	42	44	33	45	23	46	14	47	5	47	55	48	46	49	37	50	27	1	24
35		1	14	42	24	43	14	44	4	44	54	45	40	46	34	47	24	48	14	49	4	49	54	1	21
36		1	11	41	55	42	44	43	34	44	23	45	33	46	2	46	52	47	41	48	30	49	20	1	18

Tables for reducing the true Altitudes of the Sun, Moon, and Stars, to their apparent Altitudes.

TAB. XI. Reduction of the Moon's Altitude.

Arg. Moon's Horizontal Parallax.

True Altitude.	TAB. X. Reduction of Sun's Altitude.		53′		54′		55′		56′		57′		58′		59′		60′		61′		62′		TABLE XII. Red. of a Star's Altitude.	
°	′	″	°	′	°	′	°	′	°	′	°	′	°	′	°	′	°	′	°	′	°	′	′	″
37	1	9	41	25	42	14	43	3	43	51	44	41	45	29	46	18	47	6	47	56	48	45	1	16
38	1	6	40	54	41	42	42	30	43	19	44	7	44	55	45	43	46	31	47	19	48	8	1	13
39	1	3	40	23	41	10	41	57	42	45	43	33	44	20	45	8	45	56	46	43	47	31	1	10
40	1	1	39	50	40	37	41	24	42	11	42	50	43	44	44	31	45	19	46	5	46	52	1	8
41		58	39	17	40	3	40	49	41	35	42	21	43	8	43	54	44	40	45	27	46	13	1	5
42		56	38	43	39	28	40	14	40	59	41	45	42	30	43	16	44	1	44	47	45	33	1	3
43		54	38	7	38	52	39	36	40	22	41	7	41	51	42	36	43	21	44	6	44	51	1	1
44		53	37	31	38	16	38	59	39	43	40	27	41	11	41	55	42	40	43	24	44	9		59
45		51	36	54	37	38	38	21	39	4	39	48	40	31	41	14	41	58	42	42	43	25		57
46		49	36	17	36	59	37	42	38	25	39	7	39	50	40	32	41	15	41	58	42	40		55
47		47	35	39	36	21	37	2	37	44	38	26	39	8	39	50	40	32	41	13	41	55		54
48		45	35	0	35	41	36	22	37	3	37	44	38	25	39	6	39	47	40	28	41	9		52
49		43	34	20	35	0	35	40	36	20	37	1	37	42	38	22	39	2	39	42	40	23		49
50		42	33	39	34	19	34	58	35	37	36	17	36	57	37	36	38	16	38	55	39	35		48
51		40	32	58	33	37	34	16	34	54	35	33	36	11	36	50	37	29	38	7	38	46		46
52		39	32	16	32	54	33	32	34	10	34	47	35	25	36	3	36	41	37	19	37	57		44
53		37	31	34	32	11	32	48	33	25	34	2	34	39	35	16	35	53	36	30	37	7		43
54		36	30	50	31	26	32	2	32	39	33	15	33	51	34	28	35	4	35	40	36	16		41
55		35	30	6	30	42	31	17	31	52	32	27	33	2	33	38	34	14	34	49	35	24		40
56		33	29	22	29	57	30	31	31	5	31	39	32	14	32	49	33	23	33	58	34	32		38
57		32	28	36	29	10	29	43	30	17	30	51	31	24	31	58	32	32	33	5	33	39		37
58		30	27	51	28	24	28	56	29	29	30	1	30	34	31	7	31	40	32	12	32	45		35
59		29	27	5	27	37	28	9	28	40	29	11	29	43	30	15	30	47	31	19	31	51		34
60		28	26	18	26	48	27	19	27	50	28	21	28	51	29	22	29	53	30	24	30	55		33
61		27	25	30	26	0	26	30	27	0	27	29	27	59	28	29	28	59	29	29	29	59		32
62		26	24	42	25	11	25	40	26	9	26	38	27	7	27	36	28	5	28	34	29	3		30
63		25	23	54	24	22	24	50	25	18	25	46	26	14	26	42	27	10	27	38	28	6		29
64		24	23	5	23	32	23	59	24	26	24	53	25	20	25	47	26	14	26	42	27	9		28
65		22	22	16	22	42	23	8	23	34	24	0	24	26	24	52	25	18	25	45	26	11		26
66		21	21	26	21	51	22	16	22	41	23	7	23	32	23	57	24	22	24	47	25	12		25
67		20	20	36	21	0	21	24	21	48	22	12	22	36	23	0	23	25	23	49	24	13		24
68		19	19	45	20	8	20	31	20	54	21	17	21	41	22	4	22	27	22	50	23	14		23
69		18	18	54	19	16	19	38	20	0	20	22	20	44	21	7	21	29	21	51	22	13		22
70		17	18	2	18	23	18	44	19	5	19	27	19	48	20	9	20	30	20	51	21	13		21
71		16	17	11	17	31	17	51	18	11	18	31	18	51	19	11	19	32	19	52	20	12		19
72		15	16	19	16	38	16	57	17	16	17	35	17	54	18	13	18	32	18	52	19	11		18
73		14	15	26	15	44	16	2	16	20	16	38	16	56	17	15	17	33	17	51	18	9		17
74		14	14	34	14	50	15	7	15	24	15	41	15	58	16	16	16	33	16	50	17	7		16
75		13	13	40	13	56	14	12	14	28	14	44	15	0	15	16	15	32	15	48	16	4		15
76		12	12	46	13	1	13	16	13	31	13	46	14	1	14	16	14	31	14	46	15	1		14
77		11	11	53	12	7	12	20	12	34	12	48	13	2	13	16	13	30	13	44	13	58		13
78		10	10	59	11	12	11	25	11	38	11	50	12	3	12	16	12	29	12	42	12	55		12
79		9	10	5	10	16	10	28	10	40	10	52	11	4	11	15	11	27	11	39	11	51		11
80		8	9	10	9	21	9	32	9	42	9	53	10	4	10	15	10	26	10	37	10	47		10
81		8	8	16	8	26	8	35	8	45	8	55	9	4	9	14	9	24	9	34	9	43		9
82		7	7	21	7	30	7	38	7	47	7	56	8	4	8	13	8	21	8	30	8	39		8
83		6	6	27	6	34	6	41	6	49	6	57	7	4	7	11	7	19	7	26	7	34		7
84		5	5	32	5	38	5	44	5	51	5	57	6	4	6	10	6	17	6	24	6	30		6
85		4	4	36	4	42	4	47	4	53	4	58	5	3	5	9	5	14	5	20	5	25		5
86		3	3	41	3	45	3	50	3	54	3	59	4	3	4	7	4	11	4	15	4	20		4
87		3	2	46	2	50	2	53	2	56	2	59	3	2	3	5	3	8	3	12	3	15		3
88		2	1	51	1	53	1	55	1	57	1	59	2	1	2	4	2	6	2	8	2	10		2
89		1	0	56	0	57	0	58	0	59	1	0	1	1	1	2	1	3	1	4	1	5		1
90		0	0	0	0	0	0	0	0	0	0	0	0	0	0	0	0	0	0	0	0	0		0

TABLE XIII. To reduce the Sun's declination to any given Meridian, and to any given time under that Meridian.

Arg. time fr. noon.		Arg. Sun's Declination North increasing; or between March and June.																Arg. long. of th.
		0°	2°	4°	6°	8°	10°	12°	14°	16°	18°	19°	20°	21°	22°	23°	23° 28'	
h.	'	'	'	'	'	'	'	'	'	'	'	'	'	'	'	'	'	°
0	0	0.0	0.0	0.0	0.0	0.0	0.0	0.0	0.0	0.0	0.0	0.0	0.0	0.0	0.0	0.0	0.0	0
0	12	0.2	0.2	0.2	0.2	0.2	0.2	0.2	0.2	0.1	0.1	0.1	0.1	0.1	0.1	0.0	0.0	3
0	24	0.4	0.4	0.4	0.4	0.4	0.4	0.3	0.3	0.3	0.3	0.2	0.2	0.2	0.1	0.1	0.0	6
0	36	0.6	0.6	0.6	0.6	0.6	0.5	0.5	0.5	0.4	0.4	0.4	0.3	0.3	0.2	0.1	0.0	9
0	48	0.8	0.8	0.8	0.8	0.7	0.7	0.7	0.6	0.6	0.5	0.5	0.4	0.4	0.3	0.2	0.0	12
1	0	1.0	1.0	1.0	0.9	0.9	0.9	0.8	0.8	0.7	0.6	0.6	0.5	0.4	0.3	0.2	0.0	15
1	12	1.2	1.2	1.2	1.1	1.1	1.1	1.0	0.9	0.9	0.8	0.7	0.6	0.5	0.4	0.2	0.0	18
1	24	1.4	1.4	1.4	1.3	1.3	1.2	1.2	1.1	1.0	0.9	0.8	0.7	0.6	0.5	0.3	0.0	21
1	36	1.6	1.6	1.6	1.5	1.5	1.4	1.3	1.3	1.2	1.0	0.9	0.8	0.7	0.6	0.3	0.0	24
1	48	1.8	1.8	1.7	1.7	1.7	1.6	1.5	1.4	1.3	1.1	1.1	0.9	0.8	0.6	0.4	0.0	27
2	0	2.0	2.0	1.9	1.9	1.8	1.8	1.7	1.6	1.4	1.3	1.2	1.0	0.9	0.7	0.4	0.0	30
2	12	2.2	2.2	2.1	2.1	2.0	1.9	1.8	1.7	1.6	1.4	1.3	1.1	1.0	0.8	0.4	0.0	33
2	24	2.4	2.3	2.3	2.3	2.2	2.1	2.0	1.9	1.7	1.5	1.4	1.2	1.1	0.8	0.5	0.0	36
2	36	2.6	2.5	2.5	2.5	2.4	2.3	2.2	2.1	1.9	1.7	1.5	1.4	1.2	0.9	0.5	0.0	39
2	48	2.8	2.7	2.7	2.7	2.6	2.5	2.4	2.2	2.0	1.8	1.6	1.5	1.2	1.0	0.6	0.0	42
3	0	3.0	2.9	2.9	2.8	2.8	2.7	2.5	2.4	2.2	1.9	1.8	1.6	1.3	1.0	0.6	0.0	45
3	12	3.2	3.1	3.1	3.0	2.9	2.8	2.7	2.5	2.3	2.0	1.9	1.7	1.4	1.1	0.6	0.0	48
3	24	3.4	3.3	3.3	3.2	3.1	3.0	2.9	2.7	2.5	2.2	2.0	1.8	1.5	1.2	0.7	0.0	51
3	36	3.6	3.5	3.5	3.4	3.3	3.2	3.0	2.8	2.6	2.3	2.1	1.9	1.6	1.3	0.7	0.0	54
3	48	3.8	3.7	3.7	3.6	3.5	3.4	3.2	3.0	2.7	2.4	2.2	2.0	1.7	1.3	0.8	0.0	57
4	0	4.0	3.9	3.9	3.8	3.7	3.5	3.4	3.2	2.9	2.5	2.3	2.1	1.8	1.4	0.8	0.0	60
4	12	4.1	4.1	4.1	4.0	3.9	3.7	3.5	3.3	3.0	2.7	2.5	2.2	1.9	1.5	0.8	0.0	63
4	24	4.3	4.3	4.3	4.2	4.1	3.9	3.7	3.5	3.2	2.8	2.6	2.3	2.0	1.5	0.9	0.0	66
4	36	4.5	4.5	4.5	4.4	4.2	4.1	3.9	3.6	3.3	2.9	2.7	2.4	2.1	1.6	0.9	0.0	69
4	48	4.7	4.7	4.7	4.6	4.4	4.3	4.0	3.8	3.5	3.1	2.8	2.5	2.1	1.7	1.0	0.0	72
5	0	4.9	4.9	4.8	4.7	4.6	4.4	4.2	4.0	3.6	3.2	2.9	2.6	2.2	1.7	1.0	0.0	75
5	12	5.1	5.1	5.0	4.9	4.8	4.6	4.4	4.1	3.8	3.3	3.0	2.7	2.3	1.8	1.0	0.0	78
5	24	5.3	5.3	5.2	5.1	5.0	4.8	4.5	4.3	3.9	3.4	3.2	2.8	2.4	1.9	1.1	0.0	81
5	36	5.5	5.5	5.4	5.3	5.2	5.0	4.7	4.4	4.0	3.6	3.3	2.9	2.5	2.0	1.1	0.0	84
5	48	5.7	5.7	5.6	5.5	5.4	5.1	4.9	4.6	4.2	3.7	3.4	3.0	2.6	2.1	1.2	0.0	87
6	0	5.9	5.9	5.8	5.7	5.5	5.3	5.0	4.7	4.3	3.8	3.5	3.1	2.7	2.1	1.2	0.0	90
6	12	6.1	6.1	6.0	5.9	5.7	5.5	5.2	4.9	4.5	3.9	3.6	3.2	2.8	2.2	1.2	0.0	93
6	24	6.3	6.3	6.2	6.1	5.9	5.7	5.4	5.1	4.6	4.1	3.7	3.3	2.9	2.2	1.3	0.0	96
6	36	6.5	6.5	6.4	6.3	6.1	5.9	5.6	5.2	4.8	4.2	3.9	3.4	2.9	2.3	1.3	0.0	99
6	48	6.7	6.7	6.6	6.5	6.3	6.0	5.7	5.4	4.9	4.3	4.0	3.5	3.0	2.4	1.4	0.0	102
7	0	6.9	6.9	6.8	6.6	6.5	6.2	5.9	5.5	5.0	4.5	4.1	3.6	3.1	2.4	1.4	0.0	105
7	12	7.1	7.1	7.0	6.8	6.6	6.4	6.1	5.7	5.2	4.6	4.2	3.7	3.2	2.5	1.4	0.0	108
7	24	7.3	7.2	7.2	7.0	6.8	6.6	6.3	5.9	5.3	4.7	4.3	3.9	3.3	2.6	1.5	0.0	111
7	36	7.5	7.4	7.4	7.2	7.0	6.7	6.4	6.0	5.5	4.8	4.4	4.0	3.4	2.6	1.5	0.0	114
7	48	7.7	7.6	7.6	7.5	7.2	6.9	6.6	6.2	5.6	5.0	4.6	4.1	3.5	2.7	1.6	0.0	117
8	0	7.9	7.8	7.8	7.7	7.4	7.1	6.8	6.3	5.8	5.1	4.7	4.2	3.6	2.8	1.6	0.0	120
8	12	8.1	8.0	8.0	7.8	7.6	7.3	6.9	6.5	5.9	5.2	4.8	4.3	3.7	2.9	1.6	0.0	123
8	24	8.3	8.2	8.2	8.0	7.7	7.5	7.1	6.6	6.1	5.3	4.9	4.4	3.7	2.9	1.7	0.0	126
8	36	8.5	8.4	8.3	8.2	7.9	7.6	7.3	6.8	6.2	5.5	5.0	4.5	3.8	3.0	1.7	0.0	129
8	48	8.7	8.6	8.5	8.4	8.1	7.8	7.4	7.0	6.4	5.6	5.1	4.6	3.9	3.1	1.8	0.0	132
9	0	8.9	8.8	8.7	8.5	8.3	8.0	7.6	7.1	6.5	5.7	5.3	4.7	4.0	3.1	1.8	0.0	135
9	12	9.1	9.0	8.9	8.7	8.5	8.2	7.8	7.3	6.6	5.8	5.4	4.8	4.1	3.2	1.8	0.0	138
9	24	9.3	9.2	9.1	8.9	8.7	8.3	7.9	7.4	6.8	6.0	5.5	4.9	4.2	3.3	1.9	0.0	141
9	36	9.5	9.4	9.3	9.1	8.9	8.5	8.1	7.6	6.9	6.1	5.6	5.0	4.3	3.3	1.9	0.0	144
9	48	9.7	9.6	9.5	9.3	9.0	[illegible]	8.3	7.8	7.1	6.2	5.7	5.1	4.4	3.4	2.0	0.0	147
10	0	9.9	9.8	9.7	9.5	9.2	8.9	8.5	7.9	7.2	6.4	5.8	5.2	4.5	3.5	2.0	0.0	150
10	12	10.1	10.0	9.9	9.7	9.4	9.1	8.6	8.1	7.4	6.5	6.0	5.3	4.5	3.6	2.0	0.0	153
10	24	10.3	10.2	10.1	9.9	9.6	9.2	8.8	8.2	7.5	6.6	6.1	5.4	4.6	3.6	2.1	0.0	156
10	36	10.5	10.4	[illegible]	10.1	9.8	9.4	9.0	8.4	7.6	6.7	6.2	5.5	4.7	3.7	2.1	0.0	159
10	48	10.7	10.6	10.5	10.3	10.0	9.6	9.1	8.5	7.8	6.9	6.3	5.6	4.8	3.8	2.2	0.0	162
11	0	10.9	10.8	10.7	10.4	10.1	9.8	9.3	8.7	7.9	7.0	6.4	5.7	4.9	3.8	2.2	0.0	165
11	12	11.1	11.0	10.9	10.6	10.3	9.9	9.5	8.9	8.1	7.1	6.5	5.8	5.0	3.9	2.2	0.0	168
11	24	11.3	11.2	11.0	10.8	10.5	10.1	9.6	9.0	8.2	7.2	6.7	5.9	5.1	4.0	2.3	0.0	171
11	36	11.5	11.4	11.2	10.0	10.7	10.3	9.8	9.2	8.4	7.4	6.8	6.0	5.2	4.0	2.3	0.0	174
11	48	11.6	11.6	11.4	11.2	10.9	10.5	10.0	9.3	8.5	7.5	[illegible]	6.1	5.3	4.1	2.4	0.0	177
12	0	11.8	11.8	11.6	11.4	11.3	10.6	10.1	9.5	8.7	7.7	7.0	6.2	5.3	4.2	2.4	0.0	180

Tables for adapting the former table to any other time of the year.

Arg. time fr. noon.	TAB. XIV Arg. Decl. of Sun North decr. or between June and Sept. 0°	2°	4°	6°	8°	10°	12°	14°	16°	18°	19°	20°	21°	22°	23°	23°.28	Arg. Long. of Ship.
	—	—	—	—	—	—	—	—	—	—	—	—	—	—	—	—	
	′	′	′	′	′	′	′	′	′	′	′	′	′	′	′	′	
0h.	0.0	0.0	0.0	0.0	0.0	0.0	0.0	0.0	0.0	0.0	0.0	0.0	0.0	0.0	0.0	0.0	0°
1	0.0	0.0	0.0	0.0	0.0	0.0	0.0	0.0	0.0	0.0	0.0	0.0	0.0	0.0	0.0	0.0	15
2	0.0	0.0	0.0	0.0	0.0	0.0	0.0	0.0	0.0	0.0	0.0	0.0	0.0	0.0	0.0	0.0	30
3	0.0	0.0	0.0	0.0	0.0	0.0	0.0	0.0	0.0	0.0	0.0	0.0	0.0	0.0	0.0	0.0	45
4	0.0	0.0	0.0	0.0	0.0	0.0	0.0	0.0	0.0	0.0	0.0	0.0	0.0	0.0	0.0	0.0	60
5	0.1	0.1	0.1	0.1	0.0	0.0	0.0	0.0	0.0	0.0	0.0	0.0	0.0	0.0	0.0	0.0	75
6	0.1	0.1	0.1	0.1	0.1	0.1	0.1	0.0	0.0	0.0	0.0	0.0	0.0	0.0	0.0	0.0	90
7	0.1	0.1	0.1	0.1	0.1	0.1	0.1	0.1	0.0	0.0	0.0	0.0	0.0	0.0	0.0	0.0	105
8	0.1	0.1	0.1	0.1	0.1	0.1	0.1	0.1	0.0	0.0	0.0	0.0	0.0	0.0	0.0	0.0	120
9	0.1	0.1	0.1	0.1	0.1	0.1	0.1	0.1	0.0	0.0	0.0	0.0	0.0	0.0	0.0	0.0	135
10	0.1	0.1	0.1	0.1	0.1	0.1	0.1	0.1	0.1	0.0	0.0	0.0	0.0	0.0	0.0	0.0	150
11	0.1	0.1	0.1	0.1	0.1	0.1	0.1	0.1	0.1	0.0	0.0	0.0	0.0	0.0	0.0	0.0	165
12	0.1	0.1	0.1	0.1	0.1	0.1	0.1	0.1	0.1	0.1	0.0	0.0	0.0	0.0	0.0	0.0	180

Arg. time fr. noon.	TAB. XV. Arg. Decl. of Sun South incr. or between Sept. and Dec. 0°	2°	4°	6°	8°	10°	12°	14°	16°	18°	19°	20°	21°	22°	23°	23°.28	Arg. Long. of Ship.
	—	—	+	+	+	+	+	+	+	+	+	+	+	+	+	+	
	′	′	′	′	′	′	′	′	′	′	′	′	′	′	′		
0h.	0.0	0.0	0.0	0.0	0.0	0.0	0.0	0.0	0.0	0.0	0.0	0.0	0.0	0.0	0.0	0.0	0°
1	0.0	0.0	0.0	0.0	0.0	0.0	0.0	0.0	0.0	0.0	0.0	0.0	0.0	0.0	0.0	0.0	15
2	0.0	0.0	0.0	0.0	0.0	0.0	0.1	0.1	0.1	0.1	0.1	0.1	0.1	0.0	0.0	0.0	30
3	0.0	0.0	0.0	0.0	0.0	0.1	0.1	0.1	0.1	0.1	0.1	0.1	0.1	0.1	0.0	0.0	45
4	0.0	0.0	0.0	0.0	0.0	0.1	0.1	0.1	0.1	0.1	0.1	0.1	0.1	0.1	0.1	0.0	60
5	0.1	0.0	0.0	0.0	0.1	0.1	0.1	0.1	0.1	0.1	0.1	0.1	0.1	0.1	0.1	0.0	75
6	0.1	0.0	0.0	0.0	0.1	0.1	0.2	0.2	0.2	0.2	0.2	0.2	0.2	0.1	0.1	0.0	90
7	0.1	0.0	0.0	0.0	0.1	0.2	0.2	0.2	0.2	0.2	0.2	0.2	0.2	0.2	0.1	0.0	105
8	0.1	0.0	0.0	0.1	0.1	0.2	0.2	0.2	0.2	0.2	0.2	0.2	0.2	0.2	0.1	0.0	120
9	0.1	0.0	0.0	0.1	0.1	0.2	0.2	0.2	0.3	0.3	0.3	0.3	0.2	0.2	0.1	0.0	135
10	0.1	0.1	0.0	0.1	0.1	0.2	0.3	0.3	0.3	0.3	0.3	0.3	0.3	0.2	0.1	0.0	150
11	0.1	0.1	0.0	0.1	0.1	0.3	0.3	0.3	0.3	0.3	0.3	0.3	0.3	0.2	0.1	0.0	165
12	0.1	0.1	0.0	0.1	0.1	0.3	0.3	0.3	0.3	0.3	0.3	0.3	0.3	0.3	0.2	0.0	180

Arg. time fr. noon.	TAB. XVI. Arg. Decl. of Sun South decr. or between Dec. and March. 0°	2°	4°	6°	8°	10°	12°	14°	16°	18°	19°	20°	21°	22°	23°	23°.28	Arg. Long. of Ship.
	+	+	+	+	+	+	+	+	+	+	+	+	+	+	+	+	
	′	′	′	′	′	′	′	′	′	′	′	′	′	′	′	′	
0h.	0.0	0.0	0.0	0.0	0.0	0.0	0.0	0.0	0.0	0.0	0.0	0.0	0.0	0.0	0.0	0.0	0°
1	0.0	0.0	0.0	0.0	0.0	0.0	0.0	0.0	0.0	0.0	0.0	0.0	0.0	0.0	0.0	0.0	15
2	0.0	0.0	0.0	0.0	0.0	0.1	0.1	0.1	0.1	0.1	0.1	0.1	0.1	0.0	0.0	0.0	30
3	0.0	0.0	0.0	0.0	0.1	0.1	0.1	0.1	0.1	0.1	0.1	0.1	0.1	0.0	0.0	0.0	45
4	0.0	0.0	0.0	0.1	0.1	0.1	0.1	0.1	0.1	0.1	0.1	0.1	0.1	0.1	0.0	0.0	60
5	0.0	0.0	0.1	0.1	0.1	0.1	0.2	0.2	0.2	0.2	0.2	0.2	0.1	0.1	0.1	0.0	75
6	0.0	0.0	0.1	0.1	0.1	0.2	0.2	0.2	0.2	0.2	0.2	0.2	0.2	0.1	0.1	0.0	90
7	0.0	0.0	0.1	0.1	0.2	0.2	0.2	0.2	0.2	0.2	0.2	0.2	0.2	0.1	0.1	0.0	105
8	0.0	0.0	0.1	0.1	0.2	0.2	0.2	0.3	0.3	0.3	0.3	0.2	0.2	0.1	0.1	0.0	120
9	0.0	0.0	0.1	0.1	0.2	0.2	0.3	0.3	0.3	0.3	0.3	0.3	0.2	0.1	0.1	0.0	135
10	0.0	0.0	0.1	0.2	0.2	0.3	0.3	0.3	0.3	0.3	0.3	0.3	0.3	0.2	0.1	0.0	150
11	0.0	0.0	0.1	0.2	0.2	0.3	0.3	0.4	0.4	0.4	0.4	0.3	0.3	0.2	0.2	0.0	165
12	0.0	0.1	0.1	0.2	0.3	0.3	0.4	0.4	0.4	0.4	0.4	0.4	0.3	0.2	0.2	0.0	180

TABLE XVII.

For ſelecting a proper Star, from which the Moon's Diſtance is to be obſerved.

Diff. of Lat.	Diff. of Long.		Diff. of Lat.	Diff. of Long.		Diff. of Lat.	Diff. of Long.		Diff. of Lat.	Diff. of Long.		Diff. of Lat.	Diff. of Long.		Diff. of Lat.	Diff. of Long.	
°	°		°	°	′	°	°	′	°	°	′	°	°	′	°	°	′
0	10		5	10	0	10	18	34	15	28	57	20	41	6	25	57	21
1	10		6	10	57	11	20	33	16	31	12	21	43	54	26	61	46
2	10		7	12	49	12	22	35	17	33	31	22	46	52	27	66	58
3	10		8	14	42	13	24	39	18	35	56	23	50	3	28	73	48
4	10		9	16	37	14	26	46	19	38	28	24	53	32			

TABLE XVIII. To reduce the Right Afcenfion of the Sun to any given Meridian; and to any given Time under that Meridian.

Arg. time from noon.	Argument. Right Afcenfion of the Sun.											Arg. longitude of Ship.
	0 h	1 h.	2 h.	3 h.	4 h.	5 h.	6 h.	7 h.	8 h.	9 h.	10 h.	11 h.	
h. ′	′ ″	′ ″	′ ″	′ ″	′ ″	′ ″	′ ″	′ ″	′ ″	′ ″	′ ″	′ ″	°
0 0	0 0	0 0	0 0	0 0	0 0	0 0	0 0	0 0	0 0	0 0	0 0	0 0	0
0 12	0 2	0 2	0 2	0 2	0 2	0 2	0 2	0 2	0 2	0 2	0 2	0 2	3
0 24	0 4	0 4	0 4	0 4	0 4	0 4	0 4	0 4	0 4	0 4	0 4	0 4	6
0 36	0 6	0 6	0 6	0 6	0 6	0 6	0 6	0 6	0 6	0 6	0 6	0 5	9
0 48	0 7	0 7	0 7	0 8	0 8	0 8	0 8	0 8	0 8	0 8	0 7	0 7	12
1 0	0 9	0 9	0 9	0 10	0 10	0 10	0 10	0 10	0 10	0 10	0 9	0 9	15
1 12	0 11	0 11	0 11	0 12	0 12	0 12	0 12	0 12	0 12	0 11	0 11	0 11	18
1 24	0 13	0 13	0 13	0 14	0 14	0 14	0 14	0 14	0 14	0 13	0 13	0 13	21
1 36	0 15	0 15	0 15	0 15	0 16	0 16	0 17	0 16	0 16	0 15	0 15	0 14	24
1 48	0 16	0 16	0 17	0 17	0 18	0 18	0 19	0 18	0 18	0 17	0 17	0 16	27
2 0	0 18	0 18	0 19	0 19	0 20	0 21	0 21	0 21	0 20	0 19	0 18	0 18	30
2 12	0 20	0 20	0 21	0 21	0 22	0 23	0 23	0 23	0 22	0 21	0 20	0 20	33
2 24	0 22	0 22	0 22	0 23	0 24	0 25	0 25	0 25	0 24	0 23	0 22	0 22	36
2 36	0 24	0 24	0 24	0 25	0 26	0 27	0 27	0 27	0 26	0 25	0 24	0 23	39
2 48	0 25	0 25	0 26	0 27	0 28	0 29	0 29	0 29	0 28	0 27	0 26	0 25	42
3 0	0 27	0 27	0 28	0 29	0 30	0 31	0 31	0 31	0 30	0 29	0 28	0 27	45
3 12	0 29	0 29	0 30	0 31	0 32	0 33	0 33	0 33	0 32	0 31	0 30	0 29	48
3 24	0 31	0 31	0 32	0 33	0 34	0 35	0 35	0 35	0 34	0 33	0 32	0 31	51
3 36	0 33	0 33	0 34	0 35	0 36	0 37	0 37	0 37	0 36	0 35	0 33	0 32	54
3 48	0 35	0 35	0 36	0 37	0 38	0 39	0 39	0 39	0 38	0 36	0 35	0 34	57
4 0	0 36	0 36	0 37	0 39	0 40	0 41	0 42	0 41	0 40	0 38	0 37	0 36	60
4 12	0 38	0 38	0 39	0 41	0 42	0 43	0 44	0 43	0 42	0 40	0 39	0 38	63
4 24	0 40	0 40	0 41	0 43	0 44	0 45	0 46	0 45	0 44	0 42	0 41	0 40	66
4 36	0 42	0 42	0 43	0 45	0 46	0 47	0 48	0 47	0 46	0 44	0 42	0 41	69
4 48	0 44	0 44	0 45	0 46	0 48	0 49	0 50	0 49	0 48	0 46	0 44	0 43	72
5 0	0 45	0 46	0 47	0 48	0 50	0 51	0 52	0 51	0 50	0 48	0 46	0 45	75
5 12	0 47	0 47	0 49	0 50	0 52	0 54	0 54	0 53	0 52	0 50	0 48	0 47	78
5 24	0 49	0 49	0 50	0 52	0 54	0 56	0 56	0 55	0 54	0 52	0 50	0 49	81
5 36	0 51	0 51	0 52	0 54	0 56	0 58	0 58	0 58	0 56	0 54	0 52	0 50	84
5 48	0 53	0 53	0 54	0 56	0 58	1 0	1 0	1 0	0 58	0 56	0 54	0 52	87
6 0	0 54	0 55	0 56	0 58	1 0	1 2	1 2	1 2	1 0	0 58	0 55	0 54	90
6 12	0 56	0 56	0 58	1 0	1 2	1 4	1 4	1 4	1 2	1 0	0 57	0 56	93
6 24	0 58	0 58	1 0	1 2	1 4	1 6	1 6	1 6	1 4	1 2	0 59	0 58	96
6 36	1 0	1 0	1 2	1 4	1 6	1 8	1 9	1 8	1 6	1 3	1 1	0 59	99
6 48	1 2	1 2	1 3	1 6	1 8	1 10	1 11	1 10	1 8	1 5	1 3	1 1	102
7 0	1 4	1 4	1 5	1 8	1 10	1 12	1 13	1 12	1 10	1 7	1 5	1 3	105
7 12	1 5	1 6	1 7	1 10	1 12	1 14	1 15	1 14	1 12	1 9	1 7	1 5	108
7 24	1 7	1 7	1 9	1 12	1 14	1 16	1 17	1 16	1 14	1 11	1 8	1 7	111
7 36	1 9	1 9	1 11	1 14	1 16	1 18	1 19	1 18	1 16	1 13	1 10	1 9	114
7 48	1 11	1 11	1 13	1 16	1 18	1 20	1 21	1 20	1 18	1 15	1 12	1 10	117
8 0	1 13	1 13	1 15	1 17	1 20	1 22	1 23	1 22	1 20	1 17	1 14	1 12	120
8 12	1 14	1 15	1 17	1 19	1 22	1 23	1 25	1 24	1 22	1 19	1 16	1 14	123
8 24	1 16	1 16	1 18	1 21	1 24	1 27	1 27	1 26	1 24	1 21	1 18	1 16	126
8 36	1 18	1 18	1 20	1 23	1 26	1 29	1 29	1 28	1 26	1 23	1 19	1 18	129
8 48	1 20	1 20	1 22	1 25	1 28	1 31	1 31	1 30	1 28	1 25	1 21	1 19	132
9 0	1 22	1 21	1 24	1 27	1 30	1 33	1 34	1 32	1 30	1 26	1 23	1 21	135
9 12	1 24	1 24	1 26	1 29	1 32	1 35	1 36	1 35	1 32	1 28	1 25	1 23	138
9 24	1 25	1 26	1 28	1 31	1 34	1 37	1 38	1 37	1 34	1 30	1 27	1 25	141
9 36	1 27	1 27	1 30	1 33	1 36	1 39	1 40	1 39	1 36	1 32	1 29	1 27	144
9 48	1 29	1 29	1 32	1 35	1 38	1 41	1 42	1 41	1 38	1 34	1 31	1 28	147
10 0	1 31	1 31	1 33	1 37	1 40	1 43	1 44	1 43	1 40	1 36	1 32	1 30	150
10 12	1 33	1 33	1 35	1 39	1 42	1 45	1 46	1 45	1 42	1 38	1 34	1 32	153
10 24	1 34	1 35	1 37	1 41	1 44	1 47	1 48	1 47	1 44	1 40	1 36	1 34	156
10 36	1 36	1 37	1 39	1 43	1 46	1 49	1 50	1 49	1 46	1 42	1 38	1 36	159
10 48	1 38	1 38	1 41	1 45	1 48	1 51	1 52	1 51	1 48	1 44	1 40	1 37	162
11 0	1 40	1 40	1 43	1 46	1 50	1 53	1 54	1 53	1 50	1 46	1 42	1 39	165
11 12	1 42	1 42	1 45	1 48	1 52	1 55	1 56	1 55	1 52	1 48	1 44	1 41	168
11 24	1 44	1 44	1 46	1 50	1 54	1 57	1 59	1 57	1 54	1 49	1 45	1 43	171
11 36	1 45	1 46	1 48	1 52	1 57	2 0	2 1	1 59	1 56	1 51	1 47	1 45	174
11 48	1 47	1 48	1 50	1 54	1 59	2 2	2 3	2 1	1 58	1 53	1 49	1 46	177
12 0	1 49	1 49	1 52	1 56	2 1	2 4	2 5	2 3	2 0	1 55	1 51	1 48	180

TABLE XVIII. To reduce the Right Aſcenſion of the Sun to any given Meridian; and to any given Time under that Meridian.

Arg. time from Noon.		Argument. Right Aſcenſion of the Sun. 12 h.		13 h.		14 h.		15 h.		16 h.		17 h.		18 h.		19 h.		20 h.		21 h.		22 h.		23 h.		Arg. longitude of ship.
h.	'	'	"	'	"	'	"	'	"	'	"	'	"	'	"	'	"	'	"	'	"	'	"	'	"	°
0	0	0	0	0	0	0	0	0	0	0	0	0	0	0	0	0	0	0	0	0	0	0	0	0	0	0
0	12	0	2	0	2	0	2	0	2	0	2	0	2	0	2	0	2	0	2	0	2	0	2	0	2	3
0	24	0	4	0	4	0	4	0	4	0	4	0	4	0	5	0	4	0	4	0	4	0	4	0	4	6
0	36	0	5	0	5	0	6	0	6	0	6	0	7	0	7	0	7	0	6	0	6	0	6	0	6	9
0	48	0	7	0	7	0	8	0	8	0	8	0	9	0	9	0	9	0	8	0	8	0	8	0	7	12
1	0	0	9	0	9	0	10	0	10	0	11	0	11	0	11	0	11	0	11	0	10	0	10	0	9	15
1	12	0	11	0	11	0	11	0	12	0	13	0	13	0	13	0	13	0	13	0	12	0	12	0	11	18
1	24	0	13	0	13	0	13	0	14	0	15	0	15	0	16	0	15	0	15	0	14	0	14	0	13	21
1	36	0	14	0	15	0	15	0	16	0	17	0	17	0	18	0	18	0	17	0	16	0	15	0	15	24
1	48	0	16	0	17	0	17	0	18	0	19	0	20	0	20	0	20	0	19	0	18	0	17	0	17	27
2	0	0	18	0	18	0	19	0	20	0	21	0	22	0	22	0	22	0	21	0	20	0	19	0	19	30
2	12	0	20	0	20	0	21	0	21	0	23	0	24	0	24	0	24	0	23	0	22	0	21	0	21	33
2	24	0	22	0	22	0	23	0	24	0	25	0	26	0	27	0	26	0	26	0	24	0	23	0	23	36
2	36	0	23	0	24	0	25	0	26	0	27	0	28	0	29	0	29	0	28	0	26	0	25	0	24	39
2	48	0	25	0	26	0	27	0	28	0	30	0	31	0	31	0	31	0	30	0	28	0	27	0	26	42
3	0	0	27	0	28	0	29	0	30	0	32	0	33	0	33	0	33	0	32	0	30	0	29	0	28	45
3	12	0	29	0	29	0	31	0	32	0	34	0	35	0	35	0	35	0	34	0	32	0	31	0	30	48
3	24	0	31	0	31	0	33	0	34	0	36	0	37	0	38	0	37	0	36	0	35	0	33	0	32	51
3	36	0	32	0	33	0	34	0	36	0	38	0	39	0	40	0	40	0	38	0	37	0	35	0	34	54
3	48	0	34	0	35	0	36	0	38	0	40	0	42	0	42	0	42	0	40	0	39	0	37	0	36	57
4	0	0	36	0	37	0	38	0	40	0	42	0	44	0	44	0	44	0	41	0	41	0	39	0	37	60
4	12	0	38	0	39	0	40	0	42	0	44	0	46	0	47	0	46	0	45	0	43	0	41	0	39	63
4	24	0	40	0	40	0	42	0	44	0	47	0	48	0	49	0	48	0	47	0	45	0	43	0	41	66
4	36	0	41	0	42	0	44	0	46	0	49	0	50	0	51	0	51	0	49	0	47	0	44	0	43	69
4	48	0	43	0	44	0	46	0	48	0	51	0	53	0	53	0	53	0	51	0	49	0	46	0	45	72
5	0	0	45	0	46	0	48	0	50	0	53	0	55	0	56	0	55	0	53	0	51	0	48	0	47	75
5	12	0	47	0	48	0	50	0	52	0	55	0	57	0	58	0	57	0	55	0	53	0	50	0	49	78
5	24	0	49	0	50	0	52	0	54	0	57	0	59	1	0	0	59	0	58	0	55	0	52	0	51	81
5	36	0	50	0	51	0	54	0	56	0	59	1	1	1	2	1	2	1	0	0	57	0	54	0	52	84
5	48	0	52	0	53	0	56	0	59	1	1	1	4	1	4	1	4	1	2	0	59	0	56	0	54	87
6	0	0	54	0	55	0	58	1	1	1	4	1	6	1	7	1	6	1	4	1	1	0	58	0	56	90
6	12	0	56	0	57	0	59	1	3	1	6	1	8	1	9	1	8	1	6	1	6	1	0	0	58	93
6	24	0	58	0	59	1	2	1	5	1	8	1	10	1	11	1	10	1	8	1	8	1	2	1	0	96
6	36	0	59	1	1	1	3	1	7	1	10	1	12	1	13	1	13	1	10	1	7	1	4	1	2	99
6	48	1	1	1	2	1	5	1	9	1	12	1	15	1	16	1	15	1	12	1	9	1	6	1	4	102
7	0	1	3	1	4	1	7	1	11	1	14	1	17	1	18	1	17	1	14	1	11	1	8	1	6	105
7	12	1	5	1	6	1	9	1	13	1	16	1	19	1	20	1	19	1	17	1	13	1	10	1	7	108
7	24	1	7	1	8	1	11	1	15	1	18	1	21	1	22	1	21	1	19	1	15	1	12	1	9	111
7	36	1	8	1	10	1	13	1	17	1	20	1	23	1	24	1	24	1	21	1	17	1	14	1	11	114
7	48	1	10	1	12	1	15	1	19	1	23	1	25	1	27	1	26	1	23	1	19	1	16	1	13	117
8	0	1	12	1	13	1	17	1	21	1	25	1	28	1	29	1	28	1	25	1	21	1	17	1	15	120
8	12	1	14	1	15	1	19	1	23	1	27	1	30	1	31	1	30	1	27	1	23	1	19	1	17	123
8	24	1	16	1	17	1	21	1	25	1	29	1	32	1	33	1	32	1	29	1	25	1	21	1	19	126
8	36	1	17	1	19	1	22	1	27	1	31	1	34	1	36	1	35	1	32	1	27	1	23	1	21	129
8	48	1	19	1	21	1	24	1	29	1	33	1	36	1	38	1	37	1	34	1	29	1	25	1	22	132
9	0	1	21	1	23	1	26	1	31	1	35	1	39	1	40	1	39	1	36	1	31	1	27	1	24	135
9	12	1	23	1	25	1	28	1	33	1	37	1	41	1	42	1	41	1	38	1	33	1	29	1	26	138
9	24	1	25	1	26	1	30	1	35	1	40	1	43	1	44	1	43	1	40	1	36	1	31	1	28	141
9	36	1	26	1	28	1	32	1	37	1	42	1	45	1	47	1	46	1	42	1	38	1	33	1	30	144
9	48	1	28	1	30	1	34	1	39	1	44	1	47	1	49	1	46	1	44	1	40	1	35	1	32	147
10	0	1	30	1	32	1	36	1	41	1	46	1	50	1	51	1	50	1	46	1	42	1	37	1	34	150
10	12	1	32	1	34	1	38	1	43	1	48	1	52	1	53	1	52	1	49	1	44	1	39	1	36	153
10	24	1	34	1	36	1	40	1	45	1	50	1	54	1	56	1	54	1	51	1	46	1	41	1	38	156
10	36	1	35	1	37	1	42	1	47	1	52	1	56	1	58	1	57	1	53	1	48	1	43	1	39	159
10	48	1	37	1	39	1	44	1	49	1	54	1	58	2	0	1	59	1	55	1	50	1	45	1	41	162
11	0	1	39	1	41	1	46	1	51	1	57	2	1	2	2	2	1	1	57	1	52	1	46	1	43	165
11	12	1	41	1	43	1	47	1	53	1	59	2	3	2	4	2	3	1	59	1	54	1	48	1	45	168
11	24	1	43	1	45	1	49	1	55	2	1	2	5	2	7	2	5	2	1	1	56	1	50	1	47	171
11	36	1	44	1	47	1	51	1	57	2	3	2	7	2	9	2	8	2	4	1	58	1	52	1	49	174
11	48	1	46	1	48	1	53	1	59	2	5	2	9	2	11	2	10	2	6	2	0	1	54	1	51	177
12	0	1	48	1	40	1	55	2	1	2	7	2	12	2	13	2	12	2	8	2	2	1	56	1	53	180

TABLE XIX. To reduce the Declination of the Moon to any given Meridian; and to any given Time under that Meridian.

Arg. time fr. noon.	Argument	Variation of the Moon's Declination in twelve Hours.																	Arg. Log.
	′ 5	′ 10	′ 15	′ 20	′ 25	′ 30	′ 35	′ 40	′ 45	′ 50	′ 55	° ′ 1 0	° ′ 1 5	° ′ 1 10	° ′ 1 15	° ′ 1 20	° ′ 1 25	° ′ 1 30	° ′ 1 35	
h. ′	′	′	′	′	′	′	′	′	′	′	′	′	′	′	′	′	′	′	′	°
0 0	0	0	0	0	0	0	0	0	0	0	0	0	0	0	0	0	0	0	0	0
0 12	0	0	0	0	0	0	1	1	1	1	1	1	1	1	1	1	1	1	2	3
0 24	0	0	0	1	1	1	1	1	1	2	2	2	2	2	2	3	3	3	3	6
0 36	0	0	1	1	1	1	2	2	2	2	3	3	3	3	4	4	4	4	5	9
0 48	0	1	1	1	2	2	2	3	3	3	4	4	4	5	5	5	6	6	6	12
1 0	0	1	1	2	2	2	3	3	4	4	5	5	5	6	6	7	7	7	8	15
1 12	0	1	1	2	2	3	3	4	4	5	5	6	6	7	7	8	8	9	9	18
1 24	1	1	2	2	3	3	4	5	5	6	6	7	8	8	9	9	10	10	11	21
1 36	1	1	2	3	3	4	5	5	6	7	7	8	9	9	10	11	11	12	13	24
1 48	1	1	2	3	4	4	5	6	7	7	8	9	10	10	11	12	13	13	14	27
2 0	1	2	2	3	4	5	6	7	7	8	9	10	11	12	12	13	14	15	16	30
2 12	1	2	3	4	5	5	6	7	8	9	10	11	12	13	14	15	16	16	17	33
2 24	1	2	3	4	5	6	7	8	9	10	11	12	13	14	15	16	17	18	19	36
2 36	1	2	3	4	5	6	8	9	10	11	12	13	14	15	16	17	18	19	21	39
2 48	1	2	3	5	6	7	8	9	10	12	13	14	15	16	17	19	20	21	22	42
3 0	1	2	4	5	6	7	9	10	11	12	14	15	16	17	19	20	21	22	24	45
3 12	1	3	4	5	7	8	9	11	12	13	15	16	17	19	20	21	23	24	25	48
3 24	1	3	4	6	7	8	10	11	13	14	16	17	18	20	21	23	24	25	27	51
3 36	1	3	4	6	7	9	10	12	13	15	16	18	19	21	22	24	25	27	28	54
3 48	2	3	5	6	8	9	11	13	14	16	17	19	21	22	24	25	27	28	30	57
4 0	2	3	5	7	8	10	12	13	15	17	18	20	22	23	25	27	28	30	32	60
4 12	2	3	5	7	9	10	12	14	16	17	19	21	23	24	26	28	30	31	33	63
4 24	2	4	5	7	9	11	13	15	16	18	20	22	24	26	27	29	31	33	35	66
4 36	2	4	6	8	10	11	13	15	17	19	21	23	25	27	29	31	33	34	36	69
4 48	2	4	6	8	10	12	14	16	18	20	22	24	26	28	30	32	34	36	38	72
5 0	2	4	6	8	10	12	15	17	19	21	23	25	27	29	31	33	35	37	40	75
5 12	2	4	6	9	11	13	15	17	19	22	24	26	28	30	32	35	37	39	41	78
5 24	2	4	7	9	11	13	16	18	20	22	25	27	29	31	34	36	38	40	43	81
5 36	2	5	7	9	12	14	16	19	21	23	26	28	30	33	35	37	40	42	44	84
5 48	2	5	7	10	12	14	17	19	22	24	27	29	31	34	36	39	41	43	46	87
6 0	2	5	7	10	12	15	17	20	22	25	27	30	32	35	37	40	42	45	47	90
6 12	3	5	8	10	13	15	18	21	23	26	28	31	34	36	39	41	44	46	49	93
6 24	3	5	8	11	13	16	19	21	24	27	29	32	35	37	40	43	45	48	51	96
6 36	3	5	8	11	14	16	19	22	25	27	30	33	36	38	41	44	47	49	52	99
6 48	3	6	8	11	14	17	20	23	25	28	31	34	37	40	42	45	48	51	54	102
7 0	3	6	9	12	15	17	20	23	26	29	32	35	38	41	44	47	50	52	55	105
7 12	3	6	9	12	15	18	21	24	27	30	33	36	39	42	45	48	51	54	57	108
7 24	3	6	9	12	15	18	22	25	28	31	34	37	40	43	46	49	52	55	59	111
7 36	3	6	9	13	16	19	22	25	28	32	35	38	41	44	47	51	54	57	60	114
7 48	3	6	10	13	16	19	23	26	29	32	36	39	42	45	49	52	55	58	62	117
8 0	3	7	10	13	17	20	23	27	30	33	37	40	43	47	50	53	57	60	63	120
8 12	3	7	10	14	17	20	24	27	31	34	38	41	44	48	51	55	58	61	65	123
8 24	3	7	10	14	17	21	24	28	31	35	38	42	45	49	52	56	59	63	66	126
8 36	4	7	11	14	18	21	25	29	32	36	39	43	47	50	54	57	61	64	68	129
8 48	4	7	11	15	18	22	26	29	33	37	40	44	48	51	55	59	62	66	70	132
9 0	4	7	11	15	19	22	26	30	34	37	41	45	49	52	56	60	64	67	71	135
9 12	4	8	11	15	19	23	27	31	34	38	42	46	50	54	57	61	65	69	73	138
9 24	4	8	12	16	20	23	27	31	35	39	43	47	51	55	59	63	67	70	74	141
9 36	4	8	12	16	20	24	28	32	36	40	44	48	52	56	60	64	68	72	76	144
9 48	4	8	12	16	20	24	29	33	37	41	45	49	53	57	61	65	69	73	78	147
10 0	4	8	12	17	21	25	29	33	37	42	46	50	54	58	62	67	71	75	79	150
10 12	4	8	13	17	21	25	30	34	38	42	47	51	55	59	64	68	72	76	81	153
10 24	4	9	13	17	22	26	30	35	39	43	48	52	56	61	65	69	74	78	82	156
10 36	4	9	13	18	22	26	31	35	40	44	49	53	57	62	66	71	75	79	84	159
10 48	4	9	13	18	22	27	31	36	40	45	49	54	58	63	67	72	76	81	85	162
11 0	5	9	14	18	23	27	32	37	41	46	50	55	60	64	69	73	78	82	87	165
11 12	5	9	14	19	23	28	33	37	42	47	51	56	61	65	70	75	79	84	89	168
11 24	5	9	14	19	24	28	33	38	43	47	52	57	62	66	71	76	81	85	90	171
11 36	5	10	14	19	24	29	34	39	44	48	53	58	63	68	72	77	82	87	92	174
11 48	5	10	15	20	25	29	34	39	44	49	54	59	64	69	74	79	84	88	93	177
12 0	5	10	15	20	25	30	35	40	45	50	55	60	65	70	75	80	85	90	95	180

TABLE XIX. To reduce the Declination of the Moon to any given Meridian; and to any given Time under that Meridian.

Arg. time fr. noon.		Argument. Variation of the Moon's Declination in twelve Hours.													Arg. Log.
		° ' 1 40	° ' 1 45	° ' 1 50	° ' 1 55	° ' 2 0	° ' 2 5	° ' 2 10	° ' 2 15	° ' 2 20	° ' 2 25	° ' 2 30	° ' 2 35	° ' 2 40	
h.	'	° '	° '	° '	° '	° '	° '	° '	° '	° '	° '	° '	° '	° '	°
0	0	0 0	0 0	0 0	0 0	0 0	0 0	0 0	0 0	0 0	0 0	0 0	0 0	0 0	0
0	12	0 2	0 2	0 2	0 2	0 2	0 2	0 2	0 2	0 2	0 2	0 2	0 3	0 3	3
0	24	0 3	0 3	0 4	0 4	0 4	0 4	0 4	0 4	0 5	0 5	0 5	0 5	0 5	6
0	36	0 5	0 5	0 5	0 6	0 6	0 6	0 6	0 7	0 7	0 7	0 7	0 8	0 8	9
0	48	0 7	0 7	0 7	0 8	0 8	0 8	0 9	0 9	0 9	0 10	0 10	0 10	0 11	12
1	0	0 8	0 9	0 9	0 10	0 10	0 10	0 11	0 11	0 12	0 12	0 12	0 13	0 13	15
1	12	0 10	0 10	0 11	0 11	0 12	0 12	0 13	0 13	0 14	0 14	0 15	0 15	0 16	18
1	24	0 12	0 12	0 13	0 13	0 14	0 15	0 15	0 16	0 16	0 17	0 17	0 18	0 19	21
1	36	0 13	0 14	0 15	0 15	0 16	0 17	0 17	0 18	0 19	0 19	0 20	0 21	0 21	24
1	48	0 15	0 16	0 16	0 17	0 18	0 19	0 19	0 20	0 21	0 22	0 22	0 23	0 24	27
2	0	0 17	0 17	0 18	0 19	0 20	0 21	0 22	0 22	0 23	0 24	0 25	0 26	0 27	30
2	12	0 18	0 19	0 20	0 21	0 22	0 23	0 24	0 25	0 26	0 27	0 27	0 28	0 29	33
2	24	0 20	0 21	0 22	0 23	0 24	0 25	0 26	0 27	0 28	0 29	0 30	0 31	0 32	36
2	36	0 22	0 23	0 24	0 25	0 26	0 27	0 28	0 29	0 30	0 31	0 32	0 34	0 35	39
2	48	0 23	0 24	0 26	0 27	0 28	0 29	0 30	0 31	0 33	0 34	0 35	0 36	0 37	42
3	0	0 25	0 26	0 27	0 29	0 30	0 31	0 32	0 34	0 35	0 36	0 37	0 39	0 40	45
3	12	0 27	0 28	0 29	0 31	0 32	0 33	0 35	0 36	0 37	0 39	0 40	0 41	0 43	48
3	24	0 28	0 30	0 31	0 33	0 34	0 35	0 37	0 38	0 40	0 41	0 42	0 44	0 45	51
3	36	0 30	0 31	0 [illegible]	0 34	0 36	0 37	0 39	0 40	0 42	0 43	0 45	0 46	0 48	54
3	48	0 32	0 33	0 35	0 36	0 38	0 40	0 41	0 43	0 44	0 46	0 47	0 49	0 51	57
4	0	0 33	0 35	0 37	0 38	0 40	0 42	0 43	0 45	0 47	0 48	0 50	0 52	0 53	60
4	12	0 35	0 37	0 38	0 40	0 42	0 44	0 45	0 47	0 49	0 51	0 52	0 54	0 56	63
4	24	0 37	0 38	0 40	0 42	0 44	0 46	0 48	0 49	0 51	0 53	0 55	0 57	0 59	66
4	36	0 38	0 40	0 42	0 44	0 46	0 48	0 50	0 52	0 54	0 56	0 57	0 59	1 1	69
4	48	0 40	0 42	0 44	0 46	0 48	0 50	0 52	0 54	0 56	0 58	1 0	1 2	1 4	72
5	0	0 42	0 44	0 46	0 48	0 50	0 52	0 54	0 56	0 58	1 0	1 2	1 5	1 7	75
5	12	0 43	0 45	0 48	0 50	0 52	0 54	0 56	0 58	1 1	1 3	1 5	1 7	1 9	78
5	24	0 45	0 47	0 49	0 52	0 54	0 56	0 58	1 1	1 2	1 5	1 7	1 10	1 12	81
5	36	0 47	0 49	0 51	0 54	0 56	0 58	1 1	1 3	1 5	1 8	1 10	1 12	1 15	84
5	48	0 48	0 51	0 53	0 56	0 58	1 0	1 3	1 5	1 8	1 10	1 12	1 15	1 17	87
6	0	0 50	0 52	0 55	0 57	1 0	1 2	1 5	1 7	1 10	1 12	1 15	1 17	1 20	90
6	12	0 52	0 54	0 57	0 59	1 2	1 5	1 7	1 10	1 12	1 15	1 17	1 20	1 23	93
6	24	0 53	0 56	0 59	1 1	1 4	1 7	1 9	1 12	1 15	1 17	1 20	1 23	1 25	96
6	36	0 55	0 58	1 0	1 3	1 6	1 9	1 11	1 14	1 17	1 20	1 22	1 25	1 28	99
6	48	0 57	0 59	1 2	1 5	1 8	1 11	1 14	1 16	1 19	1 22	1 25	1 28	1 31	102
7	0	0 58	1 1	1 4	1 7	1 10	1 13	1 16	1 19	1 22	1 25	1 27	1 30	1 33	105
7	12	1 0	1 3	1 6	1 9	1 12	1 15	1 18	1 21	1 24	1 27	1 30	1 33	1 36	108
7	24	1 2	1 5	1 8	1 11	1 14	1 17	1 20	1 23	1 26	1 29	1 32	1 36	1 39	111
7	36	1 3	1 6	1 10	1 13	1 16	1 19	1 22	1 25	1 29	1 32	1 35	1 38	1 41	114
7	48	1 5	1 8	1 11	1 15	1 18	1 21	1 24	1 28	1 31	1 34	1 37	1 41	1 44	117
8	0	1 7	1 10	1 13	1 17	1 20	1 23	1 27	1 30	1 33	1 37	1 40	1 43	1 47	120
8	12	1 8	1 12	1 15	1 19	1 22	1 25	1 29	1 32	1 36	1 39	1 42	1 46	1 49	123
8	24	1 10	1 13	1 17	1 20	1 24	1 27	1 31	1 34	1 38	1 41	1 45	1 48	1 52	126
8	36	1 12	1 15	1 19	1 22	1 26	1 30	1 33	1 37	1 40	1 44	1 47	1 51	1 55	129
8	48	1 13	1 17	1 21	1 24	1 28	1 32	1 35	1 39	1 43	1 46	1 50	1 54	1 57	132
9	0	1 15	1 19	1 22	1 26	1 30	1 34	1 37	1 41	1 45	1 49	1 52	1 56	2 0	135
9	12	1 17	1 20	1 24	1 28	1 32	1 36	1 40	1 43	1 47	1 51	1 55	1 59	2 3	138
9	24	1 18	1 22	1 26	1 30	1 34	1 38	1 42	1 46	1 50	1 54	1 57	2 1	2 5	141
9	36	1 20	1 24	1 28	1 32	1 36	1 40	1 44	1 48	1 52	1 56	2 0	2 4	2 8	144
9	48	1 22	1 26	1 30	1 34	1 38	1 42	1 46	1 50	1 54	1 58	2 2	2 7	2 11	147
10	0	1 23	1 27	1 32	1 36	1 40	1 44	1 48	1 52	1 57	2 1	2 5	2 9	2 13	150
10	12	1 25	1 29	1 33	1 38	1 42	1 46	1 50	1 55	1 59	2 3	2 7	2 12	2 16	153
10	24	1 27	1 31	1 35	1 40	1 44	1 48	1 53	1 57	2 1	2 6	2 10	2 14	2 19	156
10	36	1 28	1 33	1 37	1 42	1 46	1 50	1 55	1 59	2 4	2 8	2 12	2 17	2 21	159
10	48	1 30	1 34	1 39	1 43	1 48	1 52	1 57	2 1	2 6	2 10	2 15	2 19	2 24	162
11	0	1 32	1 36	1 41	1 45	1 50	1 55	1 59	2 4	2 8	2 13	2 17	2 22	2 27	165
11	12	1 33	1 38	1 43	1 47	1 52	1 57	2 1	2 6	2 11	2 15	2 20	2 25	2 29	168
11	24	1 35	1 40	1 44	1 49	1 54	1 59	2 3	2 8	2 13	2 18	2 22	2 27	2 32	171
11	36	1 37	1 41	1 46	1 51	1 56	2 [illegible]	2 6	2 10	2 15	2 20	2 25	2 30	2 35	174
11	48	1 38	1 43	1 48	1 53	1 58	2 3	2 8	2 13	2 18	2 23	2 27	2 32	2 37	177
12	0	1 40	1 45	1 50	1 55	2 0	2 5	2 10	2 15	2 20	2 25	2 30	2 35	2 40	180

TABLE XIX. To reduce the Declination of the Moon to any given Meridian; and to any given Time under that Meridian.

Arg. time fr. noon. h.	′	Argument. Variation of the Moon's Declination in twelve Hours. 2° 45′	2° 50′	2° 55′	3° 0′	3° 5′	3° 10′	3° 15′	3° 20′	3° 25′	3° 30′	3° 35′	3° 40′	3° 45′	Arg. Long. °
		° ′	° ′	° ′	° ′	° ′	° ′	° ′	° ′	° ′	° ′	° ′	° ′	° ′	
0	0	0 0	0 0	0 0	0 0	0 0	0 0	0 0	0 0	0 0	0 0	0 0	0 0	0 0	0
0	12	0 3	0 3	0 3	0 3	0 3	0 3	0 3	0 3	0 3	0 3	0 4	0 4	0 4	3
0	24	0 5	0 6	0 6	0 6	0 6	0 6	0 6	0 7	0 7	0 7	0 7	0 7	0 7	6
0	36	0 8	0 8	0 9	0 9	0 9	0 9	0 10	0 10	0 10	0 10	0 11	0 11	0 11	9
0	48	0 11	0 11	0 12	0 12	0 12	0 13	0 13	0 13	0 14	0 14	0 14	0 15	0 15	12
1	0	0 14	0 14	0 15	0 15	0 15	0 16	0 16	0 17	0 17	0 17	0 18	0 18	0 19	15
1	12	0 16	0 17	0 17	0 18	0 18	0 19	0 19	0 20	0 20	0 21	0 21	0 22	0 22	18
1	24	0 19	0 20	0 20	0 21	0 22	0 22	0 23	0 23	0 24	0 24	0 25	0 26	0 26	21
1	36	0 22	0 23	0 23	0 24	0 25	0 25	0 26	0 27	0 27	0 28	0 29	0 29	0 30	24
1	48	0 25	0 25	0 26	0 27	0 28	0 28	0 29	0 30	0 31	0 31	0 32	0 33	0 34	27
2	0	0 27	0 28	0 29	0 30	0 31	0 32	0 32	0 33	0 34	0 35	0 36	0 37	0 37	30
2	12	0 30	0 31	0 32	0 33	0 34	0 35	0 36	0 37	0 38	0 38	0 39	0 40	0 41	33
2	24	0 33	0 34	0 35	0 36	0 37	0 38	0 39	0 40	0 41	0 42	0 43	0 44	0 45	36
2	36	0 36	0 37	0 38	0 39	0 40	0 41	0 42	0 43	0 44	0 45	0 47	0 48	0 49	39
2	48	0 38	0 40	0 41	0 42	0 43	0 44	0 45	0 47	0 48	0 49	0 50	0 51	0 52	42
3	0	0 41	0 42	0 44	0 45	0 46	0 47	0 49	0 50	0 51	0 52	0 54	0 55	0 56	45
3	12	0 44	0 45	0 47	0 48	0 49	0 51	0 52	0 53	0 55	0 56	0 57	0 59	1 0	48
3	24	0 47	0 48	0 50	0 51	0 52	0 54	0 55	0 57	0 58	0 59	1 1	1 2	1 4	51
3	36	0 49	0 51	0 52	0 54	0 55	0 57	0 58	1 0	1 1	1 3	1 4	1 6	1 7	54
3	48	0 52	0 54	0 55	0 57	0 59	1 0	1 2	1 3	1 5	1 6	1 8	1 10	1 11	57
4	0	0 55	0 57	0 58	1 0	1 2	1 3	1 5	1 7	1 8	1 10	1 12	1 13	1 15	60
4	12	0 58	0 59	1 1	1 3	1 5	1 6	1 8	1 10	1 12	1 13	1 15	1 17	1 19	63
4	24	1 0	1 2	1 4	1 6	1 8	1 10	1 11	1 13	1 15	1 17	1 19	1 21	1 22	66
4	36	1 3	1 5	1 7	1 9	1 11	1 13	1 15	1 17	1 19	1 20	1 22	1 24	1 26	69
4	48	1 6	1 8	1 10	1 12	1 14	1 16	1 18	1 20	1 22	1 24	1 26	1 28	1 30	72
5	0	1 9	1 11	1 13	1 15	1 17	1 19	1 21	1 23	1 25	1 27	1 29	1 32	1 34	75
5	12	1 11	1 14	1 16	1 18	1 20	1 22	1 24	1 27	1 29	1 31	1 33	1 35	1 37	78
5	24	1 14	1 16	1 19	1 21	1 23	1 25	1 28	1 30	1 32	1 34	1 37	1 39	1 41	81
5	36	1 17	1 19	1 22	1 24	1 26	1 29	1 31	1 33	1 36	1 38	1 40	1 43	1 45	84
5	48	1 20	1 22	1 25	1 27	1 29	1 32	1 34	1 37	1 39	1 41	1 44	1 46	1 49	87
6	0	1 22	1 25	1 27	1 30	1 32	1 35	1 37	1 40	1 42	1 45	1 47	1 50	1 52	90
6	12	1 25	1 28	1 30	1 33	1 36	1 38	1 41	1 43	1 46	1 48	1 51	1 54	1 56	93
6	24	1 28	1 31	1 33	1 36	1 39	1 41	1 44	1 47	1 49	1 52	1 55	1 57	2 0	96
6	36	1 31	1 33	1 36	1 39	1 42	1 44	1 47	1 50	1 53	1 55	1 58	2 1	2 4	99
6	48	1 33	1 36	1 39	1 42	1 45	1 48	1 50	1 53	1 56	1 59	2 2	2 5	2 7	102
7	0	1 36	1 39	1 42	1 45	1 48	1 51	1 54	1 57	2 0	2 2	2 5	2 8	2 11	105
7	12	1 39	1 42	1 45	1 48	1 51	1 54	1 47	2 0	2 3	2 6	2 9	2 12	2 15	108
7	24	1 42	1 45	1 48	1 51	1 54	1 57	2 0	2 3	2 6	2 9	2 13	2 16	2 19	111
7	36	1 44	1 48	1 51	1 54	1 57	2 0	2 3	2 7	2 10	2 13	2 16	2 19	2 22	114
7	48	1 47	1 50	1 54	1 57	2 0	2 3	2 7	2 10	2 13	2 16	2 20	2 23	2 26	117
8	0	1 50	1 53	1 57	2 0	2 3	2 7	2 10	2 13	2 17	2 20	2 23	2 27	2 30	120
8	12	1 53	1 56	2 0	2 3	2 6	2 10	2 13	2 17	2 20	2 23	2 27	2 30	2 34	123
8	24	1 55	1 59	2 3	2 6	2 9	2 13	2 16	2 20	2 23	2 27	2 30	2 34	2 37	126
8	36	1 58	2 2	2 5	2 9	2 12	2 16	2 20	2 23	2 27	2 30	2 34	2 38	2 41	129
8	48	2 0	2 5	2 8	2 12	2 16	2 19	2 23	2 27	2 30	2 34	2 38	2 41	2 45	132
9	0	2 4	2 7	2 11	2 15	2 19	2 23	2 26	2 30	2 34	2 37	2 41	2 45	2 49	135
9	12	2 6	2 10	2 14	2 18	2 22	2 26	2 29	2 33	2 37	2 41	2 45	2 49	2 52	138
9	24	2 9	2 13	2 17	2 21	2 25	2 29	2 33	2 37	2 41	2 44	2 48	2 52	2 56	141
9	36	2 12	2 16	2 20	2 24	2 28	2 32	2 36	2 40	2 44	2 48	2 52	2 56	3 0	144
9	48	2 15	2 19	2 23	2 27	2 31	2 35	2 39	2 43	2 47	2 51	2 56	3 0	3 4	147
10	0	2 17	2 22	2 26	2 30	2 34	2 38	2 42	2 47	2 51	2 55	2 59	3 3	3 7	150
10	12	2 20	2 25	2 29	2 33	2 37	2 41	2 46	2 50	2 54	2 58	3 3	3 7	3 11	153
10	24	2 23	2 27	2 32	2 36	2 40	2 45	2 49	2 53	2 58	3 2	3 6	3 11	3 15	156
10	36	2 26	2 30	2 35	2 39	2 43	2 48	2 52	2 57	3 1	3 5	3 10	3 14	3 19	159
10	48	2 28	2 33	2 37	2 42	2 46	2 51	2 55	3 0	3 4	3 9	3 13	3 18	3 22	162
11	0	2 31	2 36	2 40	2 45	2 49	2 54	2 59	3 3	3 8	3 12	3 17	3 22	3 26	165
11	12	2 34	2 39	2 43	2 48	2 53	2 57	3 2	3 7	3 11	3 16	3 21	3 25	3 30	168
11	24	2 37	2 41	2 46	2 51	2 56	3 0	3 5	3 10	3 15	3 19	3 24	3 29	3 34	171
11	36	2 39	2 44	2 49	2 54	2 59	3 4	3 8	3 13	3 18	3 23	3 28	3 33	3 37	174
11	48	2 42	2 47	2 52	2 57	3 2	3 7	3 12	3 17	3 22	3 26	3 31	3 36	3 41	177
12	0	2 45	2 50	2 55	3 0	3 5	3 10	3 15	3 20	3 25	3 30	3 35	3 40	3 45	180

TABLE XX.

To reduce the Time of the Moon's Passage over the Meridian of Greenwich, to the Time of its Passage over any other Meridian.

Arg. time fr. moon's south.	Arg. Daily Var. of the Moon's passing the Meridian. 40'	42'	44'	46'	48'	50'	52'	54'	56'	58'	60'	62'	64'	66'	Arg. Long.
0h. 0'	0'	0'	0'	0'	0'	0'	0'	0'	0'	0'	0'	0'	0'	0'	0°
0 20	1	1	1	1	1	1	1	1	1	1	1	1	1	1	5
0 40	1	1	1	1	1	1	1	1	1	2	2	2	2	2	10
1 0	2	2	2	2	2	2	2	2	2	2	2	2	3	3	15
1 20	2	2	2	2	3	3	3	3	3	3	3	3	3	4	20
1 40	3	3	3	3	[illegible]	3	3	4	4	4	4	4	4	4	25
2 0	3	3	4	4	[illegible]	4	4	4	4	5	5	5	5	5	30
2 20	4	4	4	4	5	5	5	5	5	5	6	6	6	6	35
2 40	4	4	5	5	5	5	6	6	6	6	6	7	7	7	40
3 0	5	5	5	6	6	6	6	7	7	7	7	7	8	8	45
3 20	5	6	6	6	6	7	7	7	7	8	8	8	9	9	50
3 40	6	6	7	7	7	7	8	8	8	9	9	9	9	10	55
4 0	6	7	7	7	8	8	8	9	9	9	10	10	10	11	60
4 20	7	7	8	8	8	9	9	9	10	10	10	11	11	11	65
4 40	7	8	8	9	9	9	10	10	10	11	11	12	12	12	70
5 0	8	9	9	9	10	10	10	11	11	12	12	12	13	13	75
5 20	9	9	9	10	10	11	11	12	12	12	13	13	14	14	80
5 40	9	10	10	11	11	11	12	12	13	13	14	14	14	15	85
6 0	10	10	11	11	12	12	13	13	13	14	14	15	15	16	90
6 20	10	11	11	12	12	13	13	14	14	15	15	16	16	17	95
6 40	11	11	12	12	13	13	14	14	15	15	16	17	17	18	100
7 0	11	12	12	13	14	14	15	15	16	16	17	17	18	18	105
7 20	12	12	13	14	14	15	15	16	16	17	18	18	19	19	110
7 40	12	13	14	14	15	15	16	17	17	18	18	19	20	20	115
8 0	13	14	14	15	15	16	17	17	18	19	19	20	20	21	120
8 20	13	14	15	15	16	17	17	18	19	19	20	21	21	22	125
8 40	14	15	15	16	17	17	18	19	19	20	21	21	22	23	130
9 0	14	15	16	17	17	18	19	20	20	21	22	22	23	24	135
9 20	15	16	17	17	18	19	20	20	21	22	22	23	24	25	140
9 40	15	16	17	18	19	19	20	21	22	22	23	24	25	25	145
10 0	16	17	18	19	19	20	21	22	22	23	24	25	26	26	150
10 20	16	18	18	19	20	21	22	22	23	24	25	26	26	27	155
10 40	17	18	19	20	21	21	22	23	24	25	26	26	27	28	160
11 0	17	19	20	20	21	22	23	24	25	26	26	27	28	29	165
11 20	18	19	20	21	22	23	24	25	25	26	27	28	29	30	170
11 40	18	20	21	22	23	23	24	25	26	27	28	29	30	31	175
12 0	19	20	21	22	23	24	25	26	27	28	29	30	31	32	180

TABLE XXI.

Error of Observation, arising from an Inclination in the Planes of the Central Mirror of the Circular Instrument; that Inclination being assumed equal to One Minute.

Arg obser. angle.	Obser. to the Right.	Obser. to the Left.	Cross Obser.	Arg. Obser angle.	Obser. to the Right.		Obser. to the Left.		Cross Obser.		Arg. Obser. angle.	Obser. to the Right.		Obser. to the Left.		Cross Obser.	
°	"	"	"	°	'	"	'	"	'	"	°	'	"	'	"	'	"
0	0	0	0	60	0	33	0	6	0	14	100	2	1	0	33	0	44
10	3	1	2	65	0	38	0	8	0	15	105	2	23	0	39	0	58
20	6	2	4	70	0	47	0	10	0	18	110	2	50	0	46	1	8
30	10	1	6	75	0	55	0	13	0	21	115	3	23	0	54	1	23
40	16	0	8	80	1	4	0	16	0	24	120	4	4	1	4	1	39
45	19	1	9	85	1	15	0	19	0	28	125	4	55	1	15	1	30
50	23	2	11	90	1	28	0	23	0	32	130	5	58	1	28	2	23
55	28	4	12	95	1	43	0	28	0	37							

TABLE XXII. Error of Observation, arising from an Error in the Line of Collimation.

Arg. obser. angle.	Argument. Error of the Line of Collimation. 10′	20′	24′	28′	30′	32′	34′	36′	38′	40′	42′	44′	46′	48′	50′	52′	54′	56′	58′	1° 0′
°	″	″	″	″	″	″	″	″	″	″	″	″	′ ″	′ ″	′ ″	′ ″	′ ″	′ ″	′ ″	′ ″
5	0	0	0	1	1	1	1	1	1	1	1	1	0 2	0 2	0 2	0 2	0 2	0 2	0 3	0 3
10	0	1	1	1	1	2	2	2	2	2	3	3	0 3	0 4	0 4	0 4	0 4	0 5	0 5	0 6
12	0	1	1	1	2	2	2	2	3	3	3	4	0 4	0 4	0 5	0 5	0 5	0 6	0 6	0 7
14	0	1	1	2	2	2	2	3	3	3	4	4	0 5	0 5	0 5	0 6	0 6	0 7	0 7	0 8
16	0	1	1	2	2	3	3	3	4	4	4	5	0 5	0 6	0 6	0 7	0 7	0 8	0 8	0 9
18	0	1	2	2	3	3	3	4	4	4	5	5	0 6	0 6	0 7	0 7	0 8	0 9	0 9	0 10
20	0	1	2	2	3	3	4	4	4	5	5	6	0 7	0 7	0 8	0 8	0 9	0 10	0 10	0 11
22	0	1	2	3	3	3	4	4	5	5	6	7	0 7	0 8	0 8	0 9	0 10	0 11	0 11	0 12
24	0	1	2	3	3	4	4	5	5	6	7	7	0 8	0 9	0 9	0 10	0 11	0 12	0 12	0 13
26	0	2	2	3	4	4	5	5	6	6	7	8	0 9	0 9	0 10	0 11	0 12	0 13	0 14	0 15
28	0	2	3	3	4	4	5	6	6	7	8	8	0 9	0 10	0 11	0 12	0 13	0 14	0 15	0 16
30	1	2	3	4	4	5	5	6	7	7	8	9	0 10	0 11	0 12	0 13	0 14	0 15	0 16	0 17
32	1	2	3	4	5	5	6	6	7	8	9	10	0 11	0 12	0 13	0 14	0 15	0 16	0 17	0 18
34	1	2	3	4	5	5	6	7	8	9	9	10	0 11	0 12	0 13	0 14	0 16	0 17	0 18	0 19
36	1	2	3	4	5	6	7	7	8	9	10	11	0 12	0 13	0 14	0 15	0 17	0 18	0 19	0 20
38	1	2	4	5	6	6	7	8	9	10	11	12	0 13	0 14	0 15	0 16	0 18	0 19	0 20	0 22
40	1	3	4	5	6	6	7	8	9	10	11	12	0 13	0 15	0 16	0 17	0 19	0 20	0 21	0 23
42	1	3	4	5	6	7	8	9	10	11	12	13	0 14	0 15	0 17	0 18	0 20	0 21	0 22	0 24
44	1	3	4	6	7	7	8	9	10	11	12	14	0 15	0 16	0 18	0 19	0 21	0 22	0 24	0 25
46	1	3	4	6	7	8	9	10	11	12	13	14	0 16	0 17	0 18	0 20	0 22	0 23	0 25	0 27
48	1	3	5	6	7	8	9	10	11	12	14	15	0 16	0 18	0 19	0 21	0 23	0 24	0 26	0 28
50	1	3	5	6	8	8	9	11	12	13	14	16	0 17	0 19	0 20	0 22	0 24	0 25	0 27	0 29
52	1	3	5	7	8	9	10	11	12	14	15	16	0 18	0 20	0 21	0 23	0 25	0 27	0 29	0 31
54	1	4	5	7	8	9	10	12	13	14	16	17	0 19	0 20	0 22	0 24	0 26	0 28	0 30	0 32
56	1	4	5	7	9	10	11	12	13	15	16	18	0 20	0 21	0 23	0 25	0 27	0 29	0 31	0 33
58	1	4	6	8	9	10	11	13	14	15	17	19	0 20	0 22	0 24	0 26	0 28	0 30	0 32	0 35
60	1	4	6	8	9	10	12	13	15	16	18	19	0 21	0 23	0 25	0 27	0 29	0 31	0 34	0 36
62	1	4	6	8	10	11	12	14	15	17	18	20	0 22	0 24	0 26	0 28	0 31	0 33	0 35	0 38
64	1	4	6	9	10	11	13	14	16	17	19	21	0 23	0 25	0 27	0 29	0 32	0 34	0 37	0 39
66	1	4	7	9	10	12	13	15	16	18	20	22	0 24	0 26	0 28	0 31	0 33	0 35	0 38	0 41
68	1	4	7	9	11	12	14	15	17	19	21	23	0 25	0 27	0 29	0 32	0 34	0 37	0 40	0 43
70	1	5	7	10	11	13	14	16	18	20	22	24	0 26	0 28	0 31	0 33	0 36	0 38	0 41	0 44
72	1	5	7	10	12	13	15	16	18	20	22	25	0 27	0 29	0 32	0 34	0 37	0 40	0 43	0 46
74	2	5	8	10	12	13	15	17	19	21	23	25	0 28	0 30	0 33	0 35	0 38	0 41	0 44	0 47
76	2	5	8	11	13	14	16	18	20	22	24	26	0 29	0 31	0 34	0 37	0 40	0 43	0 46	0 49
78	2	6	8	11	13	14	16	18	20	23	25	27	0 30	0 33	0 35	0 38	0 41	0 44	0 47	0 51
80	2	6	9	12	13	15	17	19	21	23	26	28	0 31	0 34	0 37	0 40	0 43	0 46	0 49	0 53
82	2	6	9	12	14	16	18	20	22	24	27	29	0 32	0 35	0 38	0 41	0 44	0 47	0 51	0 55
84	2	6	9	12	14	16	18	20	23	25	28	30	0 33	0 36	0 39	0 42	0 46	0 49	0 53	0 57
86	2	7	10	13	15	17	19	21	24	26	29	32	0 34	0 37	0 41	0 44	0 47	0 51	0 55	0 59
88	2	7	10	13	15	17	20	22	24	27	30	33	0 36	0 39	0 42	0 46	0 49	0 53	0 57	1 1
90	2	7	10	14	16	18	20	23	25	28	31	34	0 37	0 40	0 44	0 47	0 51	0 55	0 59	1 3
92	2	7	11	14	17	18	21	23	26	29	32	35	0 38	0 42	0 45	0 49	0 53	0 57	1 1	1 5
94	2	8	11	15	17	19	22	24	27	30	33	36	0 40	0 43	0 47	0 51	0 55	0 59	1 3	1 8
96	2	8	11	15	18	20	22	25	28	31	34	37	0 41	0 45	0 49	0 52	0 57	1 1	1 5	1 10
98	2	8	12	16	18	21	23	26	29	32	35	39	0 42	0 46	0 50	0 54	0 59	1 3	1 7	1 12
100	2	8	12	17	19	21	24	27	30	33	37	40	0 44	0 48	0 52	0 56	1 1	1 5	1 10	1 15
102	2	9	13	17	20	22	25	28	31	34	38	42	0 46	0 50	0 54	0 58	1 3	1 7	1 12	1 18
104	3	9	13	18	[illegible]	23	26	29	32	36	39	43	0 47	0 51	0 56	1 0	1 5	1 10	1 15	1 21
106	3	9	14	18	21	24	27	30	34	37	41	45	0 49	0 53	0 58	1 3	1 8	1 13	1 18	1 24
108	3	10	14	19	22	25	28	31	35	38	42	46	0 51	0 55	1 0	1 5	1 10	1 15	1 21	1 27
110	3	10	15	20	23	26	29	32	36	40	44	48	0 53	0 57	1 [illegible]	1 7	1 13	1 18	1 24	1 30
112	3	10	15	[illegible]	23	27	30	34	37	41	46	50	0 55	1 0	1 [illegible]	1 10	1 15	1 21	1 27	1 33
114	3	11	16	21	24	28	31	35	39	43	48	52	0 57	1 2	1 7	1 13	1 18	1 24	1 30	1 37
116	3	11	16	22	25	29	32	36	40	45	49	54	0 59	1 4	1 10	1 16	1 22	1 28	1 [illegible]	1 41
118	3	12	17	23	26	30	34	38	42	46	51	56	1 1	1 7	1 13	1 19	1 [illegible]	1 33	1 38	1 45
120	3	12	18	24	27	31	35	39	44	48	53	58	1 4	1 10	1 16	1 22	1 [illegible]	1 35	1 [illegible]	1 49

TABLE XXII. Error of Observation, arising from an Error in the Line of Collimation.

Arg. obser. angle.	Argument Error of the Line of Collimation.													
	1 2	1 4	1 6	1 8	1 10	1 12	1 14	1 16	1 18	1 20	1 22	1 24	1 26	1 28
5	0 3	0 3	0 3	0 4	0 4	0 4	0 4	0 4	0 5	0 5	0 5	0 5	0 6	0 6
10	0 6	0 6	0 7	0 7	0 7	0 8	0 8	0 9	0 9	0 10	0 10	0 11	0 11	0 12
12	0 7	0 7	0 8	0 8	0 9	0 9	0 10	0 11	0 11	0 12	0 13	0 13	0 14	0 14
14	0 8	0 9	0 9	0 10	0 11	0 11	0 12	0 12	0 13	0 14	0 14	0 15	0 16	0 16
16	0 9	0 10	0 11	0 11	0 12	0 13	0 13	0 14	0 15	0 16	0 16	0 17	0 18	0 19
18	0 11	0 11	0 12	0 13	0 14	0 14	0 15	0 16	0 17	0 18	0 19	0 19	0 20	0 21
20	0 12	0 13	0 13	0 14	0 15	0 16	0 17	0 18	0 19	0 20	0 21	0 22	0 23	0 24
22	0 13	0 14	0 15	0 16	0 17	0 18	0 19	0 20	0 21	0 22	0 23	0 24	0 25	0 26
24	0 14	0 15	0 16	0 17	0 18	0 19	0 20	0 21	0 23	0 24	0 25	0 26	0 27	0 29
26	0 16	0 16	0 18	0 19	0 20	0 21	0 22	0 23	0 25	0 26	0 27	0 28	0 30	0 31
28	0 17	0 18	0 19	0 20	0 21	0 23	0 24	0 25	0 27	0 28	0 29	0 31	0 32	0 34
30	0 18	0 19	0 20	0 22	0 23	0 24	0 26	0 27	0 28	0 30	0 31	0 33	0 35	0 36
32	0 19	0 20	0 22	0 23	0 25	0 26	0 27	0 29	0 30	0 32	0 34	0 35	0 37	0 39
34	0 21	0 21	0 23	0 25	0 26	0 28	0 29	0 31	0 33	0 34	0 36	0 38	0 39	0 41
36	0 22	0 23	0 25	0 26	0 28	0 29	0 31	0 33	0 35	0 36	0 38	0 40	0 42	0 44
38	0 23	0 25	0 26	0 28	0 29	0 31	0 33	0 35	0 37	0 38	0 40	0 42	0 44	0 47
40	0 24	0 26	0 28	0 29	0 31	0 33	0 35	0 37	0 39	0 40	0 43	0 45	0 47	0 49
42	0 26	0 27	0 29	0 31	0 33	0 35	0 37	0 39	0 41	0 43	0 45	0 47	0 50	0 52
44	0 27	0 29	0 31	0 32	0 35	0 37	0 39	0 41	0 43	0 45	0 47	0 50	0 52	0 54
46	0 28	0 30	0 31	0 34	0 36	0 38	0 41	0 43	0 45	0 47	0 50	0 52	0 55	0 57
48	0 30	0 32	0 34	0 36	0 38	0 40	0 43	0 45	0 47	0 50	0 52	0 55	0 57	1 0
50	0 31	0 33	0 35	0 38	0 40	0 42	0 45	0 47	0 50	0 52	0 55	0 57	1 0	1 3
52	0 33	0 35	0 37	0 39	0 42	0 44	0 47	0 49	0 52	0 54	0 57	1 0	1 3	1 6
54	0 34	0 36	0 39	0 41	0 44	0 46	0 49	0 51	0 54	0 57	1 0	1 3	1 6	1 9
56	0 36	0 38	0 41	0 43	0 45	0 48	0 51	0 54	0 57	0 59	1 2	1 5	1 9	1 12
58	0 37	0 39	0 42	0 45	0 47	0 50	0 53	0 56	0 59	1 2	1 5	1 8	1 11	1 15
60	0 39	0 41	0 44	0 47	0 49	0 52	0 55	0 58	1 1	1 4	1 8	1 11	1 14	1 18
62	0 40	0 43	0 46	0 49	0 51	0 54	0 57	1 0	1 4	1 7	1 10	1 14	1 17	1 21
64	0 42	0 45	0 48	0 50	0 53	0 56	1 0	1 3	1 7	1 10	1 13	1 17	1 21	1 24
66	0 44	0 46	0 50	0 52	0 55	0 59	1 2	1 5	1 9	1 12	1 16	1 20	1 24	1 28
68	0 45	0 48	0 52	0 55	0 58	1 1	1 4	1 8	1 12	1 15	1 19	1 23	1 27	1 31
70	0 47	0 50	0 53	0 57	1 0	1 3	1 7	1 11	1 14	1 18	1 22	1 26	1 30	1 34
72	0 49	0 52	0 55	0 59	1 2	1 6	1 9	1 13	1 17	1 21	1 25	1 29	1 34	1 38
74	0 51	0 54	0 57	1 1	1 4	1 8	1 12	1 16	1 20	1 24	1 28	1 33	1 37	1 42
76	0 52	0 56	1 0	1 3	1 7	1 11	1 15	1 19	1 23	1 27	1 32	1 36	1 41	1 46
78	0 54	0 58	1 2	1 5	1 9	1 13	1 17	1 22	1 26	1 30	1 35	1 40	1 45	1 49
80	0 56	1 0	1 4	1 8	1 12	1 16	1 20	1 25	1 29	1 33	1 38	1 43	1 48	1 53
82	0 58	1 2	1 6	1 10	1 14	1 19	1 23	1 28	1 32	1 37	1 42	1 47	1 52	1 57
84	1 1	1 4	1 8	1 13	1 17	1 22	1 26	1 31	1 36	1 40	1 46	1 51	1 56	2 1
86	1 3	1 7	1 11	1 15	1 20	1 25	1 29	1 34	1 39	1 44	1 50	1 55	2 0	2 6
88	1 5	1 9	1 14	1 18	1 23	1 27	1 32	1 38	1 43	1 48	1 53	1 59	2 4	2 10
90	1 7	1 11	1 16	1 21	1 26	1 30	1 36	1 41	1 46	1 52	1 57	2 3	2 9	2 15
92	1 10	1 14	1 19	1 24	1 29	1 34	1 39	1 44	1 50	1 56	2 2	2 7	2 14	2 20
94	1 12	1 17	1 22	1 27	1 32	1 37	1 43	1 48	1 54	2 0	2 6	2 12	2 18	2 25
96	1 15	1 19	1 25	1 30	1 35	1 40	1 46	1 52	1 58	2 4	2 10	2 17	2 23	2 30
98	1 17	1 22	1 28	1 33	1 38	1 44	1 50	1 56	2 2	2 8	2 15	2 22	2 28	2 35
100	1 20	1 25	1 31	1 36	1 42	1 48	1 54	2 0	2 7	2 13	2 20	2 27	2 34	2 41
102	1 23	1 28	1 34	1 40	1 46	1 52	1 58	2 5	2 11	2 18	2 25	2 32	2 39	2 47
104	1 26	1 31	1 37	1 43	1 50	1 56	2 3	2 9	2 16	2 23	2 30	2 38	2 45	2 53
106	1 29	1 35	1 41	1 47	1 54	2 0	2 7	2 14	2 21	2 28	2 36	2 43	2 51	2 59
108	1 31	1 38	1 45	1 51	1 58	2 5	2 12	2 19	2 26	2 34	2 42	2 49	2 58	3 6
110	1 35	1 42	1 49	1 55	2 2	2 9	2 17	2 24	2 32	2 39	2 48	2 56	3 4	3 13
112	1 39	1 46	1 53	2 0	2 7	2 14	2 21	2 29	2 37	2 45	2 54	3 3	3 11	3 20
114	1 43	1 50	1 57	2 4	2 12	2 19	2 27	2 35	2 43	2 52	3 1	3 10	3 19	3 28
116	1 48	1 54	2 2	2 9	2 17	2 25	2 33	2 41	2 50	2 58	3 8	3 17	3 27	3 36
118	1 52	1 59	2 7	2 14	2 22	2 31	2 39	2 48	2 57	3 6	3 15	3 25	3 35	3 45
120	1 56	2 4	2 12	2 20	2 28	2 37	2 46	2 55	3 4	3 14	3 23	3 33	3 44	3 54

XXII. Error of Obſervation, ariſing from an the Line of Collimation.

Argument. Error of the Line of Collimation.											
0	1 32	1 34	1 36	1 38	1 40	1 42	1 44	1 46	1 48	1 50	1 52
″	′ ″	′ ″	′ ″	′ ″	′ ″	′ ″	′ ″	′ ″	′ ″	′ ″	′ ″
6	0 6	0 7	0 7	0 7	0 8	0 8	0 8	0 9	0 9	0 9	0 9
2	0 13	0 13	0 14	0 15	0 15	0 16	0 17	0 17	0 18	0 18	0 19
5	0 15	0 16	0 17	0 18	0 18	0 19	0 20	0 20	0 21	0 [illegible]	0 23
7	0 18	0 19	0 20	0 20	0 21	0 22	0 23	0 24	0 25	0 26	0 27
0	0 21	0 22	0 23	0 23	0 24	0 25	0 27	0 28	0 29	0 30	0 31
2	0 23	0 24	0 25	0 26	0 28	0 29	0 30	0 31	0 32	0 33	0 34
5	0 26	0 27	0 28	0 30	0 31	0 32	0 33	0 35	0 36	0 37	0 38
7	0 29	0 30	0 31	0 33	0 34	0 35	0 37	0 38	0 39	0 41	0 42

TABLE XXIII. For computing the Equation of equal Altitudes.

PART FIRST.

Arg. Lat. of Place.	Argument. Interval between the Observations.													
	h. ′ 3 20	h. ′ 4 0	h. ′ 4 40	h. ′ 5 20	h. ′ 6 0	h. ′ 6 40	h. ′ 7 20	h. ′ 8 0	h. ′ 8 40	h. ′ 9 20	h. ′ 10 0	h. ′ 10 40	h. ′ 11 20	h. ′ 12 0
0°	0	0	0	0	0	0	0	0	0	0	0	0	0	0
1	19	19	20	20	21	21	22	22	23	24	25	26	28	29
2	38	39	39	40	41	42	43	45	46	48	50	52	55	58
3	57	58	59	60	62	63	65	67	69	72	75	79	83	87
4	77	78	79	80	82	84	87	90	93	96	100	103	110	116
5	96	97	99	100	103	106	108	112	116	120	126	132	138	145
6	115	116	118	120	123	127	130	134	139	144	151	158	166	174
7	134	136	138	140	144	148	152	157	162	168	176	184	194	204
8	154	156	158	161	165	169	174	180	186	193	202	211	221	233
9	173	176	179	182	186	191	196	203	21[illegible]	218	228	238	250	263
10	193	196	200	203	207	213	219	226	234	243	254	265	279	293
11	213	216	220	224	228	234	240	248	256	268	280	290	306	323
12	232	236	240	245	250	257	264	273	282	293	306	320	336	354
13	253	257	261	266	272	279	287	296	307	318	332	347	365	385
14	273	277	282	287	294	301	310	320	331	344	358	375	394	415
15	293	298	303	309	316	324	333	344	356	370	389	403	423	446
16	314	319	324	330	338	347	357	368	381	396	411	431	453	478
17	335	340	345	352	360	369	380	392	406	422	440	460	483	509
18	356	361	367	374	383	393	404	417	432	448	467	489	513	541
19	377	383	389	397	406	416	428	436	457	475	495	518	544	574
20	399	404	411	419	429	440	453	467	483	502	523	548	575	607
21	421	427	434	442	452	464	477	492	510	530	552	577	607	640
22	443	449	456	466	476	488	502	518	537	557	581	608	638	673
23	465	472	480	489	500	513	528	544	564	585	610	638	671	707
24	488	495	503	513	523	538	554	571	591	614	643	670	704	742
25	511	518	527	537	550	564	580	598	619	643	670	701	737	777
26	534	542	551	562	575	589	606	626	648	673	701	734	771	813
27	558	566	576	587	600	616	634	654	677	702	733	766	805	849
28	582	591	601	613	627	643	661	682	706	733	765	800	840	886
29	607	616	626	639	653	670	689	711	736	765	797	834	876	924
30	632	641	652	665	680	698	718	741	767	796	830	868	912	962
31	658	668	679	692	708	726	747	771	798	829	864	904	949	1001
32	685	694	706	72[illegible]	736	755	777	802	830	862	898	940	987	1041
33	711	722	734	748	763	785	826	833	863	896	934	977	1026	1082
34	739	750	762	777	793	815	839	865	896	930	970	1015	1066	1124
35	767	778	791	807	825	846	871	898	932	966	1007	1053	1106	1167
36	796	807	821	837	856	878	903	932	965	1002	1045	1093	1148	1211
37	825	837	852	868	888	911	937	968	1011	1039	1083	1133	1191	1256
38	856	868	883	900	921	944	971	1002	1038	1078	1123	1173	1235	1302
39	887	900	915	933	954	979	1007	1039	1075	1117	1164	1218	1280	1350
40	919	932	948	967	989	1014	1043	1077	1114	1157	1206	1262	1326	1398
41	952	966	982	1008	1024	1051	1088	1115	1155	1199	1250	1308	1374	1449
42	986	1000	1078	1038	1061	1088	1127	1135	1196	1242	1295	1354	1423	1500
43	1022	1036	1054	1075	1099	1127	116[illegible]	1197	1238	1287	1341	1403	1474	1554
44	1058	1073	1091	1113	1138	1167	1201	1239	1282	1332	1387	1453	1526	1609
45	1096	1111	1130	1152	1178	1209	1243	1283	1328	1380	1438	1504	1580	1666
46	1134	1151	1170	1193	1220	1252	1288	1329	1375	1429	1489	1558	1636	1726
47	1175	1192	1212	1236	1264	1296	1335	1376	1424	1479	1542	1613	1699	1787
48	1217	1234	1255	1280	1309	1343	1381	1425	1475	1532	1597	1671	1755	1851
49	1260	1278	1300	1326	1356	1390	1437	1476	1528	1587	1654	1730	1818	1917
50	1306	1324	1347	1373	1404	1440	1482	1529	1583	1644	1714	1793	1883	1986
51	1353	1372	1396	1423	1455	1492	1535	1584	1640	1703	1775	1857	1951	2038
52	1402	1422	1447	1475	1508	1547	1592	1642	1700	1766	1840	1925	2023	2133
53	1434	1475	1500	1529	1564	1604	1650	1703	1762	1831	1908	1996	2097	2209
54	1508	1529	1555	1586	1622	1664	1711	1766	1828	1916	1979	2071	2175	2294
55	1565	1587	1614	1646	1683	1726	1776	1832	1897	1970	2054	2148	2257	2380
56	1624	1647	1675	1709	1747	1792	1845	1902	1969	2045	2132	2230	2343	2471
57	1687	1711	1740	1775	1815	1861	1915	1976	2045	2124	2214	2316	2433	2566
58	1753	1778	1808	1844	1886	1934	1990	2053	2125	2207	2301	2407	2529	2667
59	1823	1849	1881	1919	1962	20[illegible]	2069	2135	2210	2296	2393	2503	2630	2773
60	1898	1925	1957	1996	2041	2094	2154	2222	2300	2389	2491	2606	2737	2887

TABLE XXIV. For computing the Equation of equal Altitudes.

PART SECOND.

Arg. Declin. of the Sun.	Argument. Interval between the Observations. 3h. 20′	4h. 0′	4h. 40′	5h. 20′	6h. 0′	6h. 40′	7h. 20′	8h. 0′	8h. 40′	9h. 20′	10h. 0′	10h. 40′	11h. 20′	12h. 0′
0° 0′	0	0	0	0	0	0	0	0	0	0	0	0	0	0
0 20	6	6	5	5	5	4	4	4	3	3	2	2	1	0
0 40	12	11	11	10	10	9	8	7	6	5	4	3	2	0
1 0	17	17	16	15	14	13	12	11	10	8	6	5	2	0
1 20	23	22	22	21	19	18	17	15	13	11	9	6	3	0
1 40	29	28	27	26	24	23	21	19	16	14	11	8	4	0
2 0	35	33	32	31	29	27	25	22	19	16	13	9	5	0
2 20	40	39	38	36	34	32	29	26	23	19	15	11	6	0
2 40	46	45	43	41	39	36	33	30	26	22	17	12	6	0
3 0	52	50	48	46	44	41	37	34	29	25	19	14	7	0
3 20	58	56	54	51	49	45	42	37	33	27	22	15	8	0
3 40	64	62	59	57	53	50	46	41	36	30	24	17	9	0
4 0	69	67	65	62	58	54	50	45	39	33	26	18	10	0
4 20	75	73	70	67	63	59	54	49	42	36	28	20	10	0
4 40	81	78	76	72	68	63	58	52	46	39	30	21	11	0
5 0	87	84	81	77	73	68	62	56	49	41	32	23	12	0
5 20	92	90	86	82	78	73	67	60	52	44	35	24	13	0
5 40	98	95	92	87	83	77	71	64	56	47	37	26	14	0
6 0	104	101	97	92	87	82	75	67	59	50	39	27	14	0
6 20	110	106	102	98	92	86	79	71	62	52	41	29	15	0
6 40	116	112	108	103	97	91	83	75	65	55	43	30	16	0
7 0	122	118	113	108	102	95	87	79	69	58	46	32	17	0
7 20	128	124	119	113	107	100	92	82	72	61	48	33	18	0
7 40	134	129	124	118	112	104	96	86	75	63	50	35	18	0
8 0	139	135	130	124	117	109	100	90	79	66	52	36	19	0
8 20	145	140	135	129	122	114	104	94	82	69	54	38	20	0
8 40	151	146	141	134	127	118	109	98	85	72	57	39	21	0
9 0	157	152	146	139	132	123	113	102	89	76	59	41	22	0
9 20	163	158	152	144	137	128	118	106	92	78	61	43	23	0
9 40	169	164	158	150	142	132	122	110	95	80	63	44	23	0
10 0	175	170	163	155	147	137	126	113	99	83	65	46	24	0
10 20	181	175	168	160	152	141	130	117	102	86	67	47	25	0
10 40	187	181	174	166	157	146	134	120	106	89	70	49	26	0
11 0	193	186	179	171	162	151	138	124	109	92	72	50	27	0
11 20	199	192	185	177	167	156	142	128	112	95	75	52	27	0
11 40	205	198	191	182	172	161	146	132	116	98	77	54	28	0
12 0	211	204	197	188	177	166	152	136	120	101	79	55	29	0
12 30	220	213	205	196	184	172	158	142	125	105	82	58	30	0
13 0	229	222	214	204	192	179	164	148	130	109	86	60	32	0
13 30	238	231	222	212	200	187	171	154	135	114	89	63	33	0
14 0	247	240	231	220	208	194	178	160	140	118	93	65	34	0
14 30	257	249	240	228	215	201	184	166	145	122	96	67	36	0
15 0	266	257	248	236	223	208	191	172	150	126	100	70	37	0
15 30	275	267	257	245	231	215	197	178	156	131	103	72	38	0
16 0	285	276	265	253	239	223	204	184	161	136	106	75	39	0
16 30	294	285	274	262	247	230	211	190	166	140	110	77	41	0
17 0	303	294	283	270	255	237	218	196	172	144	114	80	42	0
17 30	313	303	292	278	263	245	225	202	176	149	118	82	43	0
18 0	322	313	301	287	271	252	232	208	182	154	121	85	45	0
18 30	332	322	310	293	279	260	239	215	188	158	124	87	46	0
19 0	342	331	319	304	287	268	246	221	193	162	128	90	47	0
19 30	351	341	328	313	295	275	253	227	198	167	132	93	49	0
20 0	361	350	337	321	303	283	260	233	204	172	135	95	50	0
20 30	371	360	346	330	312	291	267	240	210	177	139	97	51	0
21 0	381	369	355	339	320	298	274	246	215	181	142	100	53	0
21 30	391	379	365	348	328	306	281	253	221	186	146	102	54	0
22 0	401	389	374	357	337	314	288	259	227	191	150	105	55	0
22 30	411	399	384	366	345	322	295	265	232	195	154	107	57	0
23 0	421	408	393	375	354	330	303	272	238	200	158	110	58	0
23 28	431	418	401	383	362	337	310	278	244	205	162	113	60	0

TABLE XXV. Altitude to be observed, in Order to ascertain the Apparent Time with the greatest Accuracy.

Arg. Lat.	Argument. Declination of the Sun or Star.																							
	2°		4°		6°		8°		10°		12°		14°		16°		18°		20°		22°		24°	
°	°	′	°	′	°	′	°	′	°	′	°	′	°	′	°	′	°	′	°	′	°	′	°	′
0	0	0	0	0	0	0	0	0	0	0	0	0	0	0	0	0	0	0	0	0	0	0	0	0
1	30	1	14	29	9	37	7	12	5	46	4	49	4	8	3	38	3	14	2	55	2	40	2	28
2	90	0	30	1	19	30	14	31	11	36	9	40	8	18	7	17	6	29	5	52	5	21	4	55
3	41	49	48	37	30	3	22	5	17	32	14	35	12	30	10	37	9	41	8	48	8	2	7	24
4	30	1	90	0	41	52	30	5	23	41	19	36	16	45	14	40	13	3	11	46	10	44	9	53
5	23	36	53	10	16	30	38	40	30	8	24	47	21	7	18	26	16	23	14	46	13	27	12	22
6	19	30	41	52	90	0	48	40	40	9	30	11	25	36	22	17	19	46	17	48	16	12	14	53
7	16	38	34	55	59	4	61	7	44	34	35	53	30	15	26	14	23	14	20	52	18	59	17	26
8	14	31	30	5	48	41	90	0	53	16	42	1	35	7	30	20	26	46	24	1	21	48	20	1
9	12	23	26	29	41	56	62	50	64	16	48	48	40	17	34	35	30	25	27	13	24	41	22	37
10	11	35	23	41	37	0	53	17	90	0	56	38	45	52	39	3	34	11	30	30	27	37	25	16
11	10	32	21	27	33	13	46	50	65	31	69	36	52	4	43	48	38	8	33	55	30	47	27	59
12	9	40	19	36	30	12	42	2	56	39	90	0	59	35	48	58	42	17	37	26	33	42	30	44
13	8	55	18	4	27	41	38	13	50	32	67	33	68	25	54	42	46	43	41	8	36	54	33	35
14	8	18	16	45	25	36	35	7	43	53	59	15	90	0	61	22	51	31	45	1	40	11	36	30
15	7	45	15	38	23	49	32	32	42	8	53	27	69	11	69	53	56	33	49	11	43	43	39	31
16	7	16	14	40	22	17	30	20	39	4	48	59	61	23	90	0	63	6	50	12	47	22	42	39
17	6	51	13	48	20	57	28	16	36	20	45	20	55	30	70	31	71	7	58	44	51	18	45	57
18	6	29	13	3	19	46	26	46	34	11	42	17	51	32	63	8	90	0	64	37	55	34	49	26
19	6	9	12	22	18	44	25	18	32	14	39	41	48	0	57	51	71	39	72	9	60	21	53	10
20	5	51	11	46	17	48	24	1	30	31	37	27	45	2	53	43	64	40	90	0	65	54	57	13
21	5	33	11	13	16	57	22	51	28	59	35	28	42	28	50	17	59	34	72	37	73	4	61	46
22	5	21	10	44	16	12	21	49	27	37	33	43	40	14	47	22	55	35	65	56	90	0	67	5
23	5	7	10	17	15	31	20	52	26	23	32	9	38	15	44	52	52	16	61	5	73	29	73	52
24	4	55	9	53	14	53	20	1	25	17	30	45	36	30	42	40	49	27	57	15	67	6	90	0
25	4	44	9	30	14	19	19	14	24	16	29	28	34	35	40	43	46	59	54	2	62	25	74	14
26	4	34	9	9	13	48	18	31	23	22	28	19	33	30	38	38	44	50	51	18	58	44	68	7
27	4	25	8	50	13	19	17	51	22	29	27	15	32	12	37	23	42	54	48	53	55	36	63	37
28	4	16	8	33	12	52	17	15	21	43	26	17	31	1	35	57	41	10	46	47	52	57	60	3
29	4	8	8	16	12	27	16	41	20	59	25	23	29	56	34	9	39	36	44	55	50	37	57	2
30	4	0	8	1	12	4	16	10	20	19	24	34	28	56	33	57	48	11	43	10	48	32	54	26
31	3	53	7	47	11	43	15	41	19	42	23	48	28	1	32	21	36	52	41	37	46	40	52	9
32	3	47	7	34	11	23	15	14	19	8	23	6	27	10	31	20	35	41	40	12	45	0	50	8
33	3	40	7	22	11	4	14	48	18	36	22	26	26	22	30	24	34	34	38	54	43	27	48	19
34	3	35	7	10	10	46	14	25	18	8	21	50	25	38	29	31	33	33	37	43	42	4	46	40
35	3	29	6	59	10	30	14	3	17	37	21	15	24	57	28	43	32	36	36	36	40	47	45	10
36	3	24	6	49	10	15	13	42	17	11	20	43	24	18	27	58	31	43	35	35	39	36	43	48
37	3	19	6	39	10	0	13	22	16	46	20	12	23	42	27	16	30	54	34	38	38	30	42	31
38	3	15	6	31	9	47	13	4	16	23	19	44	23	9	26	36	30	8	33	45	37	29	41	22
39	3	11	6	22	9	34	12	47	16	1	19	17	22	36	25	59	29	24	32	55	36	32	40	16
40	3	7	6	14	9	21	12	30	15	40	18	52	22	7	25	24	28	44	32	9	35	39	39	15
42	2	59	5	59	8	59	11	59	15	3	18	6	21	12	24	20	27	30	30	45	34	3	37	26
44	2	53	5	46	8	39	11	33	14	29	17	25	20	23	23	23	26	25	29	30	32	38	35	50
46	2	47	5	34	8	21	11	10	13	58	16	48	19	39	22	32	25	27	28	24	31	23	34	26
48	2	41	5	23	8	5	10	48	13	31	16	15	19	0	21	46	24	34	27	27	30	16	33	11
50	2	36	5	13	7	50	10	28	13	6	15	45	18	25	21	5	23	48	26	32	29	17	32	4
52	2	32	5	5	7	37	10	11	12	44	15	18	17	53	20	29	23	6	25	44	28	23	31	5
54	2	28	4	57	7	25	9	54	12	24	14	53	17	24	19	55	22	28	25	1	27	35	30	11
56	2	25	4	50	7	14	9	40	12	6	14	32	16	58	19	25	21	53	24	22	26	52	29	23
58	2	22	4	43	7	4	9	27	11	49	14	12	16	35	18	58	21	22	23	47	26	13	28	40
60	2	19	4	37	6	56	9	15	11	34	13	54	16	15	18	33	20	54	23	16	25	38	28	1
62	2	16	4	31	6	48	9	4	11	21	13	37	15	54	18	12	20	30	22	48	25	7	27	26
64	2	14	4	27	6	41	8	53	11	9	13	23	15	37	17	53	20	7	22	22	24	38	26	54
66	2	12	4	22	6	34	8	44	10	58	13	10	15	22	17	34	19	46	22	0	24	13	26	26
68	2	9	4	19	6	28	8	37	10	48	12	58	15	8	17	18	19	28	21	39	23	50	26	1
70	2	8	4	15	6	23	8	31	10	39	12	47	14	55	17	3	19	12	21	21	23	30	25	39
72	2	6	4	12	6	19	8	25	10	31	12	38	14	44	16	51	18	54	21	5	23	12	25	19
74	2	5	4	9	6	15	8	20	10	25	12	30	14	35	16	40	18	45	20	51	22	56	25	2
76	2	4	4	7	6	11	8	15	10	19	12	23	14	27	16	30	18	34	20	39	22	43	24	47
78	2	3	4	5	6	8	8	11	10	14	12	16	14	19	16	22	18	25	20	29	22	31	24	34
80	2	2	4	4	6	5	8	8	10	10	12	11	14	14	16	15	18	17	20	20	22	22	24	24

TABLE XXVI. Correction of observed Altitude for Spheroidal Figure of the Earth.

Azim. +	Argument. Latitude of the Place of Observation. 0°	2°	4°	6°	9°	12°	15°	18°	21°	24°	28°	32°	36°	40°	45°	
°	′	′	′	′	′	′	′	′	′	′	′	′	′	′	′	°
0	0	1.1	2.2	3.1	4.6	6.1	7.5	8.8	10.0	11.1	12.4	13.5	14.2	14.7	15.0	180
5	0	1.1	2.2	3.1	4.6	6.1	7.5	8.8	10.0	11.1	12.3	13.4	14.2	14.6	14.9	175
10	0	1.1	2.1	3.0	4.5	6.0	7.4	8.7	9.8	10.9	12.2	13.3	14.0	14.5	14.8	170
15	0	1.1	2.0	3.0	4.4	5.9	7.2	8.5	9.7	10.7	12.0	13.0	13.7	14.1	14.5	165
18	0	1.0	2.0	2.9	4.4	5.8	7.1	8.4	9.5	10.6	11.8	12.8	13.5	14.0	14.3	162
21	0	1.0	2.0	2.9	4.3	5.7	7.0	8.2	9.3	10.4	11.6	12.6	13.3	13.7	14.0	159
24	0	1.0	1.9	2.8	4.2	5.6	6.8	8.0	9.1	10.1	11.3	12.3	13.0	13.4	13.7	156
27	0	1.0	1.9	2.8	4.1	5.4	6.7	7.8	8.9	9.9	11.0	12.0	12.7	13.1	13.4	153
30	0	0.9	1.8	2.7	4.0	5.3	6.5	7.6	8.7	9.6	10.7	11.7	12.3	12.7	13.0	150
32	0	0.9	1.8	2.6	3.9	5.2	6.4	7.5	8.5	9.4	10.5	11.4	12.0	12.5	12.7	148
34	0	0.9	1.7	2.6	3.8	5.1	6.2	7.3	8.3	9.2	10.3	11.2	11.8	12.2	12.4	146
36	0	0.9	1.7	2.5	3.7	4.9	6.1	7.1	8.1	8.9	10.0	10.9	11.5	11.9	12.1	144
38	0	0.9	1.6	2.4	3.6	4.8	5.9	6.9	7.9	8.7	9.8	10.6	11.2	11.6	11.8	142
40	0	0.8	1.6	2.4	3.5	4.7	5.7	6.7	7.7	8.5	9.5	10.3	10.9	11.3	11.5	140
42	0	0.8	1.6	2.3	3.4	4.5	5.6	6.5	7.4	8.2	9.2	10.0	10.6	10.9	11.1	138
44	0	0.8	1.5	2.2	3.3	4.4	5.4	6.3	7.2	8.0	8.9	9.7	10.2	10.6	10.8	136
46	0	0.8	1.5	2.1	3.2	4.2	5.2	6.1	6.9	7.7	8.6	9.4	9.9	10.2	10.4	134
48	0	0.7	1.4	2.1	3.1	4.1	5.0	5.9	6.7	7.4	8.3	9.0	9.5	9.8	10.0	132
50	0	0.7	1.4	2.0	3.0	3.9	4.8	5.7	6.4	7.1	8.0	8.7	9.1	9.4	9.6	130
51	0	0.7	1.3	1.9	2.9	3.8	4.7	5.5	6.3	7.0	7.8	8.5	8.9	9.2	9.4	129
52	0	0.7	1.3	1.9	2.8	3.8	4.6	5.4	6.2	6.8	7.6	8.3	8.7	9.0	9.2	128
53	0	0.7	1.3	1.9	2.8	3.7	4.5	5.3	6.0	6.7	7.5	8.1	8.5	8.8	9.0	127
54	0	0.6	1.2	1.8	2.7	3.6	4.4	5.2	5.9	6.5	7.3	7.9	8.3	8.6	8.8	126
55	0	0.6	1.2	1.8	2.6	3.5	4.3	5.0	5.7	6.4	7.1	7.7	8.1	8.4	8.6	125
56	0	0.6	1.2	1.7	2.6	3.4	4.2	4.9	5.6	6.2	6.9	7.5	7.9	8.2	8.4	124
57	0	0.6	1.1	1.7	2.5	3.3	4.1	4.8	5.4	6.0	6.7	7.3	7.7	8.0	8.2	123
58	0	0.6	1.1	1.6	2.4	3.2	4.0	4.7	5.3	5.9	6.6	7.1	7.5	7.8	7.9	122
59	0	0.6	1.1	1.6	2.4	3.1	3.9	4.5	5.1	5.7	6.4	6.9	7.3	7.6	7.7	121
60	0	0.5	1.0	1.5	2.3	3.0	3.7	4.4	5.0	5.5	6.2	6.7	7.1	7.3	7.5	120
61	0	0.5	1.0	1.5	2.2	3.0	3.6	4.3	4.8	5.4	6.0	6.5	6.9	7.1	7.3	119
62	0	0.5	1.0	1.5	2.2	2.9	3.5	4.1	4.7	5.2	5.8	6.3	6.7	6.9	7.0	118
63	0	0.5	0.9	1.4	2.1	2.8	3.4	4.0	4.5	5.0	5.6	6.1	6.4	6.7	6.8	117
64	0	0.5	0.9	1.4	2.0	2.7	3.3	3.8	4.4	4.9	5.4	5.9	6.2	6.4	6.6	116
65	0	0.5	0.9	1.3	1.9	2.6	3.2	3.7	4.2	4.7	5.2	5.7	6.0	6.2	6.3	115
66	0	0.4	0.8	1.3	1.9	2.5	3.0	3.6	4.1	4.5	5.0	5.5	5.8	6.0	6.1	114
67	0	0.4	0.8	1.2	1.8	2.4	2.9	3.4	3.9	4.3	4.8	5.3	5.6	5.7	5.9	113
68	0	0.4	0.8	1.2	1.7	2.3	2.8	3.3	3.7	4.2	4.6	5.1	5.3	5.5	5.6	112
69	0	0.4	0.7	1.1	1.6	2.2	2.7	3.1	3.6	4.0	4.4	4.8	5.1	5.3	5.4	111
70	0	0.4	0.7	1.1	1.6	2.1	2.6	3.0	3.4	3.8	4.2	4.6	4.9	5.0	5.1	110
71	0	0.4	0.7	1.0	1.5	2.0	2.4	2.9	3.3	3.6	4.0	4.4	4.6	4.8	4.9	109
72	0	0.3	0.6	1.0	1.4	1.9	2.3	2.7	3.1	3.4	3.8	4.2	4.4	4.5	4.6	108
73	0	0.3	0.6	0.9	1.3	1.8	2.2	2.6	2.9	3.2	3.6	3.9	4.1	4.3	4.4	107
74	0	0.3	0.6	0.8	1.3	1.7	2.1	2.4	2.8	3.1	3.4	3.7	3.9	4.0	4.1	106
75	0	0.3	0.5	0.8	1.2	1.6	1.9	2.3	2.6	2.8	3.2	3.5	3.7	3.8	3.9	105
76	0	0.3	0.5	0.7	1.1	1.5	1.8	2.1	2.4	2.7	3.0	3.3	3.4	3.6	3.6	104
77	0	0.2	0.5	0.7	1.0	1.4	1.7	2.0	2.2	2.5	2.8	3.0	3.2	3.3	3.4	103
78	0	0.2	0.4	0.6	1.0	1.3	1.6	1.8	2.1	2.3	2.6	2.8	2.9	3.1	3.1	102
79	0	0.2	0.4	0.6	0.9	1.2	1.4	1.7	1.9	2.1	2.4	2.6	2.7	2.8	2.9	101
80	0	0.2	0.4	0.5	0.8	1.1	1.3	1.5	1.7	1.9	2.1	2.3	2.5	2.5	2.6	100
81	0	0.2	0.3	0.5	0.7	1.0	1.2	1.4	1.5	1.7	1.9	2.1	2.2	2.3	2.3	99
82	0	0.1	0.3	0.4	0.6	0.9	1.0	1.2	1.4	1.5	1.7	1.9	2.0	2.0	2.1	98
83	0	0.1	0.3	0.4	0.6	0.7	0.9	1.1	1.2	1.3	1.5	1.6	1.7	1.8	1.8	97
84	0	0.1	0.2	0.3	0.5	0.6	0.8	0.9	1.0	1.2	1.3	1.4	1.5	1.5	1.6	96
85	0	0.1	0.2	0.3	0.4	0.5	0.6	0.8	0.9	1.0	1.1	1.2	1.2	1.3	1.3	95
86	0	0.1	0.1	0.2	0.3	0.4	0.5	0.6	0.7	0.8	0.9	0.9	1.0	1.0	1.0	94
87	0	0.1	0.1	0.2	0.2	0.3	0.4	0.5	0.5	0.6	0.6	0.7	0.7	0.8	0.8	93
88	0	0.0	0.1	0.1	0.2	0.2	0.3	0.3	0.3	0.4	0.4	0.5	0.5	0.5	0.5	92
89	0	0.0	0.0	0.1	0.1	0.1	0.1	0.1	0.2	0.2	0.2	0.2	0.3	0.3	0.3	91
90	0	0.0	0.0	0.0	0.0	0.0	0.0	0.0	0.0	0.0	0.0	0.0	0.0	0.0	0.0	90
	90°	88°	86°	84°	81°	78°	75°	72°	69°	66°	62°	58°	54°	50°	45°	Azim. −
	Argument. Latitude of the Place of Observation.															

TABLE XXVII. For computing the Apparent Time.

Arg. Latitude.	Argument. Declination of the Sun or Star.												
	0°	1°	2°	3°	4°	5°	6°	7°	8°	9°	10°	11°	12°
0°	0000	0001	0003	0006	0011	0017	0024	0032	0042	0054	0066	0082	0096
1	0001	0001	0005	0007	0011	0017	0024	0033	0043	0054	0067	0081	0097
2	0003	0003	0005	0009	0013	0019	0026	0035	0045	0056	0069	0083	0099
3	0006	0007	0009	0012	0016	0022	0030	0038	0048	0060	0072	0086	0102
4	0011	0011	0013	0016	0021	0027	0034	0043	0053	0064	0077	0091	0106
5	0017	0017	0019	0022	0027	0033	0040	0049	0059	0070	0083	0097	0112
6	0024	0024	0026	0030	0034	0040	0048	0056	0066	0078	0090	0104	0120
7	0032	0033	0035	0038	0043	0049	0056	0065	0075	0086	0099	0113	0128
8	0042	0043	0045	0048	0053	0059	0066	0075	0085	0096	0109	0123	0138
9	0054	0054	0056	0060	0064	0070	0078	0086	0096	0108	0120	0134	0150
10	0066	0067	0069	0072	0077	0081	0091	0099	0109	0120	0133	0147	0162
11	0080	0081	0083	0086	0091	0097	0104	0113	0123	0134	0147	0161	0176
12	0096	0097	0099	0102	0106	0112	0120	0128	0138	0150	0162	0176	0192
13	0113	0113	0115	0119	0123	0129	0137	0145	0155	0167	0179	0193	0209
14	0131	0132	0134	0137	0141	0147	0155	0163	0173	0185	0197	0211	0227
15	0151	0151	0153	0156	0161	0167	0174	0183	0193	0204	0217	0231	0246
16	0172	0172	0174	0177	0181	0188	0195	0204	0214	0225	0238	0252	0267
17	0194	0195	0197	0200	0205	0211	0218	0226	0236	0248	0260	0275	0290
18	0218	0219	0221	0224	0228	0234	0242	0250	0260	0271	0284	0298	0314
19	0243	0244	0246	0249	0254	0260	0267	0276	0286	0297	0310	0324	0339
20	0270	0271	0273	0276	0281	0287	0294	0303	0313	0324	0337	0351	0366
21	0298	0299	0301	0304	0309	0315	0322	0331	0341	0352	0365	0379	0394
22	0328	0329	0331	0334	0339	0345	0352	0361	0371	0382	0395	0409	0424
23	0360	0360	0362	0366	0370	0376	0384	0392	0402	0413	0426	0441	0456
24	0393	0393	0395	0399	0403	0409	0417	0425	0435	0447	0459	0473	0489
25	0427	0428	0430	0433	0438	0444	0451	0460	0470	0481	0495	0508	0523
26	0463	0464	0466	0469	0474	0480	0487	0496	0506	0517	0530	0544	0559
27	0501	0502	0504	0507	0512	0518	0525	0534	0544	0555	0568	0582	0597
28	0541	0541	0543	0547	0551	0557	0564	0573	0583	0594	0607	0621	0637
29	0582	0582	0584	0588	0592	0598	0606	0614	[illegible]	[illegible]	0648	0662	0678
30	0623	0625	0627	0631	0635	0641	0648	0657	0667	[illegible]	0691	0705	0721
31	0669	0670	0672	0675	0680	0686	0693	0702	0712	0723	0736	0751	0765
32	0716	0716	0718	0722	0726	0732	0740	0748	0758	0770	0782	0796	0812
33	0764	0765	0767	0770	0775	0781	0788	0797	0807	0818	0831	0845	0860
34	0814	0815	0817	0820	0825	0831	0838	0847	0857	0868	0881	0895	0910
35	0866	0867	0869	0872	0877	0883	0890	0899	0909	0920	0933	0943	0963
36	0920	0921	0923	0926	0931	0937	0944	0953	0963	0974	0987	1001	1016
37	0976	0977	0979	0982	0987	0993	1000	1009	1019	1030	1043	1057	1072
38	1035	1035	1037	1041	1045	1051	1058	1067	1077	1088	1101	1115	1131
39	1095	1096	1098	1101	1105	1111	1119	1127	1137	1149	1161	1175	1191
40	1157	1158	1160	1163	1165	1174	1181	1190	1200	1211	1224	1238	1253
41	1222	1223	1225	1228	1233	1239	1246	1255	1265	1276	1289	1303	1318
42	1289	1290	1292	1295	1300	1306	1313	1322	1332	1343	1356	1370	1385
43	1359	1359	1361	1365	1369	1375	1383	1391	1401	1412	1425	1439	1455
44	1431	1431	1433	1437	1441	1447	1454	1463	1473	1484	1497	1511	1527
45	1505	1506	1508	1511	1516	1522	1529	1538	1548	1559	1572	1586	1601
46	1582	1583	1585	1588	1593	1599	1606	1615	1625	1636	1649	1663	1678
47	1662	1663	1665	1668	1673	1679	1686	1695	1705	1716	1729	1743	1758
48	1745	1745	1747	1751	1755	1761	1769	1777	1787	1799	1811	1825	1841
49	1831	1831	1833	1836	1841	1847	1854	1863	1873	1884	1897	1911	1926
50	1919	1920	1922	1925	1930	1937	1943	1952	1962	1973	1986	2000	2015
51	2011	2012	2014	2017	2022	2028	2035	2044	2054	2065	2078	2092	2107
52	2107	2107	2109	2112	2117	2123	2130	2139	2149	2160	2173	2187	2202
53	2205	2206	2208	2211	2216	2222	2229	2238	2248	2259	2271	2286	2301
54	2308	2308	2310	2314	2318	2324	2332	2340	2350	2361	2374	2388	2404
55	2414	2415	2417	2420	2425	2431	2438	2447	2457	2468	2481	2495	2510
56	2524	2525	2527	2530	2535	2541	2548	2557	2567	2578	2591	2605	2620
57	2639	2640	2642	2645	2649	2655	2663	2671	2681	2693	2705	2719	2735
58	2758	2759	2760	2764	2768	2774	2782	2790	2800	2812	2824	2838	2854
59	2882	2882	2884	2888	2892	2898	2905	2914	2924	2935	2948	2962	2978
60	3010	3012	3013	3016	3021	3027	3034	3043	3053	3064	3077	3091	3106

TABLE XXVII. For computing the Apparent Time.

Arg. Latitude.	Argument. Declination of the Sun or Star.												
	13°	14°	15°	16°	17°	18°	19°	20°	21°	22°	23°	24°	25°
0°	0113	0131	0151	0172	0194	0218	0243	0270	0298	0328	0360	0393	0427
1	0113	0132	0152	0172	0195	0219	0244	0271	0299	0329	0360	0393	0428
2	0115	0134	0153	0174	0197	0221	0246	0273	0301	0331	0362	0395	0430
3	0119	0137	0156	0177	0200	0224	0249	0276	0304	0334	0366	0399	0433
4	0123	0141	0161	0182	0205	0228	0254	0281	0309	0339	0370	0403	0438
5	0129	0147	0167	0188	0211	0234	0260	0287	0315	0345	0376	0409	0444
6	0137	0155	0174	0195	0218	0242	0267	0294	0322	0352	0384	0417	0451
7	0145	0163	0183	0204	0226	0250	0276	0303	0331	0361	0392	0425	0460
8	0155	0173	0193	0214	0236	0260	0286	0313	0341	0371	0402	0435	0470
9	0167	0185	0204	0225	0248	0272	0297	0324	0352	0382	0413	0446	0481
10	0179	0197	0217	0238	0260	0284	0310	0337	0365	0395	0426	0459	0495
11	0193	0211	0231	0252	0275	0298	0324	0351	0379	0409	0440	0473	0508
12	0209	0227	0246	0267	0290	0314	0339	0366	0394	0424	0456	0489	0523
13	0225	0244	0263	0284	0307	0331	0356	0383	0411	0441	0472	0505	0540
14	0244	0262	0281	0304	0325	0349	0374	0401	0429	0459	0491	0524	0558
15	0263	0281	0301	0322	0345	0368	0394	0421	0449	0479	0511	0543	0578
16	0284	0302	0322	0343	0366	0389	0415	0442	0470	0500	0531	0564	0599
17	0307	0325	0345	0364	0388	0412	0437	0464	0492	0522	0554	0587	0621
18	0331	0349	0368	0389	0412	0436	0461	0488	0516	0546	0578	0611	0645
19	0356	0374	0394	0415	0437	0461	0487	0513	0542	0572	0603	0636	0670
20	0381	0401	0421	0442	0464	0488	0513	0540	0569	0598	063	0663	0697
21	0411	0429	0449	0470	0492	0516	0542	0569	0597	0627	0658	0691	0726
22	0441	0459	0479	0450	0522	0546	0572	0598	0627	0657	0687	0721	0756
23	0472	0491	0510	0531	0554	0578	0603	0630	0658	0687	0719	0752	0787
24	0503	0524	0543	0564	0587	0611	0636	0663	0691	0721	0752	0783	0820
25	0540	0558	0578	0599	0621	0645	0670	0697	0726	0756	0787	0820	0854
26	0576	0594	0614	0635	0657	0681	0707	0733	0762	0792	0823	0856	0891
27	0614	0632	0652	0673	0695	0719	0744	0771	0800	0829	0861	0894	0928
28	0653	0672	0691	0712	0735	0759	0784	0811	0839	0869	0900	0933	0968
29	0695	0713	0732	0753	0776	0800	0825	0852	0880	0910	0941	0974	1009
30	0737	0756	0775	0796	0817	0843	0868	0895	0921	0953	0984	1017	1052
31	0782	0800	0820	0841	0863	0887	0913	0939	0967	0998	1029	1062	1097
32	0829	0847	0866	0887	0910	0934	0959	0986	1014	1044	1075	1108	1143
33	0877	0895	0915	0936	0958	0982	1007	1034	1063	1092	1124	1157	1192
34	0927	0945	0965	0986	1008	1032	1058	1084	1113	1143	1174	1207	1241
35	0979	0997	1017	1038	1060	1084	1110	1136	1165	1195	1226	1259	1294
36	1033	1051	1071	1092	1114	1138	1164	1191	1219	1249	1280	1313	1348
37	1089	1107	1127	1148	1170	1194	1220	1247	1275	1305	1336	1369	1404
38	1147	1166	1185	1206	1229	1253	1278	1305	1333	1363	1394	1427	1462
39	1208	1226	1245	1266	1287	1313	1338	1365	1393	1423	1455	1488	1522
40	1270	1288	1308	1329	1351	1375	1401	1428	1456	1486	1517	1550	1585
41	1335	1353	1373	1394	1416	1440	1465	1492	1521	1550	1582	1615	1649
42	1402	1420	1450	1461	1483	1507	1533	1559	1588	1618	1649	1682	1716
43	1471	1490	1509	1530	1553	1577	1602	1629	1657	1687	1718	1751	1786
44	1543	1562	1581	1602	1625	1649	1674	1701	1729	1759	1790	1823	1858
45	1618	1636	1656	1677	1699	1723	1748	1775	1804	1833	1865	1898	1932
46	1695	1713	1733	1754	1776	1800	1826	1852	1881	1911	1942	1975	2010
47	1775	1793	1813	1834	1856	1880	1905	1932	1961	1990	2022	2055	2089
48	1858	1876	1895	1916	1939	1963	1988	2015	2043	2073	2105	2138	2172
49	1943	1961	1981	2002	2025	2048	2074	2101	2129	2159	2190	2223	2258
50	2032	2050	2070	2091	2113	2137	2163	2189	2218	2248	2279	2312	2347
51	2124	2142	2162	2183	2205	2229	2255	2281	2310	2340	2371	2404	2438
52	2219	2237	2257	2278	2301	2324	2350	2377	2405	2435	2466	2499	2534
53	2318	2336	2356	2377	2399	2423	2449	2475	2504	2534	2565	2598	2633
54	2421	2439	2458	2479	2502	2526	2551	2578	2606	2636	2667	2700	2735
55	2527	2545	2565	2586	2608	2632	2657	2684	2713	2742	2764	2807	2841
56	2637	2655	2675	2696	2718	2742	2768	2794	2823	2853	2884	2917	2952
57	2752	2770	2789	2810	2833	2857	2882	2909	2937	2967	2999	3032	3066
58	2871	2889	2908	2929	2952	2976	3001	3028	3056	3086	3118	3151	3185
59	2994	3013	3032	3053	3076	3099	3125	3152	3180	3210	3241	3274	3309
60	3123	3141	3161	3182	3204	3228	3254	3280	3309	3339	3370	3403	3437

TABLE XXVIII. For computing the Apparent Time.

Arg. Latitude.	Argument. Declination of the Sun or Star.												
	0°	1°	2°	3°	4°	5°	6°	7°	8°	9°	10°	11°	12°
0°	0	0	0	0	0	0	0	0	0	0	0	0	0
1	0	3	6	9	12	15	18	21	24	28	31	34	37
2	0	6	12	18	24	30	37	43	49	55	62	68	74
3	0	9	18	27	37	46	55	64	74	83	92	102	111
4	0	12	24	37	49	61	73	86	98	111	123	136	149
5	0	15	30	46	61	76	92	107	123	139	154	170	186
6	0	18	37	55	73	92	110	129	145	166	185	204	223
7	0	21	43	64	86	107	129	151	173	194	216	239	261
8	0	24	49	74	98	123	148	173	197	223	248	273	299
9	0	28	55	83	111	139	165	194	223	251	279	308	337
10	0	31	62	93	123	154	185	216	248	279	311	343	375
11	0	34	68	102	136	170	204	239	273	308	343	378	413
12	0	37	74	111	149	186	223	261	299	337	375	413	452
13	0	40	81	121	161	202	243	283	324	366	407	449	491
14	0	43	87	131	174	218	262	306	350	395	440	485	530
15	0	47	94	140	187	234	282	329	377	424	472	521	569
16	0	50	100	150	200	251	301	353	403	452	506	557	609
17	0	53	107	160	214	267	321	375	430	484	539	594	650
18	0	57	113	170	227	284	341	399	459	515	573	632	691
19	0	60	120	180	241	301	362	423	484	545	607	669	732
20	0	63	127	191	254	318	382	447	511	576	642	707	774
21	0	67	134	201	268	336	403	471	539	608	676	746	816
22	0	70	141	212	282	353	423	496	568	640	712	785	856
23	0	74	148	222	297	371	446	521	597	672	748	825	902
24	0	78	155	233	311	389	468	547	626	705	785	865	946
25	0	81	163	244	326	408	490	573	651	739	822	906	991
26	0	85	170	256	341	427	513	579	685	772	860	948	1037
27	0	89	178	267	356	446	535	626	716	807	898	990	1083
28	0	93	186	279	372	465	559	653	747	842	937	1034	1130
29	0	97	194	290	388	485	583	681	779	878	977	1077	1178
30	0	101	202	303	404	505	607	709	811	914	1018	1122	1227
31	0	105	210	315	420	526	631	738	844	952	1059	1168	1277
32	0	109	218	327	437	547	657	767	878	990	1102	1215	1328
33	0	113	227	340	454	568	683	797	913	1029	1145	1262	1380
34	0	118	235	353	472	590	709	828	948	1068	1189	1311	1434
35	0	122	244	367	490	613	736	860	984	1109	1235	1361	1488
36	0	127	254	381	508	636	764	892	1021	1151	1281	1412	1544
37	0	131	263	396	527	659	794	925	1059	1193	1329	1465	1602
38	0	136	273	409	546	683	821	959	1098	1237	1378	1519	1661
39	0	141	283	424	566	708	851	994	1138	1283	1428	1574	1721
40	0	146	291	440	587	734	882	1030	1179	1322	1480	1635	1784
41	0	152	304	456	608	760	914	1067	1222	1377	1533	1690	1848
42	0	157	314	472	630	788	946	1106	1265	1426	1588	1750	1914
43	0	163	326	489	652	816	980	1145	1311	1477	1644	1813	1982
44	0	169	337	506	675	845	1015	1186	1357	1529	1703	1877	2053
45	0	175	349	524	699	875	1051	1228	1405	1584	1763	1944	2126
46	0	181	362	543	724	906	1088	1271	1455	1640	1826	2013	2201
47	0	187	374	562	750	938	1127	1317	1507	1698	1891	2084	2279
48	0	193	388	582	770	972	1167	1364	1561	1759	1958	2159	2361
49	0	201	401	603	804	1006	1208	1413	1617	1822	2028	2236	2445
50	0	208	416	625	833	1043	1253	1463	1675	1888	2101	2316	2533
51	0	215	431	647	863	1080	1298	1516	1735	1956	2177	2400	2625
52	0	223	447	671	895	1120	1345	1572	1799	2027	2257	2488	2721
53	0	232	463	695	928	1161	1395	1629	1865	2102	2340	2579	2821
54	0	240	481	723	962	1204	1447	1690	1934	2180	2427	2675	2926
55	0	249	497	748	993	1249	1501	1753	2007	2262	2518	2776	3036
56	0	257	518	777	1037	1297	1558	1820	2084	2348	2614	2882	3151
57	0	269	538	807	1077	1347	1618	1891	2164	2439	2715	2993	3273
58	0	279	559	839	1119	1400	1682	1965	2249	2535	2822	3111	3403
59	0	290	581	872	1164	1456	1749	2043	2339	2636	2935	3235	3537
60	0	304	605	908	1211	1515	1820	2127	2434	2743	3054	3367	3682

TABLE XXVIII. For computing the Apparent Time.

Arg. titude.	Argument. Declination of the Sun or Star.											
	13°	14°	15°	16°	17°	18°	19°	20°	21°	22°	23°	24°
0°	0	0	0	0	0	0	0	0	0	0	0	0
1	40	43	47	50	53	57	60	63	67	70	74	78
2	81	87	94	100	107	113	120	127	134	141	148	155
3	121	131	140	150	160	170	180	191	201	212	222	233
4	161	174	187	200	214	227	241	254	268	282	297	311
5	202	218	234	251	267	284	301	318	336	353	371	389
6	243	262	282	301	321	341	362	382	403	425	446	468
7	283	306	329	353	375	399	423	447	471	496	521	547
8	324	350	377	403	430	459	484	511	539	568	597	626
9	366	394	424	452	484	515	545	576	608	640	672	705
10	407	440	472	506	539	573	607	642	676	712	748	785
11	449	485	521	557	594	632	669	707	746	785	825	865
12	491	530	569	609	650	691	732	774	816	859	902	946
13	533	576	619	662	706	750	795	840	886	933	980	1028
14	576	622	668	715	762	810	858	907	957	1007	1058	1110
15	619	668	718	768	819	871	923	975	1029	1083	1137	1193
16	662	715	768	822	877	932	987	1044	1101	1158	1217	1277
17	706	762	819	877	935	993	1053	1113	1174	1235	1298	1361
18	750	810	871	932	993	1056	1119	1183	1247	1313	1379	1447
19	795	858	923	987	1053	1119	1186	1253	1322	1391	1462	1533
20	840	907	975	1044	1113	1183	1253	1325	1397	1470	1545	1620
21	886	957	1029	1101	1174	1247	1322	1397	1473	1551	1629	1709
22	933	1007	1083	1158	1235	1313	1391	1470	1551	1632	1715	1799
23	980	1058	1137	1217	1298	1379	1462	1545	1629	1715	1802	1890
24	1028	1110	1193	1277	1361	1447	1533	1602	1709	1799	1890	1982
25	1077	1163	1249	1337	1426	1515	1606	1697	1790	1884	1979	2076
26	1126	1216	1307	1399	1491	1585	1679	1775	1872	1971	2070	2171
27	1176	1270	1365	1461	1558	1656	1754	1854	1956	2059	2163	2279
28	1227	1326	1425	1525	1626	1728	1831	1935	2041	2148	2257	2367
29	1280	1382	1485	1589	1695	1802	1909	2017	2128	2240	2353	2468
30	1333	1439	1547	1655	1765	1876	1988	2101	2216	2333	2451	2570
31	1387	1498	1610	1723	1837	1952	2069	2187	2306	2428	2550	2675
32	1443	1558	1674	1792	1910	2030	2152	2274	2399	2525	2652	2782
33	1499	1619	1740	1862	1985	2110	2236	2364	2493	2624	2757	2891
34	1557	1682	1807	1934	2062	2192	2322	2455	2589	2725	2863	3003
35	1617	1746	1876	2012	2141	2275	2411	2549	2688	2829	2972	3117
36	1677	1811	1947	2083	2221	2361	2502	2644	2789	2935	3084	3235
37	1740	1879	2019	2161	2304	2448	2595	2743	2893	3045	3199	3355
38	1804	1948	2093	2240	2389	2538	2690	2844	2999	3157	3316	3478
39	1869	2019	2170	2322	2476	2631	2788	2947	3108	3272	3437	3605
40	1937	2092	2248	2406	2565	2726	2889	3054	3221	3390	3562	3736
41	2007	2167	2329	2493	2658	2824	2993	3164	3337	3512	3690	3879
42	2079	2245	2413	2582	2753	2926	3100	3277	3456	3638	3822	4009
43	2153	2325	2499	2674	2851	3030	3211	3394	3580	3768	3958	4152
44	2229	2408	2588	2769	2952	3138	3325	3515	3707	3902	4099	4299
45	2309	2493	2679	2867	3057	3249	3443	3640	3839	4040	4245	4452
46	2391	2582	2775	2969	3166	3365	3566	3769	3975	4184	4396	4610
47	2476	2674	2873	3075	3279	3484	3692	3903	4116	4333	4552	4724
48	2564	2769	2976	3185	3395	3609	3824	4042	4263	4488	4714	4945
49	2645	2868	3082	3299	3517	3738	3961	4187	4416	4648	4883	5122
50	2751	2971	3191	3417	3644	3872	4103	4338	4575	4815	5059	5306
51	2851	3079	3309	3541	3775	4012	4252	4495	4740	4989	5242	5498
52	2955	3192	3430	3670	3913	4159	4407	4659	4913	5171	5433	5699
53	3064	3309	3556	3805	4057	4312	4569	4830	5094	5362	5633	5908
54	3178	3432	3688	3947	4208	4472	4739	5100	5283	5561	5843	6128
55	3297	3561	3827	4095	4366	4640	4917	5198	5482	5770	6062	6358
56	3415	3696	3972	4251	4533	4817	5105	5396	5691	5990	6293	6601
57	3555	3839	4129	4415	4708	5003	5302	5605	5911	6221	6536	6856

TABLE XXIX. For computing the Apparent Time.

0 HOUR.

Num	M.	S.	Num	M.	S.	Num	M.	S.	Num	M.	S.	Num	M.	S.	Num	M.	S.
10000	0	0	9991	10	0	9962	20	0	9914	30	0	9848	40	0	9763	50	0
10000	0	10	9990	10	10	9961	20	10	9913	30	10	9847	40	10	9761	50	10
10000	0	20	9990	10	20	9961	20	20	9912	30	20	9845	40	20	9760	50	20
10000	0	30	9989	10	30	9960	20	30	9912	30	30	9844	40	30	9758	50	30
10000	0	40	9989	10	40	9959	20	40	9911	30	40	9843	40	40	9757	50	40
10000	0	50	9989	10	50	9959	20	50	9910	30	50	9842	40	50	9755	50	50
10000	1	0	9988	11	0	9958	21	0	9909	31	0	9840	41	0	9753	51	0
10000	1	10	9988	11	10	9957	21	10	9908	31	10	9839	41	10	9752	51	10
10000	1	20	9988	11	20	9957	21	20	9907	31	20	9838	41	20	9750	51	20
10000	1	30	9987	11	30	9956	21	30	9906	31	30	9837	41	30	9749	51	30
10000	1	40	9987	11	40	9955	21	40	9905	31	40	9835	41	40	9747	51	40
10000	1	50	9987	11	50	9955	21	50	9904	31	50	9834	41	50	9745	51	50
10000	2	0	9986	12	0	9954	22	0	9903	32	0	9832	42	0	9744	52	0
10000	2	10	9986	12	10	9953	22	10	9902	32	10	9831	42	10	9742	52	10
9999	2	20	9985	12	20	9953	22	20	9901	32	20	9830	42	20	9740	52	20
9999	2	30	9985	12	30	9952	22	30	9900	32	30	9829	42	30	9739	52	30
9999	2	40	9985	12	40	9951	22	40	9899	32	40	9827	42	40	9737	52	40
9999	2	50	9984	12	50	9950	22	50	9898	32	50	9826	42	50	9735	52	50
9999	3	0	9984	13	0	9950	23	0	9896	33	0	9824	43	0	9734	53	0
9999	3	10	9983	13	10	9949	23	10	9895	33	10	9823	43	10	9732	53	10
9999	3	20	9983	13	20	9948	23	20	9894	33	20	9822	43	20	9730	53	20
9999	3	30	9983	13	30	9947	23	30	9893	33	30	9820	43	30	9729	53	30
9999	3	40	9982	13	40	9947	23	40	9892	33	40	9819	43	40	9727	53	40
9999	3	50	9982	13	50	9946	23	50	9891	33	50	9818	43	50	9725	53	50
9998	4	0	9981	14	0	9945	24	0	9890	34	0	9816	44	0	9724	54	0
9998	4	10	9981	14	10	9944	24	10	9889	34	10	9815	44	10	9722	54	10
9998	4	20	9980	14	20	9944	24	20	9888	34	20	9813	44	20	9720	54	20
9998	4	30	9980	14	30	9943	24	30	9887	34	30	9812	44	30	9718	54	30
9998	4	40	9979	14	40	9942	24	40	9886	34	40	9811	44	40	9717	54	40
9998	4	50	9979	14	50	9941	24	50	9885	34	50	9809	44	50	9715	54	50
9998	5	0	9979	15	0	9941	25	0	9884	35	0	9808	45	0	9713	55	0
9997	5	10	9978	15	10	9940	25	10	9882	35	10	9806	45	10	9712	55	10
9997	5	20	9978	15	20	9939	25	20	9881	35	20	9805	45	20	9710	55	20
9997	5	30	9977	15	30	9938	25	30	9880	35	30	9804	45	30	9708	55	30
9997	5	40	9977	15	40	9937	25	40	9879	35	40	9802	45	40	9706	55	40
9997	5	50	9976	15	50	9937	25	50	9878	35	50	9801	45	50	9705	55	50
9997	6	0	9976	16	0	9936	26	0	9877	36	0	9799	46	0	9703	56	0
9996	6	10	9975	16	10	9935	26	10	9876	36	10	9798	46	10	9701	56	10
9996	6	20	9975	16	20	9934	26	20	9875	36	20	9796	46	20	9699	56	20
9996	6	30	9974	16	30	9933	26	30	9873	36	30	9795	46	30	9698	56	30
9996	6	40	9974	16	40	9932	26	40	9872	36	40	9793	46	40	9696	56	40
9996	6	50	9973	16	50	9932	26	50	9871	36	50	9792	46	50	9694	56	50
9995	7	0	9972	17	0	9931	27	0	9870	37	0	9790	47	0	9692	57	0
9995	7	10	9972	17	10	9930	27	10	9869	37	10	9789	47	10	9690	57	10
9995	7	20	9971	17	20	9929	27	20	9868	37	20	9787	47	20	9689	57	20
9995	7	30	9971	17	30	9928	27	30	9866	37	30	9786	47	30	9687	57	30
9994	7	40	9970	17	40	9927	27	40	9865	33	40	9784	47	40	9685	57	40
9994	7	50	9970	17	50	9926	27	50	9764	37	50	9783	47	50	9683	57	50
9994	8	0	9969	18	0	9925	28	0	9863	38	0	9781	48	0	9681	58	0
9994	8	10	9969	18	10	9925	28	10	9862	38	10	9780	48	10	9680	58	10
9993	8	20	9968	18	20	9924	28	20	9860	38	20	9778	48	20	9678	58	20
9993	8	30	9967	18	30	9923	28	30	9859	38	30	9777	48	30	9676	58	30
9993	8	40	9967	18	40	9922	28	40	9858	38	40	9775	48	40	9674	58	40
9993	8	50	9966	18	50	9921	28	50	9857	38	50	9774	48	50	9672	58	50
9992	9	0	9966	19	0	9920	29	0	9856	39	0	9772	49	0	9670	59	0
9992	9	10	9965	19	10	9919	29	10	9854	39	10	9771	49	10	9669	59	10
9992	9	20	9964	19	20	9918	29	20	9853	39	20	9769	49	20	9667	59	20
9991	9	30	9964	19	30	9917	29	30	9852	39	30	9768	49	30	9665	59	30
9991	9	40	9963	19	40	9916	29	40	9851	39	40	9766	49	40	9663	59	40
9991	9	50	9963	19	50	9915	29	50	9849	39	50	9764	49	50	9661	59	50
9991	10	0	9962	20	0	9914	30	0	9848	40	0	9763	50	0	9659	60	0

TABLE XXIX. For computing the Apparent Time.

I HOUR.

Num	M. S.	Num	M. S.	Num	M. S.	Num	M. S.	Num	M. S.	Num	M. S.
9659	0 0	9537	10 0	9397	20 0	9239	30 0	9063	40 0	8870	50 0
9657	0 10	9535	10 10	9394	20 10	9236	30 10	9060	40 10	8867	50 10
9655	0 20	9533	10 20	9392	20 20	9233	30 20	9057	40 20	8863	50 20
9654	0 30	9531	10 30	9389	20 30	9230	30 30	9054	40 30	8860	50 30
9652	0 40	9528	10 40	9387	20 40	9228	30 40	9051	40 40	8857	50 40
9650	0 50	9526	10 50	9384	20 50	9225	30 50	9048	40 50	8853	50 50
9648	1 0	9524	11 0	9382	21 0	9222	31 0	9044	41 0	8850	51 0
9646	1 10	9522	11 10	9379	21 10	9219	31 10	9041	41 10	8846	51 10
9644	1 20	9519	11 20	9377	21 20	9216	31 20	9038	41 20	8843	51 20
9642	1 30	9517	11 30	9374	21 30	9213	31 30	9035	41 30	8840	51 30
9640	1 40	9515	11 40	9372	21 40	9211	31 40	9032	41 40	8836	51 40
9638	1 50	9513	11 50	9369	21 50	9208	31 50	9029	41 50	8833	51 50
9636	2 0	9511	12 0	9367	22 0	9205	32 0	9026	42 0	8829	52 0
9634	2 10	9508	12 10	9364	22 10	9202	32 10	9023	42 10	8826	52 10
9632	2 20	9506	12 20	9362	22 20	9199	32 20	9020	42 20	8823	52 20
9630	2 30	9504	12 30	9359	22 30	9196	32 30	9016	42 30	8819	52 30
9628	2 40	9501	12 40	9356	22 40	9194	32 40	9013	42 40	8816	52 40
9626	2 50	9499	12 50	9354	22 50	9191	32 50	9010	42 50	8812	52 50
9625	3 0	9497	13 0	9351	23 0	9188	33 0	9007	43 0	8809	53 0
9623	3 10	9495	13 10	9349	23 10	9185	33 10	9004	43 10	8805	53 10
9621	3 20	9492	13 20	9346	23 20	9182	33 20	9001	43 20	8802	53 20
9619	3 30	9490	13 30	9344	23 30	9179	33 30	8997	43 30	8799	53 30
9617	3 40	9488	13 40	9341	23 40	9176	33 40	8994	43 40	8795	53 40
9615	3 50	9485	13 50	9338	23 50	9173	33 50	8991	43 50	8792	53 50
9613	4 0	9483	14 0	9336	24 0	9171	34 0	8988	44 0	8788	54 0
9611	4 10	9481	14 10	9333	24 10	9168	34 10	8985	44 10	8785	54 10
9609	4 20	9479	14 20	9331	24 20	9165	34 20	8982	44 20	8781	54 20
9607	4 30	9476	14 30	9328	24 30	9162	34 30	8978	44 30	8778	54 30
9605	4 40	9474	14 40	9325	24 40	9159	34 40	8975	44 40	8774	54 40
9602	4 50	9472	14 50	9323	24 50	9156	34 50	8972	44 50	8771	54 50
9600	5 0	9469	15 0	9320	25 0	9153	35 0	8969	45 0	8767	55 0
9598	5 10	9467	15 10	9317	25 10	9150	35 10	8965	45 10	8764	55 10
9596	5 20	9465	15 20	9315	25 20	9147	35 20	8962	45 20	8760	55 20
9594	5 30	9462	15 30	9312	25 30	9144	35 30	8959	45 30	8757	55 30
9592	5 40	9460	15 40	9309	25 40	9141	35 40	8956	45 40	8753	55 40
9590	5 50	9458	15 50	9307	25 50	9138	35 50	8952	45 50	8750	55 50
9588	6 0	9455	16 0	9304	26 0	9135	36 0	8949	46 0	8746	56 0
9586	6 10	9453	16 10	9301	26 10	9132	36 10	8946	46 10	8743	56 10
9584	6 20	9450	16 20	9299	26 20	9129	36 20	8943	46 20	8739	56 20
9582	6 30	9448	16 30	9296	26 30	9127	36 30	8939	46 30	8736	56 30
9580	6 40	9446	16 40	9293	26 40	9124	36 40	8936	46 40	8732	56 40
9578	6 50	9443	16 50	9291	26 50	9121	36 50	8933	46 50	8728	56 50
9576	7 0	9441	17 0	9288	27 0	9118	37 0	8930	47 0	8725	57 0
9574	7 10	9438	17 10	9285	27 10	9115	37 10	8926	47 10	8721	57 10
9571	7 20	9436	17 20	9283	27 20	9112	37 20	8923	47 20	8718	57 20
9569	7 30	9434	17 30	9280	27 30	9109	37 30	8920	47 30	8714	57 30
9567	7 40	9431	17 40	9277	27 40	9106	37 40	8917	47 40	8711	57 40
9565	7 50	9429	17 50	9275	27 50	9103	37 50	8913	47 50	8707	57 50
9563	8 0	9426	18 0	9272	28 0	9100	38 0	8910	48 0	8704	58 0
9561	8 10	9424	18 10	9269	28 10	9097	38 10	8907	48 10	8700	58 10
9559	8 20	9422	18 20	9266	28 20	9094	38 20	8903	48 20	8696	58 20
9557	8 30	9419	18 30	9264	28 30	9090	38 30	8900	48 30	8693	58 30
9554	8 40	9417	18 40	9261	28 40	9087	38 40	8897	48 40	8689	58 40
9552	8 50	9414	18 60	9258	28 50	9084	38 50	8893	48 50	8686	58 50
9550	9 0	9412	19 0	9255	29 0	9081	39 0	8890	49 0	8682	59 0
9548	9 10	9409	19 10	9253	29 10	9078	39 10	8887	49 10	8678	59 10
9546	9 20	9407	19 20	9250	29 20	9075	39 20	8883	49 20	8675	59 20
9544	9 30	9404	19 30	9247	29 30	9072	39 30	8880	49 30	8671	59 30
9541	9 40	9402	19 40	9244	29 40	9069	39 40	8877	49 40	8667	59 40
9539	9 50	9399	19 50	9242	29 50	9066	39 50	8873	49 50	8664	59 50
9537	10 0	9397	20 0	9239	30 0	9063	40 0	8870	50 0	8660	60 0

TABLE XXIX. For computing the Apparent Time.

II HOURS.

Num	M.	S.	Num	M.	S.	Num	M.	S.	Num	M.	S.	Num	M.	S.	Num	M.	S.
8660	0	0	8434	10	0	8191	20	0	7933	30	0	7660	40	0	7373	50	0
8657	0	10	8430	10	10	8187	20	10	7929	30	10	7656	40	10	7368	50	10
8653	0	20	8426	10	20	8183	20	20	7925	30	20	7651	40	20	7363	50	20
8649	0	30	8422	10	30	8179	20	30	7920	30	30	7646	40	30	7358	50	30
8646	0	40	8418	10	40	8175	20	40	7916	30	40	7642	40	40	7353	50	40
8642	0	50	8414	10	50	8171	20	50	7911	30	50	7637	40	50	7348	50	50
8638	1	0	8410	11	0	8166	21	0	7907	31	0	7632	41	0	7343	51	0
8635	1	10	8406	11	10	8162	21	10	7902	31	10	7628	41	10	7338	51	10
8631	1	20	8402	11	20	8158	21	20	7898	31	20	7623	41	20	7333	51	20
8627	1	30	8399	11	30	8154	21	30	7893	31	30	7618	41	30	7328	51	30
8624	1	40	8395	11	40	8150	21	40	7889	31	40	7613	41	40	7323	51	40
8620	1	50	8391	11	50	8145	21	50	7885	31	50	7609	41	50	7318	51	50
8616	2	0	8387	12	0	8141	22	0	7880	32	0	7604	42	0	7313	52	0
8612	2	10	8383	12	10	8137	22	10	7876	32	10	7599	42	10	7309	52	10
8609	2	20	8379	12	20	8133	22	20	7871	32	20	7595	42	20	7304	52	20
8605	2	30	8375	12	30	8128	22	30	7867	32	30	7590	42	30	7299	52	30
8601	2	40	8371	12	40	8124	22	40	7862	32	40	7585	42	40	7294	52	40
8598	2	50	8367	12	50	8120	22	50	7858	32	50	7580	42	50	7289	52	50
8594	3	0	8363	13	0	8116	23	0	7853	33	0	7576	43	0	7284	53	0
8590	3	10	8359	13	10	8111	23	10	7849	33	10	7571	43	10	7279	53	10
8587	3	20	8355	13	20	8107	23	20	7844	33	20	7566	43	20	7274	53	20
8583	3	30	8351	13	30	8103	23	30	7840	33	30	7561	43	30	7269	53	30
8579	3	40	8347	13	40	8099	23	40	7835	33	40	7557	43	40	7264	53	40
8575	3	50	8343	13	50	8094	23	50	7831	33	50	7552	43	50	7259	53	50
8572	4	0	8339	14	0	8090	24	0	7826	34	0	7547	44	0	7254	54	0
8568	4	10	8335	14	10	8086	24	10	7822	34	10	7542	44	10	7249	54	10
8564	4	20	8331	14	20	8082	24	20	7817	34	20	7537	44	20	7244	54	20
8560	4	30	8327	14	30	8077	24	30	7812	34	30	7533	44	30	7239	54	30
8557	4	40	8323	14	40	8073	24	40	7808	34	40	7528	44	40	7234	54	40
8553	4	50	8319	14	50	8069	24	50	7803	34	50	7523	44	50	7229	54	50
8549	5	0	8315	15	0	8064	25	0	7799	35	0	7518	45	0	7224	55	0
8545	5	10	8311	15	10	8060	25	10	7794	35	10	7514	45	10	7219	55	10
8542	5	20	8307	15	20	8056	25	20	7790	35	20	7509	45	20	7214	55	20
8538	5	30	8303	15	30	8051	25	30	7785	35	30	7504	45	30	7208	55	30
8534	5	40	8298	15	40	8047	25	40	7781	35	40	7499	45	40	7203	55	40
8530	5	50	8294	15	50	8043	25	50	7776	35	50	7494	45	50	7198	55	50
8526	6	0	8290	16	0	8039	26	0	7771	36	0	7490	46	0	7193	56	0
8523	6	10	8286	16	10	8034	26	10	7767	36	10	7485	46	10	7188	56	10
8519	6	20	8282	16	20	8030	26	20	7762	36	20	7480	46	20	7183	56	20
8515	6	30	8278	16	30	8026	26	30	7758	36	30	7475	46	30	7178	56	30
8511	6	40	8274	16	40	8021	26	40	7753	36	40	7470	46	40	7173	56	40
8507	6	50	8270	16	50	8017	26	50	7748	36	50	7465	46	50	7168	56	50
8503	7	0	8266	17	0	8012	27	0	7744	37	0	7461	47	0	7163	57	0
8499	7	10	8262	17	10	8008	27	10	7739	37	10	7456	47	10	7158	57	10
8496	7	20	8258	17	20	8004	27	20	7735	37	20	7451	47	20	7153	57	20
8492	7	30	8254	17	30	7999	27	30	7730	37	30	7446	47	30	7148	57	30
8488	7	40	8249	17	40	7995	27	40	7725	37	40	7441	47	40	7143	57	40
8484	7	50	8245	17	50	7991	27	50	7721	37	50	7436	47	50	7138	57	50
8480	8	0	8241	18	0	7986	28	0	7716	38	0	7431	48	0	7132	58	0
8477	8	10	8237	18	10	7982	28	10	7712	38	10	7427	48	10	7127	58	10
8473	8	20	8233	18	20	7978	28	20	7707	38	20	7422	48	20	7122	58	20
8469	8	30	8229	18	30	7973	28	30	7702	38	30	7417	48	30	7117	58	30
8465	8	40	8225	18	40	7969	28	40	7698	38	40	7412	48	40	7112	58	40
8461	8	50	8221	18	50	7964	28	50	7693	38	50	7407	48	50	7107	58	50
8457	9	0	8216	19	0	7960	29	0	7688	39	0	7402	49	0	7102	59	0
8453	9	10	8212	19	10	7956	29	10	7684	39	10	7397	49	10	7097	59	10
8449	9	20	8208	19	20	7951	29	20	7679	39	20	7392	49	20	7092	59	20
8446	9	30	8204	19	30	7947	29	30	7674	39	30	7387	49	30	7086	59	30
8442	9	40	8200	19	40	7942	29	40	7670	39	40	7383	49	40	7081	59	40
8438	9	50	8196	19	50	7938	29	50	7665	39	50	7378	49	50	7076	59	50
8434	10	0	8191	20	0	7933	30	0	7660	40	0	7373	50	0	7071	60	0

TABLE XXIX. For computing the Apparent Time.

III HOURS.

Num	M.	S.	Num	M.	S.	Num	M.	S.	Num	M.	S.	Num	M.	S.	Num	M.	S.
7071	0	0	6756	10	0	6428	20	0	6088	30	0	5736	40	0	5373	50	0
7066	0	10	6750	10	10	6422	20	10	6082	30	10	5730	40	10	5367	50	10
7061	0	20	6745	10	20	6417	20	20	6076	30	20	5724	40	20	5361	50	20
7056	0	30	6740	10	30	6411	20	30	6070	30	30	5718	40	30	5355	50	30
7050	0	40	6734	10	40	6406	20	40	6064	30	40	5712	40	40	5348	50	40
7045	0	50	6729	10	50	6400	20	50	6059	30	50	5706	40	50	5342	50	50
7040	1	0	6724	11	0	6394	21	0	6053	31	0	5700	41	0	5336	51	0
7035	1	10	6718	11	10	6389	21	10	6047	31	10	5694	41	10	5330	51	10
7030	1	20	6713	11	20	6383	21	20	6041	31	20	5688	41	20	5324	51	20
7025	1	30	6707	11	30	6378	21	30	6036	31	30	5682	41	30	5318	51	30
7019	1	40	6702	11	40	6372	21	40	6030	31	40	5676	41	40	5311	51	40
7014	1	50	6697	11	50	6366	21	50	6024	31	50	5670	41	50	5305	51	50
7009	2	0	6691	12	0	6361	22	0	6018	32	0	5664	42	0	5299	52	0
7004	2	10	6686	12	10	6355	22	10	6012	32	10	5658	42	10	5293	52	10
6999	2	20	6680	12	20	6350	22	20	6006	32	20	5652	42	20	5287	52	20
6993	2	30	6675	12	30	6344	22	30	6001	32	30	5646	42	30	5281	52	30
6988	2	40	6670	12	40	6338	22	40	5995	32	40	5640	42	40	5274	52	40
6983	2	50	6664	12	50	6333	22	50	5989	32	50	5634	42	50	5268	52	50
6978	3	0	6659	13	0	6327	23	0	5983	33	0	5628	43	0	5262	53	0
6973	3	10	6653	13	10	6321	23	10	5977	33	10	5622	43	10	5256	53	10
6967	3	20	6648	13	20	6316	23	20	5972	33	20	5616	43	20	5250	53	20
6962	3	30	6642	13	30	6310	23	30	5966	33	30	5610	43	30	5244	53	30
6957	3	40	6637	13	40	6304	23	40	5960	33	40	5604	43	40	5237	53	40
6952	3	50	6632	13	50	6299	23	50	5954	33	50	5598	43	50	5231	53	50
6947	4	0	6626	14	0	6293	24	0	5948	34	0	5592	44	0	5225	54	0
6941	4	10	6621	14	10	6287	24	10	5942	34	10	5586	44	10	5219	54	10
6936	4	20	6615	14	20	6281	24	20	5936	34	20	5580	44	20	5213	54	20
6931	4	30	6610	14	30	6276	24	30	5931	34	30	5574	44	30	5206	54	30
6926	4	40	6604	14	40	6271	24	40	5925	34	40	5568	44	40	5200	54	40
6920	4	50	6599	14	50	6265	24	50	5919	34	50	5562	44	50	5194	54	50
6915	5	0	6593	15	0	6259	25	0	5913	35	0	5556	45	0	5188	55	0
6910	5	10	6588	15	10	6254	25	10	5907	35	10	5550	45	10	5181	55	10
6905	5	20	6582	15	20	6248	25	20	5901	35	20	5544	45	20	5175	55	20
6899	5	30	6577	15	30	6242	25	30	5895	35	30	5538	45	30	5169	55	30
6894	5	40	6572	15	40	6236	25	40	5890	35	40	5531	45	40	5163	55	40
6889	5	50	6566	15	50	6231	25	50	5884	35	50	5525	45	50	5157	55	50
6883	6	0	6561	16	0	6225	26	0	5878	36	0	5519	46	0	5150	56	0
6878	6	10	6555	16	10	6219	26	10	5872	36	10	5513	46	10	5144	56	10
6873	6	20	6550	16	20	6214	26	20	5866	36	20	5507	46	20	5138	56	20
6868	6	30	6544	16	30	6208	26	30	5860	36	30	5501	46	30	5132	56	30
6862	6	40	6539	16	40	6202	26	40	5854	36	40	5495	46	40	5125	56	40
6857	6	50	6533	16	50	6197	26	50	5848	36	50	5489	46	50	5119	56	50
6852	7	0	6528	17	0	6191	27	0	5842	37	0	5483	47	0	5113	57	0
6846	7	10	6522	17	10	6185	27	10	5837	37	10	5477	47	10	5107	57	10
6841	7	20	6517	17	20	6179	27	20	5831	37	20	5471	47	20	5100	57	20
6836	7	30	6511	17	30	6174	27	30	5825	37	30	5465	47	30	5094	57	30
6831	7	40	6505	17	40	6168	27	40	5819	37	40	5459	47	40	5088	57	40
6825	7	50	6500	17	50	6162	27	50	5813	37	50	5452	47	50	5082	57	50
6820	8	0	6494	18	0	6157	28	0	5807	38	0	5446	48	0	5075	58	0
6815	8	10	6489	18	10	6151	28	10	5801	38	10	5440	48	10	5069	58	10
6809	8	20	6483	18	20	6145	28	20	5793	38	20	5434	48	20	5063	58	20
6804	8	30	6478	18	30	6139	28	30	5789	38	30	5428	48	30	5057	58	30
6799	8	40	6472	18	40	6134	28	40	5783	38	40	5422	48	40	5050	58	40
6793	8	50	6467	18	50	6128	28	50	5777	38	50	5416	48	50	5044	58	50
6788	9	0	6461	19	0	6122	29	0	5771	39	0	5410	49	0	5038	59	0
6783	9	10	6456	19	10	6116	29	10	5765	39	10	5404	49	10	5031	59	10
6777	9	20	6450	19	20	6111	29	20	5760	39	20	5397	49	20	5025	59	20
6772	9	30	6445	19	30	6105	29	30	5754	39	30	5391	49	30	5019	59	30
6767	9	40	6439	19	40	6099	29	40	5748	39	40	5385	49	40	5013	59	40
6761	9	50	6433	19	50	6093	29	50	5742	39	50	5379	49	50	5006	59	50
6756	10	0	6428	20	0	6088	30	0	5736	40	0	5373	50	0	5000	60	0

TABLE XXIX. For computing the Apparent Time.

IV HOURS.

Num	M.	S.	Num	M.	S.	Num	M.	S.	Num	M.	S.	Num	M.	S.	Num	M.	S.
5000	0	0	4617	10	0	4226	20	0	3827	30	0	3420	40	0	3007	50	0
4994	0	10	4611	10	10	4220	20	10	3820	30	10	3413	40	10	3000	50	10
4987	0	20	4605	10	20	4213	20	20	3813	30	20	3406	40	20	2993	50	20
4981	0	30	4598	10	30	4206	20	30	3807	30	30	3400	40	30	2986	50	30
4975	0	40	4592	10	40	4200	20	40	3800	30	40	3393	40	40	2979	50	40
4968	0	50	4585	10	50	4193	20	50	3793	30	50	3386	40	50	2972	50	50
4962	1	0	4579	11	0	4187	21	0	3786	31	0	3379	41	0	2965	51	0
4956	1	10	4572	11	10	4180	21	10	3780	31	10	3372	41	10	2958	51	10
4949	1	20	4566	11	20	4173	21	20	3773	31	20	3365	41	20	2951	51	20
4943	1	30	4559	11	30	4167	21	30	3766	31	30	3359	41	30	2945	51	30
4937	1	40	4553	11	40	4160	21	40	3759	31	40	3352	41	40	2938	51	40
4930	1	50	4546	11	50	4153	21	50	3753	31	50	3345	41	50	2931	51	50
4924	2	0	4540	12	0	4147	22	0	3746	32	0	3338	42	0	2924	52	0
4918	2	10	4533	12	10	4140	22	10	3739	32	10	3331	42	10	2917	52	10
4912	2	20	4527	12	20	4134	22	20	3733	32	20	3324	42	20	2910	52	20
4905	2	30	4520	12	30	4127	22	30	3726	32	30	3317	42	30	2903	52	30
4899	2	40	4514	12	40	4120	22	40	3719	32	40	3311	42	40	2896	52	40
4892	2	50	4507	12	50	4114	22	50	3712	32	50	3304	42	50	2889	52	50
4886	3	0	4501	13	0	4107	23	0	3706	33	0	3297	43	0	2882	53	0
4880	3	10	4494	13	10	4100	23	10	3699	33	10	3290	43	10	2875	53	10
4873	3	20	4488	13	20	4094	23	20	3692	33	20	3283	43	20	2868	53	20
4867	3	30	4481	13	30	4087	23	30	3685	33	30	3276	43	30	2861	53	30
4861	3	40	4475	13	40	4081	23	40	3678	33	40	3269	43	40	2854	53	40
4854	3	50	4468	13	50	4074	23	50	3672	33	50	3262	43	50	2847	53	50
4848	4	0	4462	14	0	4067	24	0	3665	34	0	3256	44	0	2840	54	0
4842	4	10	4455	14	10	4061	24	10	3658	34	10	3249	44	10	2833	54	10
4835	4	20	4449	14	20	4054	24	20	3651	34	20	3242	44	20	2826	54	20
4829	4	30	4442	14	30	4047	24	30	3645	34	30	3235	44	30	2819	54	30
4823	4	40	4436	14	40	4041	24	40	3638	34	40	3228	44	40	2812	54	40
4816	4	50	4429	14	50	4034	24	50	3631	34	50	3221	44	50	2805	54	50
4810	5	0	4423	15	0	4027	25	0	3624	35	0	3214	45	0	2798	55	0
4803	5	10	4416	15	10	4021	25	10	3618	35	10	3207	45	10	2791	55	10
4797	5	20	4410	15	20	4014	25	20	3611	35	20	3201	45	20	2784	55	20
4791	5	30	4403	15	30	4007	25	30	3604	35	30	3194	45	30	2777	55	30
4784	5	40	4397	15	40	4001	25	40	3597	35	40	3187	45	40	2770	55	40
4778	5	50	4390	15	50	3994	25	50	3590	35	50	3180	45	50	2763	55	50
4772	6	0	4384	16	0	3988	26	0	3584	36	0	3173	46	0	2756	56	0
4765	6	10	4377	16	10	3981	26	10	3577	36	10	3166	46	10	2749	56	10
4759	6	20	4371	16	20	3974	26	20	3570	36	20	3159	46	20	2742	56	20
4752	6	30	4364	16	30	3967	26	30	3563	36	30	3152	46	30	2735	56	30
4746	6	40	4357	16	40	3961	26	40	3556	36	40	3145	46	40	2728	56	40
4740	6	50	4351	16	50	3954	26	50	3550	36	50	3139	46	50	2721	56	50
4733	7	0	4344	17	0	3947	27	0	3543	37	0	3132	47	0	2714	57	0
4727	7	10	4338	17	10	3941	27	10	3536	37	10	3125	47	10	2707	57	10
4720	7	20	4331	17	20	3934	27	20	352[illegible]	37	20	3118	47	20	2700	57	20
4714	7	30	4325	17	30	3927	27	30	3522	37	30	3111	47	30	2693	57	30
4708	7	40	4318	17	40	3921	27	40	3516	37	40	3104	47	40	2686	57	40
4701	7	50	4312	17	50	3914	27	50	3509	37	50	3097	47	50	2679	57	50
4695	8	0	4305	18	0	3907	28	0	3502	38	0	3090	48	0	2672	58	0
4688	8	10	4298	18	10	3901	28	10	3495	38	10	3083	48	10	2665	58	10
4682	8	20	4292	18	20	3894	28	20	3488	38	20	3076	48	20	2658	58	20
4675	8	30	4285	18	30	3887	28	30	3482	38	30	3069	48	30	2651	58	30
4669	8	40	4279	18	40	3880	28	40	3475	38	40	3062	48	40	2644	58	40
4663	8	50	4272	18	50	3874	28	50	3468	38	50	3056	48	50	2637	58	50
4656	9	0	4266	19	0	3867	29	0	3461	39	0	3049	49	0	2630	59	0
4650	9	10	4259	19	10	3860	29	10	3454	39	10	3042	49	10	2623	59	10
4643	9	20	4252	19	20	3854	29	20	3447	39	20	3035	49	20	2616	59	20
4637	9	30	4246	19	30	3847	29	30	3441	39	30	3028	49	30	2609	59	30
4630	9	40	4239	19	40	384[illegible]	29	40	3434	39	40	3021	49	40	2602	59	40
4624	9	50	4233	19	50	3833	29	50	3427	39	50	3014	49	50	2595	59	50
4617	10	0	4226	20	0	3827	30	0	3420	40	0	3007	50	0	2588	60	0

TABLE XXIX. For computing the Apparent Time.

▼ HOURS.

Num	M.	S.	Num	M.	S.	Num	M.	S.	Num	M.	S.	Num	M.	S.	Num	M.	S.
2588	0	0	2164	10	0	1736	20	0	1305	30	0	872	40	0	436	50	0
2581	0	10	2157	10	10	1729	20	10	1298	30	10	864	40	10	429	50	10
2574	0	20	2150	10	20	1722	20	20	1291	30	20	857	40	20	422	50	20
2567	0	30	2143	10	30	1715	20	30	1284	30	30	850	40	30	414	50	30
2560	0	40	2136	10	40	1708	20	40	1276	30	40	843	40	40	407	50	40
2553	0	50	2129	10	50	1701	20	50	1269	30	50	835	40	50	400	50	50
2546	1	0	2122	11	0	1693	21	0	1262	31	0	828	41	0	393	51	0
2539	1	10	2115	11	10	1686	21	10	1255	31	10	821	41	10	385	51	10
2532	1	20	2108	11	20	1679	21	20	1248	31	20	814	41	20	378	51	20
2525	1	30	2100	11	30	1672	21	30	1240	31	30	806	41	30	371	51	30
2518	1	40	2093	11	40	1665	21	40	1233	31	40	799	41	40	363	51	40
2511	1	50	2086	11	50	1658	21	50	1226	31	50	792	41	50	356	51	50
2504	2	0	2079	12	0	1650	22	0	1219	32	0	785	42	0	349	52	0
2497	2	10	2072	12	10	1643	22	10	1211	32	10	777	42	10	342	52	10
2490	2	20	2065	12	20	1636	22	20	1204	32	20	770	42	20	334	52	20
2483	2	30	2058	12	30	1629	22	30	1197	32	30	763	42	30	327	52	30
2476	2	40	2051	12	40	1622	22	40	1190	32	40	756	42	40	320	52	40
2469	2	50	2044	12	50	1615	22	50	1183	32	50	748	42	50	313	52	50
2462	3	0	2036	13	0	1607	23	0	1175	33	0	741	43	0	305	53	0
2454	3	10	2029	13	10	1600	23	10	1168	33	10	734	43	10	298	53	10
2447	3	20	2022	13	20	1593	23	20	1161	33	20	727	43	20	291	53	20
2440	3	30	2015	13	30	1586	23	30	1154	33	30	719	43	30	283	53	30
2433	3	40	2008	13	40	1579	23	40	1146	33	40	712	43	40	276	53	40
2426	3	50	2001	13	50	1571	23	50	1139	33	50	705	43	50	269	53	50
2419	4	0	1994	14	0	1564	24	0	1132	34	0	698	44	0	262	54	0
2412	4	10	1987	14	10	1557	24	10	1125	34	10	690	44	10	254	54	10
2405	4	20	1979	14	20	1550	24	20	1118	34	20	683	44	20	247	54	20
2398	4	30	1972	14	30	1543	24	30	1110	34	30	676	44	30	240	54	30
2391	4	40	1965	14	40	1536	24	40	1103	34	40	668	44	40	233	54	40
2384	4	50	1958	14	50	1528	24	50	1096	34	50	661	44	50	225	54	50
2377	5	0	1951	15	0	1521	25	0	1089	35	0	654	45	0	218	55	0
2370	5	10	1944	15	10	1514	25	10	1081	35	10	647	45	10	211	55	10
2363	5	20	1937	15	20	1507	25	20	1074	35	20	639	45	20	204	55	20
2356	5	30	1929	15	30	1500	25	30	1067	35	30	632	45	30	196	55	30
2349	5	40	1922	15	40	1492	25	40	1060	35	40	625	45	40	189	55	40
2341	5	50	1915	15	50	1485	25	50	1052	35	50	618	45	50	182	55	50
2334	6	0	1908	16	0	1478	26	0	1045	36	0	610	46	0	174	56	0
2327	6	10	1901	16	10	1471	26	10	1038	36	10	603	46	10	167	56	10
2320	6	20	1894	16	20	1464	26	20	1031	36	20	596	46	20	160	56	20
2313	6	30	1887	16	30	1456	26	30	1024	36	30	589	46	30	153	56	30
2306	6	40	1879	16	40	1449	26	40	1016	36	40	581	46	40	145	56	40
2299	6	50	1872	16	50	1442	26	50	1009	36	50	574	46	50	137	56	50
2292	7	0	1865	17	0	1435	27	0	1002	37	0	567	47	0	131	57	0
2285	7	10	1858	17	10	1428	27	10	995	37	10	560	47	10	124	57	10
2278	7	20	1851	17	20	1420	27	20	987	37	20	552	47	20	116	57	20
2271	7	30	1844	17	30	1413	27	30	980	37	30	545	47	30	109	57	30
2264	7	40	1837	17	40	1406	27	40	973	37	40	538	47	40	102	57	40
2257	7	50	1829	17	50	1399	27	50	966	37	50	531	47	50	94	57	50
2249	8	0	1822	18	0	1392	28	0	958	38	0	523	48	0	87	58	0
2242	8	10	1815	18	10	1384	28	10	951	38	10	516	48	10	80	58	10
2235	8	20	1808	18	20	1377	28	20	944	38	20	509	48	20	73	58	20
2228	8	30	1801	18	30	1370	28	30	937	38	30	501	48	30	65	58	30
2221	8	40	1794	18	40	1363	28	40	929	38	40	494	48	40	58	58	40
2214	8	50	1787	18	50	1356	28	50	922	38	50	487	48	50	51	58	50
2207	9	0	1779	19	0	1348	29	0	915	39	0	480	49	0	44	59	0
2200	9	10	1772	19	10	1341	29	10	908	39	10	472	49	10	36	59	10
2193	9	20	1765	19	20	1334	29	20	900	39	20	465	49	20	29	59	20
2186	9	30	1758	19	30	1327	29	30	893	39	30	458	49	30	22	59	30
2179	9	40	1751	19	40	1320	29	40	886	39	40	451	49	40	14	59	40
2171	9	50	1744	19	50	1313	29	50	879	39	50	443	49	50	7	59	50
2164	10	0	1736	20	0	1305	30	0	872	40	0	436	50	0	[illegible]	[illegible]	0

TABLE XXX. Aug. Moon's Semid.

Arg. app. alt.	Arg. Hor. semid. of the Moon. 14′ 30″	15′ 0″	15′ 30″	16′ 0″	16′ 30″	17′ 0″
°	″	″	″	″	″	″
0	0	0	0	0	0	0
3	0	0	1	1	1	1
6	1	1	2	2	2	2
9	2	2	2	2	3	3
12	3	3	3	3	4	4
15	4	4	4	4	5	5
18	4	4	5	5	5	6
21	5	5	6	6	6	7
24	5	6	6	6	7	7
27	6	6	7	7	8	8
30	7	7	8	8	9	9
33	7	8	8	9	10	10
36	8	9	9	10	10	11
39	8	9	10	10	11	12
42	9	10	10	11	12	12
45	9	10	11	12	12	13
48	10	11	12	12	13	14
51	10	11	12	13	14	14
54	11	12	13	13	14	15
57	11	12	13	14	15	16
60	12	13	14	14	15	16
63	12	13	14	15	16	17
66	12	13	14	15	16	17
69	13	14	15	15	16	17
72	13	14	15	16	17	18
75	13	14	15	16	17	18
78	13	14	15	16	17	18
81	13	14	15	16	17	18
84	13	14	15	16	18	19
90	13	14	16	17	18	19

TABLE XXXI. Contrac. semid. Sun & Moon.

Arg. incli. of sem.	Arg. App. alt. of the Sun or Moon. 10°	11°	12°	13°	14°	16°	18°	20°	24°	32°	50°	90°
°	″	″	″	″	″	″	″	″	″	″	″	″
0	0	0	0	0	0	0	0	0	0	0	0	0
3	0	0	0	0	0	0	0	0	0	0	0	0
6	0	0	0	0	0	0	0	0	0	0	0	0
9	0	0	0	0	0	0	0	0	0	0	0	0
12	0	0	0	0	0	0	0	0	0	0	0	0
15	1	0	0	0	0	0	0	0	0	0	0	0
18	1	1	1	0	0	0	0	0	0	0	0	0
21	1	1	1	1	1	0	0	0	0	0	0	0
24	1	1	1	1	1	1	0	0	0	0	0	0
27	2	1	1	1	1	1	1	0	0	0	0	0
30	2	2	1	1	1	1	1	1	0	0	0	0
33	2	2	2	1	1	1	1	1	0	0	0	0
36	3	2	2	2	2	1	1	1	1	0	0	0
39	3	3	2	2	2	1	1	1	1	0	0	0
42	4	3	3	2	2	2	1	1	1	0	0	0
45	4	3	3	3	2	2	1	1	1	0	0	0
48	5	4	3	3	2	2	1	1	1	1	0	0
51	5	4	4	3	3	2	2	1	1	1	0	0
54	5	4	4	3	3	2	2	1	1	1	0	0
57	6	5	4	4	3	2	2	2	1	1	0	0
60	6	5	4	4	3	3	2	2	1	1	0	0
63	7	5	5	4	3	3	2	2	1	1	0	0
66	7	6	5	4	4	3	2	2	1	1	0	0
69	7	6	5	4	4	3	2	2	1	1	0	0
72	7	6	5	5	4	3	2	2	1	1	0	0
75	8	6	5	5	4	3	3	2	1	1	0	0
78	8	7	6	5	4	3	3	2	2	1	0	0
81	8	7	6	5	4	3	3	2	2	1	0	0
84	8	7	6	5	4	3	3	2	2	1	0	0
90	8	7	6	5	4	3	3	2	2	1	0	0

Corrections of the Moon's Par. arising from the Spheroidal *Figure* of the Earth.

Arg. Lat.	XXXII. Red. h. p. Arg. Hor. par. 54′	57′	60′	XXXIII. Ang. h. p. Arg. Hor. par. 54′	57′	60′	XXXIV. Cor. p. lat. Arg. Hor. par. 54′	57′	60′	XXXV. Gr. effect p. in lon. Arg. Hor. parallax. 54′	57′	60′	XXXVI. Red. of Lat.
°	″	″	″	″	″	″	″	″	″	″	″	″	′
0	0	0	0	0	0	0	0	0	0	0.0	0.0	0.0	0.0
4	0	0	0	0	0	0	2	2	2	0.8	0.9	0.9	2.1
8	0	0	0	1	1	1	4	4	4	1.5	1.6	1.7	4.1
12	1	1	1	1	1	1	5	6	6	2.3	2.5	2.6	6.1
16	1	1	1	2	2	2	7	8	8	3.2	3.4	3.5	7.9
20	2	2	2	3	4	4	9	9	10	3.9	4.1	4.3	9.6
24	2	2	3	5	5	5	11	11	12	4.6	4.9	5.1	11.1
28	3	3	3	6	7	7	12	13	13	5.3	5.6	5.9	12.4
32	4	4	5	8	8	9	14	15	15	5.9	6.3	6.6	13.3
36	5	5	5	10	10	11	15	16	17	6.7	7.0	7.4	14.2
40	6	6	6	12	12	13	17	18	18	7.2	7.6	8.0	14.7
44	7	7	8	14	14	15	18	19	20	7.8	8.3	8.7	15.9
48	8	8	9	16	16	17	19	20	21	8.4	8.8	9.3	14.9
52	9	9	10	18	18	19	20	22	23	8.8	9.3	9.8	14.5
56	10	10	11	19	20	21	21	23	24	9.4	9.9	10.4	13.9
60	10	11	12	21	22	23	22	24	25	9.7	10.2	10.8	13.0
64	11	12	13	23	24	25	23	25	26	10.1	10.6	11.2	11.8
68	12	13	13	24	25	27	24	25	27	10.4	10.9	11.4	10.5
72	13	13	14	25	27	28	25	26	27	10.6	11.2	11.8	8.9
76	13	14	14	26	28	29	25	26	28	10.8	11.4	12.0	7.1
80	14	14	15	27	29	30	25	27	28	11.0	11.6	12.2	5.2
84	14	15	15	28	29	31	26	27	29	11.2	11.8	12.4	3.2
90	14	15	16	29	30	32	26	28	29	11.4	12.0	12.6	0.0

TABLE XXXVII. Equation of Second Difference.

App. t. aft N. or M. h. ′	Argument. Second Difference of the Moon's Place. 1″	2″	3″	4″	5″	6″	7″	8″	9″	10″	15″	20″	25″	30″	40″	50″	Ap. t. aft N. or M. h. ′
	″	″	″	″	″	″	″	″	″	″	″	″	″	″	″	″	
0 0	0.0	0.0	0.0	0.0	0.0	0.0	0.0	0.0	0.0	0.0	0.0	0.0	0.0	0.0	0.0	0.0	12 0
0 10	0.0	0.0	0.0	0.0	0.0	0.0	0.0	0.0	0.1	0.1	0.1	0.1	0.2	0.2	0.3	0.3	11 50
0 20	0.0	0.0	0.0	0.0	0.1	0.1	0.1	0.1	0.1	0.1	0.2	0.3	0.3	0.4	0.5	0.7	11 40
0 30	0.0	0.0	0.1	0.1	0.1	0.1	0.1	0.2	0.2	0.2	0.3	0.4	0.5	0.6	0.8	1.0	11 30
0 40	0.0	0.1	0.1	0.1	0.1	0.2	0.2	0.2	0.2	0.3	0.4	0.5	0.7	0.8	1.0	1.3	11 20
0 50	0.0	0.1	0.1	0.1	0.2	0.2	0.2	0.3	0.3	0.3	0.5	0.6	0.8	1.0	1.3	1.6	11 10
1 0	0.0	0.1	0.1	0.1	0.2	0.2	0.3	0.3	0.3	0.4	0.6	0.8	0.9	1.1	1.5	1.9	11 0
1 10	0.0	0.1	0.1	0.2	0.2	0.3	0.3	0.3	0.4	0.4	0.7	0.9	1.1	1.3	1.8	2.2	10 50
1 20	0.0	0.1	0.1	0.2	0.2	0.3	0.3	0.4	0.4	0.5	0.7	1.0	1.2	1.5	2.0	2.5	10 40
1 30	0.0	0.1	0.2	0.2	0.3	0.3	0.4	0.4	0.5	0.5	0.8	1.1	1.4	1.6	2.2	2.7	10 30
1 40	0.1	0.1	0.2	0.2	0.3	0.4	0.4	0.5	0.5	0.6	0.9	1.2	1.5	1.8	2.4	3.0	10 20
1 50	0.1	0.1	0.2	0.3	0.3	0.4	0.4	0.5	0.6	0.6	1.0	1.3	1.6	1.9	2.6	3.2	10 10
2 0	0.1	0.1	0.2	0.3	0.3	0.4	0.5	0.6	0.6	0.7	1.0	1.4	1.7	2.1	2.8	3.5	10 0
2 10	0.1	0.1	0.2	0.3	0.4	0.4	0.5	0.6	0.7	0.7	1.1	1.5	1.8	2.2	3.0	3.7	9 50
2 20	0.1	0.2	0.2	0.3	0.4	0.5	0.5	0.6	0.7	0.8	1.2	1.6	2.0	2.3	3.1	3.9	9 40
2 30	0.1	0.2	0.2	0.3	0.4	0.5	0.6	0.7	0.7	0.8	1.2	1.6	2.1	2.5	3.3	4.1	9 30
2 40	0.1	0.2	0.3	0.3	0.4	0.5	0.6	0.7	0.8	0.9	1.3	1.7	2.2	2.6	3.5	4.3	9 20
2 50	0.1	0.2	0.3	0.4	0.4	0.5	0.6	0.7	0.8	0.9	1.3	1.8	2.2	2.7	3.6	4.5	9 10
3 0	0.1	0.2	0.3	0.4	0.5	0.6	0.7	0.7	0.8	0.9	1.4	1.9	2.3	2.8	3.8	4.7	9 0
3 10	0.1	0.2	0.3	0.4	0.5	0.6	0.7	0.8	0.9	1.0	1.5	1.9	2.4	2.9	3.9	4.9	8 50
3 20	0.1	0.2	0.3	0.4	0.5	0.6	0.7	0.8	0.9	1.0	1.5	2.0	2.5	3.0	4.0	5.0	8 40
3 30	0.1	0.2	0.3	0.4	0.5	0.6	0.7	0.8	0.9	1.0	1.5	2.1	2.6	3.1	4.1	5.2	8 30
3 40	0.1	0.2	0.3	0.4	0.5	0.6	0.7	0.8	0.9	1.1	1.6	2.1	2.6	3.2	4.2	5.3	8 20
3 50	0.1	0.2	0.3	0.4	0.5	0.6	0.8	0.9	1.0	1.1	1.6	2.2	2.7	3.3	4.3	5.4	8 10
4 0	0.1	0.2	0.3	0.4	0.6	0.7	0.8	0.9	1.0	1.1	1.7	2.2	2.8	3.3	4.4	5.6	8 0
4 20	0.1	0.2	0.3	0.5	0.6	0.7	0.8	0.9	1.0	1.2	1.7	2.3	2.9	3.5	4.6	5.8	7 40
4 40	0.1	0.2	0.4	0.5	0.6	0.7	0.8	0.9	1.0	1.2	1.8	2.4	3.0	3.6	4.8	5.9	7 20
5 0	0.1	0.2	0.4	0.5	0.6	0.7	0.8	1.0	1.1	1.2	1.8	2.4	3.0	3.6	4.9	6.1	7 0
6 0	0.1	0.2	0.4	0.5	0.6	0.7	0.9	1.0	1.1	1.3	1.9	2.5	3.1	3.8	5.0	6.3	6 0

App. t. aft N. or M. h. ′	Argument. Second Difference of the Moon's Place. 1′	2′	3′	4′	5′	6′	7′	8′	9′	10′	11′	12′	Ap. t. aft N. or M. h. ′
	″	″	″	″	″	″	″	″	″	″	″	″	
0 0	0.0	0.0	0.0	0.0	0.0	0.0	0.0	0.0	0.0	0.0	0.0	0.0	12 0
0 10	0.4	0.8	1.2	1.6	2.1	2.5	2.9	3.3	3.7	4.1	4.5	4.9	11 50
0 20	0.8	1.6	2.4	3.2	4.1	4.9	5.7	6.5	7.3	8.1	8.9	9.7	11 40
0 30	1.2	2.4	3.6	4.8	6.0	7.2	8.4	9.6	10.8	12.0	13.2	14.4	11 30
0 40	1.6	3.1	4.7	6.3	7.9	9.4	11.0	12.6	14.2	15.7	17.3	18.9	11 20
0 50	1.9	3.9	5.8	7.8	9.7	11.6	13.6	15.5	17.4	19.4	21.3	23.3	11 10
1 0	2.3	4.6	6.9	9.2	11.5	13.8	16.0	18.3	20.6	22.9	25.2	27.5	11 0
1 10	2.6	5.3	7.9	10.5	13.2	15.8	18.4	21.1	23.7	26.3	29.0	31.6	10 50
1 20	3.0	5.9	8.9	11.9	14.8	17.8	20.7	23.7	26.7	29.6	32.6	35.6	10 40
1 30	3.3	6.6	9.8	13.1	16.4	19.7	23.0	26.3	29.5	32.8	36.1	39.4	10 30
1 40	3.6	7.2	10.8	14.4	17.9	21.5	25.1	28.7	32.3	35.9	39.5	43.1	10 20
1 50	3.9	7.8	11.6	15.5	19.4	23.3	27.2	31.1	34.9	38.8	42.7	46.6	10 10
2 0	4.2	8.3	12.5	16.7	20.8	25.0	29.2	33.3	37.5	41.7	45.8	50.0	10 0
2 10	4.4	8.9	13.3	17.8	22.2	26.6	31.1	35.5	39.9	44.4	48.8	53.3	9 50
2 20	4.7	9.4	14.1	18.8	23.5	28.2	32.9	37.6	42.3	47.0	51.7	56.4	9 40
2 30	4.9	9.9	14.8	19.8	24.7	29.7	34.6	39.6	44.5	49.5	54.4	59.4	9 30
2 40	5.2	10.4	15.6	20.7	25.9	31.1	36.3	41.5	46.7	51.9	57.0	62.2	9 20
2 50	5.4	10.8	16.2	21.6	27.1	32.5	37.9	43.3	48.7	54.1	59.5	64.9	9 10
3 0	5.6	11.3	16.9	22.5	28.1	33.8	39.4	45.0	50.6	56.3	61.9	67.5	9 0
3 10	5.8	11.7	17.5	23.3	29.1	35.0	40.8	46.6	52.4	58.3	64.1	69.9	8 50
3 20	6.0	12.0	18.1	24.1	30.1	36.1	42.1	48.1	54.2	60.2	66.2	72.2	8 40
3 30	6.2	12.4	18.6	24.8	31.0	37.2	43.4	49.6	55.8	62.0	68.2	74.4	8 30
3 40	6.4	12.7	19.1	25.5	31.8	38.2	44.6	50.9	57.3	63.7	70.0	76.4	8 20
3 50	6.5	13.0	19.6	26.1	32.6	39.1	45.7	52.2	58.7	65.2	71.7	78.3	8 10
4 0	6.7	13.3	20.0	26.7	33.3	40.1	46.7	53.3	60.0	66.7	73.3	80.0	8 0
4 20	6.9	13.8	20.8	27.7	34.6	41.5	48.4	55.4	62.3	69.2	76.1	83.1	7 40
4 40	7.1	14.3	21.4	28.5	35.6	42.8	49.9	57.0	64.2	71.3	78.4	85.6	7 20
5 0	7.3	14.6	21.9	29.2	36.5	43.8	51.0	58.3	65.6	72.9	80.2	87.5	7 0
5 20	7.4	14.8	22.2	29.6	37.0	44.4	51.9	59.3	66.7	74.1	81.5	88.9	6 40
5 40	7.5	15.0	22.4	29.9	37.4	44.9	52.3	59.8	67.3	74.8	82.2	89.7	6 20
6 0	7.5	15.0	22.5	30.0	37.5	45.0	52.5	60.0	67.5	75.0	82.5	90.0	6 0

TABLE XXXVIII. Correction of approx. time, ans. to eq. of Sec. Difference.

Moon's motion in 12 hours.	Arg. Equation of Second Difference of the Moon's Longitude.																						
	1″	2″	3″	4″	5″	6″	7″	8″	9″	10″	11″	12″	13″	14″	15″	16″	17″	18″	19″	20″	22″	24″	26″	28″
h. ′	″	″	″	″	″	″	″	″	″	″	″	″	″	″	″	″	″	″	″	″	″	″	″	″
5 30	2	4	7	9	11	13	15	17	20	22	24	26	28	30	33	35	37	39	41	43	48	52	57	61
5 40	2	4	6	8	11	13	15	17	19	21	23	25	27	29	32	34	36	38	40	42	47	51	55	59
5 50	2	4	6	8	10	12	14	16	18	20	23	25	27	29	31	33	35	37	39	41	45	49	53	57
6 0	2	4	6	8	10	12	14	16	18	20	22	24	26	28	30	32	34	36	38	40	44	48	52	56
6 10	2	4	6	8	10	12	14	16	17	19	21	23	25	27	29	31	33	35	37	39	43	47	50	54
6 20	2	4	6	8	9	11	13	15	17	19	21	23	25	26	28	30	32	34	36	38	42	45	49	53
6 30	2	4	6	7	9	11	13	15	17	18	20	22	24	26	28	30	31	33	35	37	41	44	48	52
6 40	2	4	5	7	9	11	13	14	16	18	20	22	23	25	27	29	31	32	34	36	40	43	47	50
6 50	2	4	5	7	9	11	12	14	16	17	19	21	23	25	26	28	30	32	33	35	39	42	46	49
7 0	2	3	5	7	9	10	12	14	15	17	19	20	22	24	26	27	29	31	32	34	38	42	45	48
7 10	2	3	5	7	8	10	12	13	15	17	18	20	22	23	25	27	28	30	32	33	37	40	44	47
7 20	2	3	5	7	8	10	11	13	15	16	18	20	21	23	25	26	28	29	31	33	36	39	43	46
7 30	2	3	5	6	8	10	11	13	14	16	18	19	21	22	24	25	27	29	30	32	35	38	42	45
7 40	2	3	5	6	8	9	11	12	14	16	17	19	20	22	23	25	27	28	30	31	34	38	41	44

Moon's motion in 12 hours.	Arg. Equation of Second Difference of the Moon's Longitude.																	
	30″	32″	34″	36″	38″	40″	42″	44″	46″	48″	50″	55″	60″	65″	70″	75″	80″	90″
h. ′	″	″	″	″	″	″	″	″	″	″	″	″	″	″	″	″	″	″
5 30	65	70	74	78	83	87	91	96	100	104	109	120	131	142	152	163	174	196
5 40	64	68	72	76	81	85	89	93	97	101	106	116	127	138	148	159	169	190
5 50	62	66	70	74	78	82	86	90	94	98	102	113	123	134	144	154	164	185
6 0	60	64	68	72	76	80	84	88	92	96	100	110	120	130	140	150	160	180
6 10	58	62	66	70	74	78	82	86	89	93	97	107	117	126	136	146	155	175
6 20	57	60	64	68	72	76	80	83	87	91	95	104	114	123	132	142	151	170
6 30	55	59	63	66	70	74	78	81	85	88	92	101	111	120	129	138	147	166
6 40	54	57	61	65	68	72	76	79	83	86	90	99	108	117	126	135	144	162
6 50	53	56	60	63	67	70	74	77	81	84	88	97	105	114	123	131	140	158
7 0	51	55	58	62	65	69	72	75	79	82	86	94	103	111	120	129	137	154
7.30	50	54	57	60	64	67	70	74	77	80	84	92	100	109	117	126	134	150
7 20	49	52	56	59	62	65	69	72	75	79	82	90	98	106	114	123	131	147
7 30	48	51	54	58	61	64	67	70	73	77	80	88	96	104	112	120	128	144
7 40	47	50	53	56	59	63	66	68	72	75	78	86	94	102	110	117	125	141

TABLE XXXIX. Equat. semi. of moon in time.

Moon's Semidiameter	Daily var. of Moon's passing Meridi.						
	40′	44′	48′	52′	56′	60′	64′
14′ 40″	60″3	60″5	60″6	60″8	60″9	61″1	61″3
14 45	60.6	60.8	61.0	61.1	61.3	61.5	61.6
14 50	61.0	61.1	61.3	61.5	61.6	61.8	62.0
14 55	61.3	61.5	61.7	61.8	61.9	62.1	62.3
15 0	61.7	61.8	62.0	62.2	62.3	62.5	62.7
15 5	62.0	62.2	62.3	62.5	62.7	62.8	63.0
15 10	62.4	62.5	62.7	62.9	63.0	63.2	63.4
15 15	62.7	62.9	63.0	63.2	63.4	63.5	63.7
15 20	63.0	63.2	63.4	63.5	63.7	63.9	64.1
15 25	63.4	63.6	63.7	63.9	64.0	64.2	64.4
15 30	63.7	63.9	64.1	64.2	64.4	64.6	64.8
15 35	64.1	64.2	64.4	64.6	64.8	64.9	65.1
15 40	64.4	64.6	64.7	64.9	65.1	65.3	65.5
15 45	64.7	64.9	65.1	65.3	65.4	65.6	65.8
15 50	65.1	65.3	65.4	65.6	65.8	66.0	66.1
15 55	65.4	65.6	65.8	66.0	66.1	66.3	66.5
16 0	65.8	66.0	66.1	66.3	66.5	66.7	66.8
16 5	66.1	66.3	66.5	66.6	66.8	67.0	67.2
16 10	66.5	66.6	66.8	67.0	67.2	67.4	67.5
16 15	66.8	67.0	67.2	67.3	67.5	67.7	67.9
16 20	67.1	67.3	67.5	67.7	67.9	68.1	68.2
16 25	67.5	67.7	67.9	68.0	68.2	68.4	68.6
16 30	67.8	68.0	68.2	68.4	68.6	68.7	68.9
16 35	68.2	68.4	68.5	68.7	68.9	69.1	69.3
16 40	68.5	68.7	68.9	69.1	69.3	69.4	69.6
16 45	68.9	69.0	69.2	69.4	69.6	69.8	70.0
16 50	69.2	69.4	69.6	69.8	70.0	70.1	70.3

TABLE XL. Incr. moon's sem. in t.

Moon's Declination	Moon's eq. sem. in time.					
	60″	62″	64″	66″	68″	70″
3° 0′	0″1	0″1	0″1	0″1	0″1	0″1
5 0	0.2	0.2	0.2	0.2	0.3	0.3
7 0	0.5	0.5	0.5	0.5	0.5	0.5
9 0	0.8	0.8	0.8	0.8	0.8	0.9
11 0	1.1	1.2	1.2	1.2	1.3	1.3
13 0	1.5	1.6	1.7	1.7	1.8	1.8
14 0	1.8	1.9	2.0	2.0	2.1	2.1
15 0	2.1	2.2	2.2	2.3	2.4	2.5
16 0	2.4	2.5	2.6	2.7	2.7	2.8
17 0	2.7	2.8	2.9	3.0	3.1	3.2
18 0	3.1	3.2	3.3	3.4	3.5	3.6
19 0	3.4	3.6	3.7	3.8	3.9	4.0
20 0	3.8	4.0	4.1	4.2	4.3	4.4
21 0	4.3	4.4	4.5	4.7	4.8	5.0
22 0	4.7	4.9	5.0	5.2	5.3	5.5
23 0	5.1	5.3	5.5	5.7	5.9	6.0
24 0	5.7	5.9	6.0	6.2	6.4	6.6
24 30	5.9	6.1	6.3	6.5	6.7	6.9
25 0	6.2	6.4	6.6	6.8	7.0	7.2
25 30	6.4	6.7	6.9	7.1	7.3	7.5
26 0	6.7	7.0	7.2	7.4	7.6	7.8
26 30	7.0	7.3	7.5	7.8	8.0	8.2
27 0	7.3	7.6	7.8	8.1	8.3	8.6
27 30	7.6	7.9	8.2	8.4	8.6	8.9
28 0	7.9	8.2	8.5	8.7	9.0	9.3
28 30	8.2	8.5	8.8	9.1	9.4	9.6
29 0	8.6	8.5	9.2	9.5	9.8	10.0

P. part { [illegible] | P. part

TABLE XLI. Natural Versed Sines.

M	Versed sine 2 degrees. 0"	10"	20"	30"	40"	50"	
0	000609	0611	0613	0614	0616	0618	59
1	0619	0621	0623	0624	0626	0628	58
2	0630	0631	0633	0635	0637	0638	57
3	0640	0642	0643	0645	0647	0649	56
4	0650	0652	0654	0656	0657	0659	55
5	0661	0663	0665	0666	0668	0670	54
6	0672	0673	0675	0677	0679	0681	53
7	0682	0684	0686	0688	0689	0691	52
8	0693	0695	0697	0699	0700	0702	51
9	0704	0706	0708	0709	0711	0713	50
10	000715	0717	0719	0720	0722	0724	49
11	0726	0728	0730	0732	0733	0735	48
12	0737	0739	0741	0743	0745	0746	47
13	0748	0750	0752	0754	0756	0758	46
14	0760	0761	0763	0765	0767	0769	45
15	0771	0773	0775	0777	0779	0781	44
16	0782	0784	0786	0788	0790	0792	43
17	0794	0796	0798	0800	0802	0804	42
18	0806	0808	0810	0811	0813	0815	41
19	0817	0819	0821	0823	0825	0827	40
20	000829	0831	0833	0835	0837	0839	39
21	0841	0843	0845	0847	0849	0851	38
22	0853	0855	0857	0859	0861	0863	37
23	0865	0867	0869	0871	0873	0875	36
24	0877	0879	0881	0883	0885	0887	35
25	0889	0891	0893	0896	0898	0900	34
26	0902	0904	0906	0908	0810	0912	33
27	0914	0916	0918	0920	0922	0924	32
28	0927	0929	0931	0933	0935	0937	31
29	0939	0941	0943	0945	0948	0950	30
30	000952	0954	0956	0958	0960	0962	29
31	0965	0967	0969	0971	0973	0975	28
32	0977	0979	0982	0984	0986	0988	27
33	0990	0992	0995	0997	0999	1001	26
34	1003	1005	1008	1010	1012	1014	25
35	1016	1018	1021	1023	1025	1027	24
36	1029	1032	1034	1036	1038	1040	23
37	1043	1045	1047	1049	1052	1054	22
38	1056	1058	1060	1063	1065	1067	21
39	1069	1072	1074	1076	1078	1081	20
40	001083	1085	1087	1090	1092	1094	19
41	1096	1099	1101	1103	1106	1108	18
42	1110	1112	1115	1117	1119	1122	17
43	1124	1126	1128	1131	1133	1135	16
44	1138	1140	1142	1145	1147	1149	15
45	1152	1154	1156	1159	1161	1163	14
46	1166	1168	1170	1173	1175	1177	13
47	1180	1182	1184	1187	1189	1191	12
48	1194	1196	1199	1201	1203	1206	11
49	1208	1210	1213	1215	1218	1220	10
50	001222	1225	1227	1230	1232	1234	9
51	1237	1239	1242	1244	1247	1249	8
52	1251	1254	1256	1259	1261	1264	7
53	1266	1268	1271	1273	1276	1278	6
54	1281	1283	1286	1288	1290	1293	5
55	1295	1298	1300	1303	1305	1308	4
56	1310	1313	1315	1318	1320	1323	3
57	1325	1328	1330	1333	1335	1338	2
58	1340	1343	1345	1348	1350	1353	1
59	1355	1358	1360	1363	1365	1368	0
	60"	50"	40"	30"	20"	10"	M
	Co-versed sine 87 degrees.						

P. part	1"	2"	3"	4"	5"	6"	7"	8"	9"
	0	0	1	1	1	1	1	2	2

M	Versed sine 3 degrees. 0"	10"	20"	30"	40"	50"	
0	001370	1373	1376	1378	1381	1383	59
1	1386	1388	1391	1393	1396	1399	58
2	1401	1404	1406	1409	1411	1414	57
3	1417	1419	1422	1424	1427	1429	56
4	1432	1435	1437	1440	1442	1445	55
5	1448	1450	1453	1455	1458	1461	54
6	1463	1466	1469	1471	1474	1476	53
7	1479	1482	1484	1487	1490	1492	52
8	1495	1498	1500	1503	1506	1508	51
9	1511	1514	1516	1519	1522	1524	50
10	001527	1530	1532	1535	1538	1540	49
11	1543	1546	1548	1551	1554	1557	48
12	1559	1562	1565	1567	1570	1573	47
13	1575	1578	1581	1584	1586	1589	46
14	1592	1595	1597	1600	1603	1606	45
15	1608	1611	1614	1617	1619	1622	44
16	1625	1628	1630	1633	1636	1639	43
17	1641	1644	1647	1650	1653	1655	42
18	1658	1661	1664	1667	1669	1672	41
19	1675	1678	1681	1683	1686	1689	40
20	001692	1695	1697	1700	1703	1706	39
21	1709	1712	1714	1717	1720	1723	38
22	1726	1729	1732	1734	1737	1740	37
23	1743	1746	1749	1752	1754	1757	36
24	1760	1763	1766	1769	1772	1775	35
25	1777	1780	1783	1786	1789	1792	34
26	1795	1798	1801	1804	1806	1809	33
27	1812	1815	1818	1821	1824	1827	32
28	1830	1833	1836	1839	1842	1845	31
29	1847	1850	1853	1856	1859	1862	30
30	001865	1868	1871	1874	1877	1880	29
31	1883	1886	1889	1892	1895	1898	28
32	1901	1904	1907	1910	1913	1916	27
33	1919	1922	1925	1928	1931	1934	26
34	1937	1940	1943	1946	1949	1952	25
35	1955	1958	1961	1964	1967	1970	24
36	1973	1976	1979	1982	1985	1989	23
37	1997	1995	1998	2001	2004	2007	22
38	2010	2083	2016	2019	2022	2025	21
39	2028	2032	1035	2038	2041	2044	20
40	002047	2050	2053	1056	2059	2063	19
41	2066	2069	2072	2075	2078	2081	18
42	2084	2088	2091	2094	2097	2100	17
43	2103	2106	2109	2113	2116	2119	16
44	2122	2125	2128	2132	2135	2138	15
45	2141	2144	2147	2151	2154	2157	14
46	2160	2163	2167	2170	2173	2176	13
47	2179	2182	2186	2189	2192	2195	12
48	2199	2202	2205	2208	2211	2215	11
49	2218	2221	2224	2228	2231	2234	10
50	002237	2240	2244	2247	2250	2253	9
51	2257	2260	2263	2267	2270	2273	8
52	2276	2280	2283	2286	2289	2293	7
53	2296	2299	2303	2306	2309	2312	6
54	2316	2319	2322	2326	2329	2332	5
55	2336	2339	2342	2345	2349	2352	4
56	2355	2359	2362	2365	2369	2372	3
57	2375	2379	2382	2385	2389	2392	2
58	2396	2399	2402	2406	2409	2412	1
59	2416	2419	2422	2426	2429	2433	0
	60"	50"	40"	30"	20"	10"	M
	Co-versed sine 86 degrees.						

P. part	1"	2"	3"	4"	5"	6"	7"	8"	9"
	0	1	1	1	1	2	2	2	3

TABLE XLI. Natural Versed Sines.

M	Versed sine 4 degrees. 0"	10"	20"	30"	40"	50"	
0	002436	2439	2443	2446	2449	2453	59
1	2456	2460	2463	2466	2470	2473	58
2	2477	2480	2484	2487	2490	2494	57
3	2497	2501	2504	2507	2511	2514	56
4	2518	2521	2525	2528	2532	2535	55
5	2538	2542	2545	2549	2552	2556	54
6	2559	2563	2566	2570	2573	2577	53
7	2580	2584	2587	2591	2594	2597	52
8	2601	2604	2608	2612	2615	2618	51
9	2622	2625	2629	2633	2636	2640	50
10	002643	2647	2650	2654	2657	2661	49
11	2664	2668	2671	2675	2678	2682	48
12	2686	2689	2693	2696	2700	2603	47
13	2707	2710	2714	2718	2721	2725	46
14	2728	2732	2735	2739	2743	2746	45
15	2750	2753	2757	2761	2764	2768	44
16	2771	2775	2779	2782	2786	2789	43
17	2793	2797	2800	2804	2808	2811	42
18	2815	2819	2822	2826	2829	2833	41
19	2837	2840	2844	2848	2851	2855	40
20	002859	2862	2866	2870	2873	2877	39
21	2881	2884	2888	2892	2895	2899	38
22	2903	2906	2910	2914	2918	2921	37
23	2925	2929	2932	2936	2940	2944	36
24	2947	2951	2955	2958	2962	2966	35
25	2970	2973	2977	2981	2985	2988	34
26	2992	2996	3000	3003	3007	3011	33
27	3015	3018	3022	3026	3030	3033	32
28	3037	3041	3045	3049	3052	3056	31
29	3060	3064	3067	3071	3075	3079	30
30	003083	3086	3090	3094	3098	3102	29
31	3106	3109	3113	3117	3121	3125	28
32	3128	3132	3136	3140	3144	3148	27
33	3152	3155	3159	3163	3167	3171	26
34	3175	3178	3182	3186	3190	3194	25
35	3198	3202	3206	3209	3213	3217	24
36	3221	3225	3229	3233	3237	3241	23
37	3244	3248	3252	3256	3260	3264	22
38	3268	3272	3276	3280	3284	3288	21
39	3291	3295	3299	3303	3307	3311	20
40	003315	3319	3323	3327	3331	3335	19
41	3339	3343	3347	3351	3355	3359	18
42	3363	3367	3371	3375	3379	3383	17
43	3386	3390	3394	3398	3402	3406	16
44	3410	3414	3418	3422	3426	3430	15
45	3434	3439	3443	3447	3451	3455	14
46	3459	3463	3467	3471	3475	3479	13
47	3483	3487	3491	3495	3499	3503	12
48	3507	3511	3515	3519	3523	3527	11
49	3532	3536	3540	3544	3548	3552	10
50	003556	3560	3564	3568	3572	3576	9
51	3581	3585	3589	3593	3597	3601	8
52	3605	3609	3613	3618	3622	3626	7
53	3630	3634	3638	3642	3646	3651	6
54	3655	3659	3663	3667	3671	3675	5
55	3680	3684	3688	3692	3696	3700	4
56	3705	3709	3713	3717	3721	3725	3
57	3730	3734	3738	3742	3746	3751	2
58	3755	3759	3763	3767	3772	3776	1
59	3780	3784	3788	3793	3797	3801	0
	60"	50"	40"	30"	20"	10"	M
	Co-versed sine 85 degrees.						

P. part	1"	2"	3"	4"	5"	6"	7"	8"	9"
	0	1	1	1	2	2	3	3	3

M	Versed sine 5 degrees. 0"	10"	20"	30"	40"	50"	
0	003805	3810	3814	3818	3822	3826	59
1	3831	3835	3839	3843	3848	3852	58
2	3856	3860	3865	3869	3873	3877	57
3	3882	3886	3890	3895	3899	3903	56
4	3907	3912	3916	3920	3925	3929	55
5	3933	3937	3942	3946	3950	3955	54
6	3959	3963	3968	3972	3976	3981	53
7	3985	3989	3993	3998	4002	4006	52
8	4011	4015	4019	4024	4028	4033	51
9	4037	4041	4046	4050	4054	4059	50
10	004063	4067	4072	4076	4081	4085	49
11	4089	4094	4098	4102	4107	4111	48
12	4116	4120	4124	4129	4133	4138	47
13	4142	4146	4151	4155	4160	4164	46
14	4168	4173	4177	4182	4186	4191	45
15	4195	4200	4204	4208	4213	4217	44
16	4222	4226	4231	4235	4240	4244	43
17	4248	4253	4257	4262	4266	4271	42
18	4275	4280	4284	4289	4293	4298	41
19	4302	4307	4311	4316	4320	4325	40
20	004329	4334	4338	4343	4347	4352	39
21	4356	4361	4365	4370	4374	4379	38
22	4383	4388	4393	4397	4402	4406	37
23	4411	4415	4420	4424	4429	4433	36
24	4438	4443	4447	4452	4456	4461	35
25	4465	4470	4475	4479	4484	4488	34
26	4493	4498	4502	4507	4511	4516	33
27	4521	4525	4530	4534	4539	4544	32
28	4548	4553	4557	4562	4567	4571	31
29	4576	4581	4585	4590	4595	4599	30
30	004604	4608	4613	4618	4622	4627	29
31	4632	4636	4641	4646	4650	4655	28
32	4660	4664	4669	4674	4678	4683	27
33	4688	4693	4697	4702	4707	4711	26
34	4716	4721	4725	4730	4735	4740	25
35	4744	4749	4754	4758	4763	4768	24
36	4773	4777	4782	4787	4792	4796	23
37	4801	4806	4811	4815	4820	4825	22
38	4830	4834	4839	4844	4849	4853	21
39	4858	4863	4868	4872	4877	4882	20
40	004887	4892	4896	4901	4906	4911	19
41	4916	4920	4925	4930	4935	4940	18
42	4944	4949	4954	4959	4964	4969	17
43	4973	4978	4983	4988	4993	4998	16
44	5002	5007	5012	5017	5022	5027	15
45	5031	5036	5041	5046	5051	5056	14
46	5061	5066	5070	5075	5080	5085	13
47	5090	5095	5100	5105	5109	5014	12
48	5119	5124	5129	5134	5139	5144	11
49	5149	5154	5159	5163	5168	5173	10
50	005178	5183	5188	5193	5198	5203	9
51	5208	5213	5218	5223	5228	5233	8
52	5238	5243	5247	5252	5257	5262	7
53	5267	5272	5277	5282	5287	5292	6
54	5297	5302	5307	5312	5317	5322	5
55	5327	5332	5337	5342	5347	5352	4
56	5357	5362	5367	5372	5377	5382	3
57	5387	5392	5397	5402	5407	5412	2
58	5417	5423	5428	5433	5438	5443	1
59	5448	5453	5458	5463	5468	5473	0
	60"	50"	40"	30"	20"	10"	M
	Co-versed sine 84 degrees.						

P. part	1"	2"	3"	4"	5"	6"	7"	8"	9"
	0	1	1	2	2	3	3	4	4

Versed sine 6 degrees.

M	0"	10"	20"	30"	40"	50"	
0	005478	5483	5488	5493	5498	5503	59
1	5509	5514	5519	5524	5529	5534	58
2	5539	5544	5549	5554	5559	5565	57
3	5570	5575	5580	5585	5590	5595	56
4	5600	5606	5610	5616	5621	5626	55
5	5631	5636	5641	5647	5652	5657	54
6	5662	5667	5672	5678	5683	5688	53
7	5693	5698	5703	5709	5714	5719	52
8	5724	5729	5734	5740	5745	5750	51
9	5755	5760	5766	5771	5776	5781	50
10	005786	5792	5797	5802	5807	5812	49
11	5818	5823	5828	5833	5839	5844	48
12	5849	5854	5860	5865	5870	5875	47
13	5880	5886	5891	5896	5902	5907	46
14	5912	5917	5923	5928	5933	5938	45
15	5944	5949	5954	5960	5965	5970	44
16	5975	5981	5986	5991	5997	6002	43
17	6007	6012	6018	6023	6028	6034	42
18	6039	6044	6050	6055	6060	6066	41
19	6071	6076	6082	6087	6092	6098	40
20	006103	6108	6114	6119	6124	6130	39
21	6135	6141	6146	6151	6157	6162	38
22	6167	6173	6178	6184	6189	6194	37
23	6200	6205	6210	6216	6221	6227	36
24	6232	6237	6243	6248	6254	6259	35
25	6265	6270	6275	6281	6286	6292	34
26	6297	6303	6308	6313	6319	6324	33
27	6330	6335	6341	6346	6352	6357	32
28	6362	6368	6373	6379	6384	6390	31
29	6395	6401	6406	6412	6417	6423	30
30	006428	6434	6439	6445	6450	6456	29
31	6461	6467	6472	6478	6483	6489	28
32	6494	6500	6505	6511	6516	6522	27
33	6527	6533	6538	6544	6549	6555	26
34	6561	6566	6572	6577	6583	6588	25
35	6594	6599	6605	6611	6616	6622	24
36	6627	6633	6638	6644	6650	6655	23
37	6661	6666	6672	6677	6683	6689	22
38	6694	6700	6705	6711	6717	6722	21
39	6728	6734	6739	6745	6750	6756	20
40	006762	6767	6773	6779	6784	6790	19
41	6795	6801	6807	6812	6818	6824	18
42	6829	6835	6841	6846	6852	6858	17
43	6863	6869	6875	6880	6886	6892	16
44	6897	6903	6909	6914	6920	6926	15
45	6932	6937	6943	6949	6954	6960	14
46	6966	6971	6977	6983	6989	6994	13
47	7000	7006	7012	7017	7023	7029	12
48	7034	7040	7046	7052	7057	7063	11
49	7069	7075	7080	7086	7092	7098	10
50	007104	7109	7115	7121	7127	7132	9
51	7138	7144	7150	7156	7161	7167	8
52	7173	7179	7185	7190	7196	7202	7
53	7208	7214	7219	7225	7231	7237	6
54	7243	7248	7254	7260	7266	7272	5
55	7278	7283	7289	7295	7301	7307	4
56	7313	7319	7324	7330	7336	7342	3
57	7348	7354	7360	7365	7371	7377	2
58	7383	7389	7395	7401	7407	7412	1
59	7418	7424	7430	7436	7442	7448	0
	60"	50"	40"	30"	20"	10"	M

Co-versed sine 83 degrees.

P. part	1"	2"	3"	4"	5"	6"	7"	8"	9"
	1	1	2	2	3	3	4	4	5

Versed sine 7 degrees.

M	0"	10"	20"	30"	40"	50"	
0	007454	7460	7466	7472	7478	7483	59
1	7489	7495	7501	7507	7513	7519	58
2	7525	7531	7537	7543	7549	7555	57
3	7561	7567	7572	7578	7584	7590	56
4	7596	7602	7608	7614	7620	7626	55
5	7632	7638	7644	7650	7656	7662	54
6	7668	7674	7680	7686	7692	7698	53
7	7704	7710	7716	7722	7728	7734	52
8	7740	7746	7752	7758	7764	7770	51
9	7776	7682	7788	7794	7800	7807	50
10	007813	7819	7825	7831	7837	7843	49
11	7849	7855	7861	7867	7873	7879	48
12	7885	7891	7897	7904	7910	7916	47
13	7922	7928	7934	7940	7946	7952	46
14	7958	7964	7971	7977	7983	7989	45
15	7995	8001	8007	8013	8020	8026	44
16	8032	8038	8044	8050	8056	8062	43
17	8069	8075	8081	8087	8093	8099	42
18	8106	8112	8118	8124	8130	8136	41
19	8143	8149	8155	8161	8167	8173	40
20	008180	8186	8192	8198	8204	8211	39
21	8217	8223	8229	8235	8242	8248	38
22	8254	8260	8267	8273	8279	8285	37
23	8291	8298	8304	8310	8316	8323	36
24	8329	8335	8341	8348	8354	8360	35
25	8366	8373	8379	8385	8391	8398	34
26	8404	8410	8416	8423	8429	8435	33
27	8442	8448	8454	8460	8467	8473	32
28	8479	8486	8492	8498	8505	8511	31
29	8517	8524	8530	8536	8542	8549	30
30	008555	8561	8568	8574	8580	8587	29
31	8593	8599	8606	8612	8619	8625	28
32	8631	8638	8644	8650	8657	8663	27
33	8669	8676	8682	8689	8695	8701	26
34	8708	8714	8720	8727	8733	8740	25
35	8746	8752	8759	8765	8772	8778	24
36	8784	8791	8797	8804	8810	8817	23
37	8823	8829	8836	8842	8849	8855	22
38	8862	8868	8874	8881	8887	8894	21
39	8900	8907	8913	8920	8926	8933	20
40	008939	8945	8952	8958	8965	8971	19
41	8978	8984	8991	8997	9004	9010	18
42	9017	9023	9030	9036	9043	9049	17
43	9056	9062	9069	9075	9082	9088	16
44	9095	9101	9108	9114	9121	9128	15
45	9134	9141	9147	9154	9160	9167	14
46	9173	9180	9186	8193	9200	9206	13
47	9213	9219	9226	9232	9239	9246	12
48	9252	9259	9265	9272	9279	9285	11
49	9292	9298	9305	9311	9318	9325	10
50	009331	9338	9345	9351	9358	9364	9
51	9372	9378	9384	9391	9397	9404	8
52	9411	9417	9424	9431	9437	9444	7
53	9451	9457	9464	9471	9477	9484	6
54	9491	9497	9504	9511	9517	9524	5
55	9531	9537	9544	9551	9557	9564	4
56	9571	9577	9584	9591	9567	9604	3
57	9611	9618	9624	9631	9638	9644	2
58	9651	9658	9665	9671	9678	9685	1
59	9691	9698	9705	9712	9718	9725	0
	60"	50"	40"	30"	20"	10"	M

Co-versed sine 82 degrees.

P. part	1"	2"	3"	4"	5"	6"	7"	8"	9"
	1	1	2	3	3	4	4	5	6

M	Versed sine 8 degrees. 0"	10"	20"	30"	40"	50"	
0	009732	9739	9745	9752	9759	9766	59
1	9772	9779	9786	9793	9800	9806	58
2	9813	9820	9827	9833	9840	9847	57
3	9854	9861	9867	9874	9881	9888	56
4	9895	9901	9908	9915	9922	9929	55
5	9935	9942	9949	9956	9963	9970	54
6	9976	9983	9990	9997	0004	0011	53
7	010017	0024	0031	0038	0045	0052	52
8	0058	0065	0072	0079	0086	0093	51
9	0100	0107	0113	0120	0127	0134	50
10	010141	0148	0155	0162	0169	0175	49
11	0182	0189	0196	0203	0210	0217	48
12	0224	0231	0238	0245	0251	0258	47
13	0265	0272	0279	0286	0293	0300	46
14	0307	0314	0321	0328	0335	0342	45
15	0349	0356	0363	0369	0376	0383	44
16	0390	0397	0404	0411	0418	0425	43
17	0432	0439	0446	0453	0460	0467	42
18	0474	0481	0488	0495	0502	0509	41
19	0516	0523	0530	0537	0544	0551	40
20	010558	0565	0572	0579	0586	0594	39
21	0601	0608	0615	0622	0629	0636	38
22	0643	0650	0657	0664	0671	0678	37
23	0685	0692	0699	0706	0714	0721	36
24	0728	0735	0742	0749	0756	0763	35
25	0770	0777	0784	0792	0799	0806	34
26	0813	0820	0827	0834	0841	0848	33
27	0855	0863	0870	0877	0884	0891	32
28	0898	0905	0913	0920	0927	0934	31
29	0941	0948	0955	0963	0970	0977	30
30	010984	0991	0998	1006	1013	1020	29
31	1027	1034	1042	1049	1056	1063	28
32	1070	1077	1085	1092	1099	1106	27
33	1114	1121	1128	1035	1042	1150	26
34	1157	1164	1172	1178	1186	1193	25
35	1200	1207	1215	1222	1229	1236	24
36	1244	1251	1258	1265	1273	1280	23
37	1287	1294	1302	1309	1316	1324	22
38	1331	1338	1345	1353	1360	1367	21
39	1374	1382	1389	1396	1404	1411	20
40	011418	1426	1433	1440	1448	1455	19
41	1462	1469	1477	1484	1491	1499	18
42	1506	1513	1521	1528	1535	1543	17
43	1550	1558	1565	1572	1580	1587	16
44	1594	1602	1609	1616	1624	1631	15
45	1638	1646	1653	1661	1668	1675	14
46	1683	1690	1698	1705	1712	1720	13
47	1727	1735	1742	1749	1757	1764	12
48	1772	1779	1786	1794	1801	1809	11
49	1816	1824	1831	1838	1846	1853	10
50	011861	1868	1876	1883	1891	1898	9
51	1906	1913	1920	1928	1935	1943	8
52	1950	1958	1965	1973	1980	1988	7
53	1995	2003	2010	2018	2025	2033	6
54	2040	2048	2055	2063	2070	2078	5
55	2085	2093	2100	2108	2115	2123	4
56	2130	2138	2145	2153	2160	2168	3
57	2176	2183	2191	2198	2206	2213	2
58	2221	2228	2236	2243	2251	2259	1
59	2266	2274	2281	2289	22[illegible]6	2304	0
	60"	50"	40"	30"	20"	10"	
	Co-versed sine 81 degrees.						M

P. part { 1" 2" 3" 4" 5" 6" 7" 8" 9" / 1 1 2 3 4 4 5 6 6

M	Versed sine 9 degrees. 0"	10"	20"	30"	40"	50"	
0	012312	2319	2327	2334	2342	2350	59
1	2357	2365	2372	2380	2388	2395	58
2	2403	2410	2418	2426	2433	2441	57
3	2449	2456	2464	2471	2479	2487	56
4	2494	2502	2510	2517	2525	2533	55
5	2540	2548	2556	2563	2571	2579	54
6	2586	2594	2602	2609	2617	2625	53
7	2632	2640	2648	2655	2663	2671	52
8	2678	2686	2694	2701	2709	2717	51
9	2725	2732	2740	2748	2755	2763	50
10	012771	2779	2786	2794	2802	2810	49
11	2817	2825	2833	2840	2848	2856	48
12	2864	2871	2879	2887	2895	2903	47
13	2910	2918	2926	2934	2941	2949	46
14	2957	2965	2972	2980	2988	2996	45
15	3004	3011	3019	3027	3035	3043	44
16	3050	3058	3066	3074	3082	3089	43
17	3097	3105	3113	3121	3129	3136	42
18	3144	3152	3160	3168	3176	3183	41
19	3191	3199	3207	3215	3223	3231	40
20	013238	3246	3254	3262	3270	3278	39
21	3286	3294	3301	3309	3317	3325	38
22	3333	3341	3349	3357	3365	3372	37
23	3380	3388	3396	3404	3412	3420	36
24	3428	3436	3444	3452	3460	3468	35
25	3475	3483	3491	3499	3507	3515	34
26	3523	3531	3539	3547	3555	3563	33
27	3571	3579	3587	3595	3603	3611	32
28	3619	3627	3634	3642	3650	3658	31
29	3666	3674	3682	3690	3698	3706	30
30	013714	3722	3730	3738	3746	3754	29
31	3762	3770	3778	3787	3795	3803	28
32	3811	3819	3827	3835	3843	3851	27
33	3859	3867	3875	3883	3891	3899	26
34	3907	3915	3923	3931	3939	3947	25
35	3955	3964	1972	3980	3988	3996	24
36	4004	4012	4020	4028	4036	4044	23
37	4053	4061	4069	4077	4085	4093	22
38	4101	4109	4117	4126	4134	4142	21
39	4150	4158	4166	4174	4182	4191	20
40	014199	4207	4215	4223	4231	4239	19
41	4248	4256	4264	4272	4280	4288	18
42	4297	4305	4313	4321	4329	4337	17
43	4346	4354	4362	4370	4378	4387	16
44	4395	4403	4411	4419	4428	4436	15
45	4444	4452	4460	4469	4477	4485	14
46	4493	4501	4510	4518	4526	4534	13
47	4543	4551	4559	4567	4576	4584	12
48	4592	4600	4609	4617	4625	4633	11
49	4642	4650	4658	4666	4675	4683	10
50	014691	4700	4708	4716	4724	4733	9
51	4741	4749	4758	4766	4774	4783	8
52	4791	4799	4807	4816	4824	4832	7
53	4841	4849	4857	4866	4874	4882	6
54	4891	4899	4907	4916	4924	4932	5
55	4941	4949	4957	4966	4974	4983	4
56	4991	4999	5008	5016	5024	5033	3
57	5041	5049	5058	5066	5075	5083	2
58	5091	5090	5108	5117	5125	5133	1
59	5142	5150	5159	5167	5175	5184	0
	60"	50"	40"	30"	20"	10"	
	Co-versed sine 80 degrees.						M

P. part { 1" 2" 3" 4" 5" 6" 7" 8" 9" / 1 2 2 3 4 5 6 6 7

 TABLE XLI. Natural Versed Sines.

M	Versed sine 10 degrees. 0"	10"	20"	30"	40"	50"	
0	015192	5201	5209	5218	5226	5234	59
1	5243	5251	5260	5268	5277	5285	58
2	5293	5302	5310	5319	5327	5336	57
3	5344	5353	5361	5370	5378	5386	56
4	5395	5403	5412	5420	5429	5437	55
5	5446	5454	5463	5471	5480	5488	54
6	5497	5505	5514	5522	5531	5539	53
7	5548	5556	5565	5573	5582	5590	52
8	5599	5607	5616	5625	5633	5642	51
9	5650	5659	5667	5676	5684	5693	50
10	015702	5710	5719	5727	5736	5744	49
11	5753	5761	5770	5779	5787	5796	48
12	5804	5813	5822	5830	5839	5847	47
13	5856	5865	5873	5882	5890	5899	46
14	5908	5916	5925	5933	5942	5951	45
15	5959	5968	5977	5985	5994	6002	44
16	6011	6020	6028	6037	6046	6054	43
17	6063	6072	6080	6089	6098	6106	42
18	6115	6124	6132	6141	6150	6158	41
19	6167	6176	6184	6193	6202	6210	40
20	016219	6228	6237	6245	6254	6263	39
21	6271	6280	6289	6298	6306	6315	38
22	6324	6332	6341	6350	6359	6367	37
23	6376	6385	6394	6402	6411	6420	36
24	6429	6437	6446	6455	6464	6472	35
25	6481	6490	6499	6507	6516	6525	34
26	6534	6542	6551	6560	6569	6578	33
27	6586	6595	6604	6613	6622	6630	32
28	6639	6648	6657	6666	6674	6683	31
29	6692	6701	6710	6719	6727	6736	30
30	016745	6754	6763	6772	6780	6789	29
31	6798	6807	6816	6825	6834	6842	28
32	6851	6860	6869	6878	6887	6896	27
33	6904	6913	6922	6931	6940	6949	26
34	6958	6967	6976	6985	6993	7002	25
35	7011	7020	7029	7038	7047	7056	24
36	7065	7074	7082	7091	7100	7109	23
37	7118	7127	7136	7145	7154	7163	22
38	7172	7181	7190	7199	7208	7217	21
39	7226	7235	7243	7252	7261	7270	20
40	017279	7288	7297	7306	7315	7324	19
41	7333	7342	7351	7360	7369	7378	18
42	7387	7396	7405	7414	7423	7432	17
43	7441	7450	7459	7468	7477	7486	16
44	7495	7504	7513	7522	7532	7541	15
45	7550	7559	7568	7577	7586	7595	14
46	7604	7613	7622	7631	7640	7649	13
47	7658	7667	7676	7686	7695	7704	12
48	7713	7722	7731	7740	7749	7758	11
49	7767	7776	7785	7795	7804	7813	10
50	017822	7831	7840	7849	7858	7868	9
51	7877	7886	7895	7904	7913	7922	8
52	7931	7941	7950	7959	7968	7977	7
53	7986	7995	8005	8014	8023	8032	6
54	8042	8050	8060	8069	8078	8087	5
55	8096	8106	8115	8124	8133	8142	4
56	8151	8161	8170	8179	8188	8197	3
57	8207	8216	8225	8234	8244	8253	2
58	8262	8271	8280	8290	8299	8308	1
59	8317	8327	8336	8345	8354	8364	0
	60"	50"	40"	30"	20"	10"	
	Co-versed sine 79 degrees.						M

P. part { 1" 2" 3" 4" 5" 6" 7" 8" 9" / 1 2 3 4 4 5 6 7 8 }

M	Versed sine 41 degrees. 0"	10"	20"	30"	40"	50"	
0	018373	8382	8391	8401	8410	8419	59
1	8428	8438	8447	8456	8465	8475	58
2	8484	8493	8503	8512	8521	8530	57
3	8540	8549	8558	8568	8577	8586	56
4	8595	8605	8614	8623	8633	8642	55
5	8651	8661	8670	8679	8689	8698	54
6	8707	8717	8726	8735	8745	8754	53
7	8763	8773	8782	8791	8801	8810	52
8	8819	8829	8838	8848	8857	8866	51
9	8876	8885	8894	8904	8913	8923	50
10	018932	8941	8951	8960	8970	8979	49
11	8988	8998	9007	9017	9026	9035	48
12	9045	9054	9064	9073	9082	9092	47
13	9101	9111	9120	9130	9139	9149	46
14	9158	9167	9177	9186	9196	9205	45
15	9215	9224	9234	9243	9253	9262	44
16	9272	9281	9290	9300	9309	9319	43
17	9328	9338	9347	9357	9366	9376	42
18	9385	9395	9404	9414	9423	9433	41
19	9442	9452	9461	9471	9480	9490	40
20	019500	9509	9519	9528	9538	9547	39
21	9557	9566	9576	9585	9595	9604	38
22	9614	9624	9633	9643	9652	9662	37
23	9671	9681	9691	9700	9710	9719	36
24	9729	9738	9748	9757	9767	9777	35
25	9786	9796	9806	9815	9825	9834	34
26	9844	9854	9863	9873	9882	9892	33
27	9902	9911	9921	9931	9940	9950	32
28	9959	9969	9979	9988	9998	0008	31
29	020017	0027	0037	0046	0056	0066	30
30	020075	0085	0095	0104	0114	0124	29
31	0133	0143	0153	0162	0172	0182	28
32	0191	0201	0211	0221	0230	0240	27
33	0250	0259	0269	0279	0288	0298	26
34	0308	0318	0327	0337	0347	0357	25
35	0366	0376	0386	0396	0405	0415	24
36	0425	0435	0444	0454	0464	0474	23
37	0483	0493	0503	0513	0522	0532	22
38	0542	0552	0561	0571	0581	0591	21
39	0601	0610	0620	0630	0640	0650	20
40	020659	0669	0679	0689	0699	0708	19
41	0718	0728	0738	0748	0758	0767	18
42	0777	0787	0797	0807	0817	0826	17
43	0836	0846	0856	0866	0876	0885	16
44	0895	0905	0915	0925	0935	0945	15
45	0955	0964	0974	0984	0994	1004	14
46	1014	1024	1034	1043	1053	1063	13
47	1073	1083	1093	1103	1113	1123	12
48	1133	1143	1152	1162	1172	1182	11
49	1192	1202	1212	1222	1232	1242	10
50	021252	1262	1272	1282	1292	1301	9
51	1311	1321	1331	1341	1351	1361	8
52	1371	1381	1391	1401	1411	1421	7
53	1431	1441	1451	1461	1471	1481	6
54	1491	1501	1511	1521	1531	1541	5
55	1551	1561	1572	1581	1591	1601	4
56	1611	1621	1631	1641	1651	1661	3
57	1671	1681	1691	1701	1712	1722	2
58	1732	1742	1752	1762	1772	1782	1
59	1792	1802	1812	1822	1832	1842	0
	60"	50"	40"	30"	20"	10"	
	Co-versed sine 78 degrees.						M

P. part { 1" 2" 3" 4" 5" 6" 7" 8" 9" / 1 2 3 4 5 6 7 8 9 }

4529	4540	4551	4501	45
4594	4604	4615	4626	46
4658	4668	4679	4690	47
4722	4733	4743	4754	47
4786	4797	4808	4818	48
4851	4861	4872	4883	48
4915	4926	4937	4947	49
4980	4990	5001	5012	50
5044	5055	5066	5077	50
5109	5120	5131	5141	51
5174	5185	5196	5206	52
5239	5250	5260	5271	52
5304	5315	5325	5336	53
5369	5380	5391	5401	54
5434	5445	5456	5467	54
5499	5510	5521	5532	55
5565	5575	5586	5597	56
60"	50"	40"	30"	20

TABLE XLI. Natural Versed Sines.

M	Versed sine 14 degrees. 0"	10"	20"	30"	40"	50"	
0	029704	9716	9728	9739	9751	9763	59
1	9775	9786	9798	9810	9822	9833	58
2	9845	9857	9869	9881	9892	9904	57
3	9916	9928	9939	9951	9963	9975	56
4	9987	9998	0010	0022	0034	0045	55
5	030057	0069	0081	0093	0104	0116	54
6	0128	0140	0152	0163	0175	0187	53
7	0199	0211	0222	0234	0246	0258	52
8	0270	0282	0294	0305	0317	0329	51
9	0341	0353	0365	0377	0388	0400	50
10	030412	0424	0436	0448	0460	0471	49
11	0483	0495	0507	0519	0531	0543	48
12	0555	0567	0578	0590	0602	0614	47
13	0626	0638	0650	0662	0674	0686	46
14	0698	0709	0721	0733	0745	0757	45
15	0769	0781	0793	0805	0817	0829	44
16	0841	0853	0865	0877	0889	0901	43
17	0913	0924	0936	0948	0960	0972	42
18	0984	0996	1008	1020	1032	1044	41
19	1056	1068	1080	1092	1104	1116	40
20	031128	1140	1152	1164	1176	1188	39
21	1200	1212	1224	1236	1248	1260	38
22	1272	1284	1296	1308	1320	1332	37
23	1344	1357	1369	1381	1393	1405	36
24	1417	1429	1441	1453	1465	1477	35
25	1489	1501	1513	1525	1538	1550	34
26	1562	1574	1586	1598	1610	1622	33
27	1634	1646	1658	1671	1683	1695	32
28	1707	1719	1731	1743	1755	1767	31
29	1780	1792	1804	1816	1828	1840	30
30	031852	1864	1877	1889	1901	1913	29
31	1925	1937	1950	1962	1974	1986	28
32	1998	2010	2023	2035	2047	2059	27
33	2071	2083	2096	2108	2120	2132	26
34	2144	2157	2169	2181	2193	2205	25
35	2218	2230	2242	2254	2266	2279	24
36	2291	2303	2315	2328	2340	2352	23
37	2364	2376	2389	2401	2413	2425	22
38	2438	2450	2462	2474	2487	2499	21
39	2511	2523	2536	2548	2560	2573	20
40	032585	2597	2609	2621	2634	2646	19
41	2658	2671	2683	2695	2708	2720	18
42	2732	2745	2757	2769	2781	2794	17
43	2806	2818	2831	2843	2855	2868	16
44	2880	2892	2905	2917	2929	2942	15
45	2954	2966	2979	2991	3003	3016	14
46	3028	3041	3053	3065	3078	3090	13
47	3102	3115	3127	3139	3152	3164	12
48	3177	3189	3201	3214	3226	3239	11
49	3251	3263	3276	3288	3301	3313	10
50	033325	3338	3350	3363	3375	3387	9
51	3400	3412	3424	3437	3450	3462	8
52	3474	3487	3499	3512	3524	3537	7
53	3549	3562	3574	3587	3599	3611	6
54	3624	3636	3649	3661	3674	3686	5
55	3699	3711	3724	3736	3749	3761	4
56	3774	3786	3799	3801	3814	3826	3
57	3849	3861	3874	3886	3899	3911	2
58	3924	3936	3949	3961	3974	3986	1
59	3999	4011	4024	4037	4049	4062	0
	60"	50"	40"	30"	20"	10"	
	Co-versed sine 75 degrees.						M

P. part { 1" 2" 3" 4" 5" 6" 7" 8" 9" / 1 2 4 5 6 7 8 10 11

M	Versed sine 15 degrees. 0"	10"	20"	30"	40"	50"	
0	034074	4087	4099	4112	4124	4137	59
1	4150	4162	4175	4187	4200	4212	58
2	4225	4237	4250	4263	4275	4288	57
3	4300	4313	4326	4338	4351	4363	56
4	4376	4389	4401	4414	4426	4439	55
5	4452	4464	4477	4489	4502	4515	54
6	4527	4540	4553	4565	4578	4591	53
7	4603	4616	4628	4641	4654	4666	52
8	4679	4692	4704	4717	4730	4742	51
9	4755	4768	4780	4793	4806	4818	50
10	034831	4844	4857	4869	4882	4895	49
11	4907	4920	4933	4945	4958	4971	48
12	4984	4996	5009	5022	5034	5047	47
13	5060	5073	5085	5098	5111	5123	46
14	5136	5149	5162	5174	5187	5200	45
15	5213	5225	5238	5251	5264	5276	44
16	5289	5302	5315	5328	5340	5353	43
17	5366	5379	5391	5404	5417	5430	42
18	5443	5455	5468	5481	5494	5507	41
19	5519	5532	5545	5558	5571	5583	40
20	035596	5609	5622	5635	5648	5660	39
21	5673	5686	5699	5712	5725	5737	38
22	5750	5763	5776	5789	5802	5815	37
23	5827	5840	5853	5866	5879	5892	36
24	5905	5917	5930	5943	5956	5969	35
25	5982	5995	6008	6021	6033	6046	34
26	6059	6072	6085	6098	6111	6124	33
27	6137	6150	6163	6175	6188	6201	32
28	6214	6227	6240	6253	6266	6279	31
29	6292	6305	6318	6331	6344	6357	30
30	036370	6383	6395	6408	6421	6434	29
31	6447	6460	6473	6486	6499	6512	28
32	6525	6538	6551	6564	6577	6590	27
33	6603	6616	6629	6642	6655	6668	26
34	6681	6694	6707	6720	6733	6746	25
35	6759	6772	6785	6798	6811	6824	24
36	6837	6850	6864	6877	6890	6903	23
37	6916	6929	6942	6955	6968	6981	22
38	6994	7007	7020	7033	7046	7059	21
39	7072	7086	7099	7112	7125	7138	20
40	037151	7164	7177	7190	7203	7216	19
41	7230	7243	7256	7269	7282	7295	18
42	7308	7321	7334	7348	7361	7374	17
43	7387	7400	7413	7426	7440	7453	16
44	7466	7479	7492	7505	7518	7532	15
45	7545	7558	7572	7584	7597	7611	14
46	7624	7637	7650	7663	7676	7690	13
47	7703	7716	7729	7742	7756	7769	12
48	7782	7795	7808	7822	7835	7848	11
49	7861	7874	7888	7901	7914	7927	10
50	037941	7954	7967	7980	7994	8007	9
51	8020	8033	8046	8060	8073	8086	8
52	8099	8113	8126	8139	8153	8166	7
53	8179	8192	8206	8219	8232	8245	6
54	8259	8272	8285	8299	8312	8325	5
55	8338	8352	8365	8378	8392	8405	4
56	8418	8432	8445	8458	8471	8485	3
57	8498	8511	8525	8538	8551	8565	2
58	8578	8591	8605	8618	8631	8645	1
59	8658	8672	8685	8698	8712	8725	0
	60"	50"	40"	30"	20"	10"	
	Co-versed sine 74 degrees.						M

P. part { 1" 2" 3" 4" 5" 6" 7" 8" 9" / 1 3 4 5 6 8 9 10 12

M	Versed sine 16 degrees. 0"	10"	20"	30"	40"	50"	
0	038738	8752	8765	8778	8792	8805	59
1	8819	8832	8845	8859	8872	8885	58
2	8899	8912	8926	8939	8952	8966	57
3	8979	8993	9006	9019	9033	9046	56
4	9060	9073	9087	9100	9113	9127	55
5	9140	9154	9167	9181	9194	9207	54
6	9221	9234	9248	9261	9275	9288	53
7	9302	9315	9328	9342	9355	9369	52
8	9382	9396	9409	9423	9436	9450	51
9	9463	9477	9490	9504	9517	9531	50
10	039544	9558	9571	9585	9598	9612	49
11	9625	9639	9652	9666	9679	9693	48
12	9706	9720	9733	9747	9760	9774	47
13	9788	9801	9815	9828	9842	9855	46
14	9869	9882	9896	9909	9923	9937	45
15	9950	9964	9977	9991	0004	0018	44
16	040031	0045	0059	0072	0086	0100	43
17	0113	0127	0140	0154	0167	0181	42
18	0195	0208	0222	0236	0249	0263	41
19	0276	0290	0304	0317	0331	0345	40
20	040358	0372	0385	0399	0413	0426	39
21	0440	0454	0467	0481	0495	0508	38
22	0522	0536	0549	0563	0577	0590	37
23	0604	0618	0631	0645	0659	0672	36
24	0686	0770	0713	0727	0741	0754	35
25	0768	0782	0796	0809	0823	0837	34
26	0850	0864	0878	0892	0905	0919	33
27	0933	0947	0960	0974	0988	1001	32
28	1015	1029	1043	1056	1070	1084	31
29	1098	1111	1125	1139	1153	1166	30
30	041180	1194	1208	1222	1235	1249	29
31	1263	1277	1290	1304	1318	1332	28
32	1346	1359	1373	1387	1401	1415	27
33	1428	1442	1456	1470	1484	1498	26
34	1511	1525	1539	1553	1567	1581	25
35	1594	1608	1622	1636	1650	1664	24
36	1677	1691	1705	1719	1733	1747	23
37	1761	1774	1788	1802	1816	1830	22
38	1844	1858	1872	1885	1899	1913	21
39	1927	1941	1955	1969	1983	1997	20
40	042010	2024	2038	2052	2066	2080	19
41	2094	2108	2122	2136	2150	2164	18
42	2178	2191	2205	2219	2233	2247	17
43	2261	2275	2289	2303	2317	2331	16
44	2345	2359	2373	2387	2401	2415	15
45	2429	2443	2457	2471	2485	2499	14
46	2513	2526	2540	2554	2568	2582	13
47	2596	2610	2624	2638	2652	2666	12
48	2681	2695	2709	2723	2737	2751	11
49	2765	2779	2793	2807	2821	2835	10
50	042849	2863	2877	2891	2905	2919	9
51	2933	2947	2961	2975	2989	3003	8
52	3017	3032	3046	3060	3074	3088	7
53	3102	3116	3130	3144	3158	3172	6
54	3186	3201	3215	3229	3243	3257	5
55	3271	3285	3299	3313	3327	3342	4
56	3356	3370	3384	3398	3412	3426	3
57	3440	3455	3469	3483	3497	3511	2
58	3525	3539	3554	3568	3582	3596	1
59	3610	3624	3639	3653	3667	3681	0
	60"	50"	40"	30"	20"	10"	
	Co-versed sine 73 degrees.						M

P. part	1"	2"	3"	4"	5"	6"	7"	8"	9"
	1	3	4	6	7	8	10	11	12

M	Versed sine 17 degrees. 0"	10"	20"	30"	40"	50"	
0	043695	3709	3724	3738	3752	3766	59
1	3780	3795	3809	3823	3837	3851	58
2	3866	3880	3894	3908	3922	3937	57
3	3951	3965	3979	3993	4008	4021	56
4	4036	4050	4065	4079	4093	4107	55
5	4121	4136	4150	4164	4178	4193	54
6	4207	4222	4236	4250	4264	4278	53
7	4293	4307	4321	4335	4350	4364	52
8	4378	4392	4407	4421	4435	4450	51
9	4464	4478	4493	4507	4521	4535	50
10	044550	4564	4578	4593	4607	4621	49
11	4636	4650	4664	4679	4693	4707	48
12	4722	4736	4750	4765	4779	4793	47
13	4808	4822	4836	4851	4865	4879	46
14	4894	4908	4923	4937	4951	4966	45
15	4980	4994	5009	5023	5038	5052	44
16	5066	5081	5095	5110	5124	5138	43
17	5153	5167	5182	5196	5210	5225	42
18	5239	5254	5268	5282	5297	5311	41
19	5326	5340	5355	5369	5383	5398	40
20	045412	5427	5441	5456	5470	5485	39
21	5499	5514	5528	5542	5557	5571	38
22	5586	5600	5615	5629	5644	5658	37
23	5673	5687	5702	5716	5731	5745	36
24	5760	5774	5789	5803	5818	5832	35
25	5847	5861	5876	5890	5905	5919	34
26	5934	5948	5963	5977	5992	6006	33
27	6021	6036	6050	6065	6079	6094	32
28	6108	6123	6137	6152	6166	6181	31
29	6196	6210	6225	6239	6254	6268	30
30	046283	6298	6312	6327	6341	6356	29
31	6371	6385	6400	6414	6429	6444	28
32	6458	6473	6487	6502	6517	6531	27
33	6546	6560	6575	6590	6604	6619	26
34	6634	6648	6663	6677	6692	6707	25
35	6721	6736	6751	6765	6780	6795	24
36	6809	6824	6839	6853	6868	6883	23
37	6897	6912	6927	6941	6956	6971	22
38	6985	7000	7015	7029	7044	7059	21
39	7074	7088	7103	7118	7132	7147	20
40	047162	7177	7191	7206	7221	7235	19
41	7250	7265	7280	7294	7309	7324	18
42	7339	7353	7368	7383	7397	7412	17
43	7427	7442	7457	7471	7486	7501	16
44	7516	7530	7545	7560	7575	7589	15
45	7604	7619	7634	7649	7663	7678	14
46	7693	7708	7723	7737	7752	7767	13
47	7782	7797	7811	7826	7841	7856	12
48	7871	7885	7900	7915	7930	7945	11
49	7960	7974	7989	8004	8019	8034	10
50	048049	8063	8078	8093	8108	8123	9
51	8138	8153	8167	8182	8197	8212	8
52	8227	8242	8257	8272	8286	8301	7
53	8316	8331	8346	8361	8376	8391	6
54	8406	8420	8435	8450	8465	8480	5
55	8495	8510	8525	8540	8555	8570	4
56	8585	8599	8614	8629	8644	8659	3
57	8674	8689	8704	8719	8734	8749	2
58	8764	8779	8794	8809	8824	8839	1
59	8854	8869	8884	8899	8914	8929	0
	60"	50"	40"	30"	20"	10"	
	Co-versed sine 72 degrees.						M

P. part	1"	2"	3"	4"	5"	6"	7"	8"	9"
	1	3	4	6	7	9	10	11	13

TABLE XLI. Natural Versed Sines.

	Versed sine 18 degrees.						
M	0"	10"	20"	30"	40"	50"	
0	048943	8958	8973	8988	9003	9018	59
1	9033	9048	9063	9078	9093	9108	58
2	9123	9138	9153	9168	9183	9198	57
3	9214	9229	9244	9259	9274	9289	56
4	9304	9319	9334	9349	9364	9379	55
5	9394	9409	9424	9439	9454	9469	54
6	9484	9499	9514	9529	9545	9560	53
7	9575	9590	9605	9620	9635	9650	52
8	9665	9680	9695	9710	9726	9741	51
9	9756	9771	9786	9801	9816	9831	50
10	049846	9862	9877	9892	9907	9922	49
11	9937	9952	9967	9982	9998	0013	48
12	050028	0043	0058	0073	0089	0104	47
13	0119	0134	0149	0164	0179	0195	46
14	0210	0225	0240	0255	0271	0286	45
15	0301	0316	0331	0346	0362	0377	44
16	0392	0407	0422	0438	0453	0468	43
17	0483	0498	0514	0529	0544	0559	42
18	0575	0590	0605	0620	0635	0651	41
19	0666	0681	0696	0712	0727	0742	40
20	050757	0773	0788	0803	0818	0834	39
21	0849	0864	0879	0895	0910	0925	38
22	0941	0956	0971	0986	1001	1017	37
23	1032	1048	1063	1078	1093	1109	36
24	1124	1139	1155	1170	1185	1201	35
25	1216	1231	1246	1262	1277	1292	34
26	1308	1323	1338	1354	1369	1384	33
27	1400	1415	1430	1446	1461	1477	32
28	1492	1507	1523	1538	1553	1569	31
29	1584	1599	1615	1630	1646	1661	30
30	051676	1692	1707	1723	1738	1753	29
31	1769	1784	1799	1815	1830	1846	28
32	1861	1877	1892	1907	1923	1938	27
33	1954	1969	1984	2000	2015	2031	26
34	2046	2062	2077	2093	2108	2123	25
35	2139	2154	2170	2185	2201	2216	24
36	2232	2247	2263	2278	2293	2309	23
37	2324	2340	2355	2371	2386	2402	22
38	2417	2433	2448	2464	2479	2495	21
39	2510	2526	2541	2557	2572	2588	20
40	052603	2619	2634	2650	2665	2681	19
41	2696	2712	2728	2743	2759	2774	18
42	2790	2805	2821	2836	2852	2867	17
43	2883	2899	2914	2930	2945	2961	16
44	2976	2992	3008	3023	3039	3054	15
45	3070	3085	3101	3117	3132	3148	14
46	3163	3179	3195	3210	3226	3241	13
47	3257	3273	3288	3304	3319	3335	12
48	3351	3366	3382	3398	3413	3429	11
49	3445	3460	3476	3491	3507	3523	10
50	053538	3554	3570	3585	3601	3617	9
51	3632	3648	3664	3679	3695	3711	8
52	3726	3742	3758	3773	3789	3805	7
53	3820	3836	3852	3868	3883	3899	6
54	3915	3930	3946	3962	3977	3993	5
55	4009	4025	4040	4056	4072	4088	4
56	4103	4119	4135	4150	4166	4182	3
57	4198	4213	4229	4245	4261	4276	2
58	4292	4308	4324	4339	4355	4371	1
59	4387	4403	4418	4434	4450	4466	0
	60"	50"	40"	30"	20"	10"	
	Co-versed sine 71 degrees.						M

P. part	1"	2"	3"	4"	5"	6"	7"	8"	9"
	2	3	5	6	8	9	11	12	14

	Versed sine 19 degrees.						
M	0"	10"	20"	30"	40"	50"	
0	054481	4497	4513	4529	4545	4560	59
1	4576	4592	4608	4624	4639	4655	58
2	4671	4687	4703	4718	4734	4750	57
3	4766	4782	4798	4813	4829	4845	56
4	4861	4877	4893	4908	4924	4940	55
5	4956	4972	4988	5004	5019	5035	54
6	5051	5067	5083	5099	5115	5130	53
7	5146	5162	5178	5194	5210	5226	52
8	5242	5258	5273	5289	5305	5321	51
9	5337	5353	5369	5385	5401	5417	50
10	055432	5448	5464	5480	5496	5512	49
11	5528	5544	5560	5576	5592	5608	48
12	5624	5640	5656	5671	5687	5703	47
13	5719	5735	5751	5767	5783	5799	46
14	5815	5831	5847	5863	5879	5895	45
15	5911	5927	5943	5959	5975	5991	44
16	6007	6023	6039	6055	6071	6087	43
17	6103	6119	6135	6151	6167	6183	42
18	6199	6215	6231	6247	6263	6279	41
19	6295	6311	6327	6343	6359	6375	40
20	056391	6408	6424	6440	6456	6472	39
21	6488	6504	6520	6536	6552	6568	38
22	6584	6600	6616	6632	6649	6665	37
23	6681	6697	6713	6729	6745	6761	36
24	6777	6793	6810	6826	6842	6858	35
25	6874	6890	6906	6922	6938	6955	34
26	6971	6987	7003	7019	7035	7051	33
27	7068	7084	7100	7116	7132	7148	32
28	7164	7181	7197	7213	7229	7245	31
29	7261	7278	7294	7310	7326	7342	30
30	057359	7375	7391	7407	7423	7439	29
31	7456	7472	7488	7504	7520	7537	28
32	7553	7569	7585	7602	7618	7634	27
33	7650	7666	7683	7699	7715	7731	26
34	7748	7764	7780	7796	7813	7829	25
35	7845	7861	7878	7894	7910	7926	24
36	7943	7959	7975	7991	8008	8024	23
37	8040	8056	8073	8089	8105	8122	22
38	8138	8154	8170	8187	8203	8219	21
39	8236	8252	8268	8285	8301	8317	20
40	058334	8350	8366	8382	8399	8415	19
41	8431	8448	8464	8480	8497	8513	18
42	8529	8546	8562	8578	8595	8611	17
43	8628	8644	8660	8677	8693	8709	16
44	8726	8742	8758	8775	8791	8808	15
45	8824	8840	8857	8873	8890	8906	14
46	8922	8939	8955	8972	8988	9004	13
47	9021	9037	9054	9070	9086	9103	12
48	9119	9136	9152	9169	9185	9201	11
49	9218	9234	9251	9267	9284	9300	10
50	059316	9333	9349	9366	9382	9399	9
51	9415	9432	9448	9465	9481	9498	8
52	9514	9530	9547	9563	9580	9596	7
53	9613	9629	9646	9662	9679	9695	6
54	9712	9728	9745	9761	9778	9794	5
55	9811	9827	9844	9860	9877	9894	4
56	9910	9927	9943	9960	9976	9993	3
57	060009	0026	0042	0059	0075	0092	2
58	0109	0125	0142	0158	0175	0191	1
59	0208	0224	0241	0258	0274	0291	0
	60"	50"	40"	30"	20"	10"	
	Co-versed sine 70 degrees.						M

P. part	1"	2"	3"	4"	5"	6"	7"	8"	9"
	2	3	5	6	8	10	11	13	15

Versed sine 20 degrees.

M	0"	10"	20"	30"	40"	50"	
0	060307	0324	0341	0357	0374	0390	59
1	0407	0424	0440	0457	0473	0490	58
2	0507	0523	0540	0556	0573	0590	57
3	0606	0623	0639	0656	0673	0689	56
4	0706	0723	0739	0756	0773	0789	55
5	0806	0822	0839	0856	0872	0889	54
6	0906	0922	0939	0956	0972	0989	53
7	1006	1022	1039	1056	1072	1089	52
8	1106	1123	1139	1156	1173	1189	51
9	1206	1223	1239	1256	1273	1290	50
10	061306	1323	1340	1356	1373	1390	49
11	1407	1423	1440	1457	1474	1490	48
12	1507	1524	1540	1557	1574	1591	47
13	1607	1624	1641	1658	1674	1691	46
14	1708	1725	1742	1758	1775	1792	45
15	1809	1825	1842	1859	1876	1893	44
16	1909	1926	1943	1960	1977	1993	43
17	2010	2027	2044	2061	2077	2094	42
18	2111	2128	2145	2162	2178	2195	41
19	2212	2229	2246	2263	2279	2296	40
20	062313	2330	2347	2364	2380	2397	39
21	2414	2431	2448	2465	2482	2499	38
22	2515	2532	2549	2566	2583	2600	37
23	2617	2634	2650	2667	2684	2701	36
24	2718	2735	2752	2769	2786	2803	35
25	2819	2836	2853	2870	2887	2904	34
26	2921	2938	2955	2972	2989	3006	33
27	3023	3039	3056	3073	3090	3107	32
28	3124	3141	3158	3175	3192	3209	31
29	3226	3243	3260	3277	3294	3311	30
30	063328	3345	3362	3379	3396	3413	29
31	3430	3447	3464	3481	3498	3515	28
32	3532	3549	3566	3583	3600	3617	27
33	3634	3651	3668	3685	3702	3719	26
34	3736	3753	3770	3787	3804	3821	25
35	3838	3855	3872	3889	3906	3923	24
36	3940	3958	3975	3992	4009	4026	23
37	4043	4060	4077	4094	4111	4128	22
38	4145	4162	4179	4197	4214	4231	21
39	4248	4265	4282	4299	4316	4333	20
40	064350	4368	4385	4402	4419	4436	19
41	4453	4470	4487	4505	4522	4539	18
42	4556	4573	4590	4607	4625	4642	17
43	4659	4676	4693	4710	4727	4745	16
44	4762	4779	4796	4813	4830	4848	15
45	4865	4882	4899	4916	4934	4951	14
46	4968	4985	5002	5019	5037	5054	13
47	5071	5088	5105	5123	5140	5157	12
48	5174	5192	5209	5226	5243	5260	11
49	5278	5295	5312	5329	5347	5364	10
50	065381	5398	5416	5433	5450	5467	9
51	5485	5502	5519	5536	5554	5571	8
52	5588	5605	5623	5640	5657	5675	7
53	5692	5709	5726	5744	5761	5778	6
54	5796	5813	5830	5847	5865	5882	5
55	5899	5917	5934	5951	5969	5986	4
56	6003	6021	6038	6055	6073	6090	3
57	6107	6125	6142	6159	6177	6194	2
58	6211	6229	6246	6263	6281	6298	1
59	6315	6333	6350	6367	6385	6402	0
	60"	50"	40"	30"	20"	10"	M

Co-versed sine 69 degrees.

P. part	1"	2"	3"	4"	5"	6"	7"	8"	9"
	2	3	5	7	8	10	12	14	15

Versed sine 21 degrees.

M	0"	10"	20"	30"	40"	50"	
0	066420	6437	6454	6472	6489	6506	59
1	6524	6541	6559	6576	6593	6611	58
2	6628	6646	6663	6680	6698	6715	57
3	6733	6750	6768	6785	6802	6820	56
4	6837	6855	6872	6889	6907	6924	55
5	6942	6959	6977	6994	7012	7029	54
6	7046	7064	7081	7099	7116	7134	53
7	7151	7169	7186	7204	7221	7239	52
8	7256	7274	7291	7309	7326	7343	51
9	7361	7378	7396	7413	7431	7448	50
10	067466	7483	7501	7518	7536	7554	49
11	7571	7589	7606	7624	7641	7659	48
12	7676	7694	7711	7729	7746	7764	47
13	7781	7799	7817	7834	7852	7869	46
14	7887	7904	7922	7939	7957	7975	45
15	7992	8010	8027	8045	8062	8080	44
16	8098	8115	8133	8150	8168	8186	43
17	8203	8221	8238	8256	8274	8291	42
18	8309	8326	8344	8362	8379	8397	41
19	8414	8432	8450	8467	8485	8503	40
20	068520	8538	8556	8573	8591	8608	39
21	8626	8644	8661	8679	8697	8714	38
22	8732	8750	8767	8785	8803	8820	37
23	8838	8856	8873	8891	8909	8926	36
24	8944	8962	8980	8997	9015	9033	35
25	9050	9068	9086	9103	9121	9139	34
26	9157	9174	9192	9210	9227	9245	33
27	9263	9281	9298	9316	9334	9352	32
28	9369	9387	9405	9423	9440	9458	31
29	9476	9494	9511	9529	9547	9565	30
30	069582	9600	9618	9636	9654	9671	29
31	9689	9707	9725	9742	9760	9778	28
32	9796	9814	9831	9849	9867	9885	27
33	9903	9920	9938	9956	9974	9992	26
34	070010	0027	0045	0063	0081	0099	25
35	0116	0134	0152	0170	0188	0206	24
36	0224	0241	0259	0277	0295	0313	23
37	0331	0348	0366	0384	0401	0420	22
38	0438	0456	0474	0491	0509	0527	21
39	0545	0563	0581	0599	0617	0635	20
40	070652	0670	0688	0706	0724	0742	19
41	0760	0778	0796	0814	0832	0850	18
42	0867	0885	0903	0921	0939	0957	17
43	0975	0993	1011	1029	1047	1065	16
44	1083	1101	1119	1137	1155	1172	15
45	1190	1208	1226	1244	1262	1280	14
46	1298	1316	1334	1352	1370	1388	13
47	1406	1424	1442	1460	1478	1496	12
48	1514	1532	1550	1568	1586	1604	11
49	1622	1640	1658	1676	1694	1712	10
50	071730	1748	1766	1784	1803	1821	9
51	1839	1857	1875	1893	1911	1929	8
52	1947	1965	1983	2001	2019	2037	7
53	2055	2073	2091	2110	2128	2146	6
54	2164	2182	2200	2218	2236	2254	5
55	2272	2290	2308	2327	2345	2363	4
56	2381	2399	2417	2435	2453	2471	3
57	2490	2508	2526	2544	2562	2580	2
58	2598	2617	2635	2653	2671	2689	1
59	2707	2725	2744	2762	2780	2798	0
	60"	50"	40"	30"	20"	10"	M

Co-versed sine 68 degrees.

P. part	1"	2"	3"	4"	5"	6"	7"	8"	9"
	2	4	5	7	9	11	12	14	16

 TABLE XLI. Natural Versed Sines.

M	Versed sine 22 degrees. 0"	10"	20"	30"	40"	50"	
0	072516	2834	2852	2871	2889	2907	59
1	2925	2943	2961	2980	2998	3016	58
2	3034	3052	3071	3089	3107	3125	57
3	3143	3162	3180	3198	3216	3234	56
4	3253	3271	3289	3307	3326	3344	55
5	3362	3380	3398	3417	3435	3453	54
6	3471	3490	3508	3526	3544	3563	53
7	3581	3599	3617	3636	3654	3672	52
8	3690	3709	3727	3745	3763	3782	51
9	3800	3818	3837	3855	3873	3891	50
10	073910	3928	3946	3965	3983	4001	49
11	4020	4038	4056	4074	4093	4111	48
12	4129	4148	4166	4184	4203	4221	47
13	4239	4258	4276	4294	4313	4331	46
14	4349	4368	4386	4404	4423	4441	45
15	4459	4478	4496	4515	4533	4551	44
16	4570	4588	4606	4625	4643	4662	43
17	4680	4698	4717	4735	4753	4772	42
18	4790	4809	4827	4845	4864	4882	41
19	4901	4919	4938	4956	4974	4993	40
20	075011	5030	5048	5066	5085	5103	39
21	5122	5140	5159	5177	5196	5214	38
22	5232	5251	5269	5288	5306	5325	37
23	5343	5362	5380	5399	5417	5435	36
24	5454	5472	5491	5509	5528	5546	35
25	5565	5583	5602	5620	5639	5657	34
26	5676	5694	5713	5731	5750	5768	33
27	5787	5805	5824	5842	5861	5879	32
28	5898	5916	5935	5954	5972	5991	31
29	6009	6028	6046	6065	6083	6102	30
30	076120	6139	6158	6176	6195	6213	29
31	6232	6250	6269	6288	6306	6325	28
32	6343	6362	6380	6399	6418	6436	27
33	6455	6473	6492	6511	6529	6548	26
34	6566	6585	6604	6622	6641	6659	25
35	6678	6697	6715	6734	6753	6771	24
36	6790	6808	6827	6846	6864	6883	23
37	6902	6920	6939	6958	6976	6995	22
38	7014	7032	7051	7069	7088	7107	21
39	7126	7144	7163	7182	7200	7219	20
40	077238	7256	7275	7294	7312	7331	19
41	7350	7368	7387	7406	7424	7443	18
42	7462	7481	7499	7518	7537	7555	17
43	7574	7593	7612	7630	7649	7668	16
44	7687	7705	7724	7743	7762	7780	15
45	7799	7818	7837	7855	7874	7893	14
46	7912	7930	7949	7968	7987	8005	13
47	8024	8043	8062	8080	8099	8118	12
48	8137	8156	8174	8193	8212	8231	11
49	8250	8268	8287	8306	8325	8344	10
50	078362	8381	8400	8419	8438	8457	9
51	8475	8494	8513	8532	8551	8570	8
52	8588	8607	8626	8645	8664	8683	7
53	8701	8720	8739	8758	8777	8796	6
54	8815	8833	8852	8871	8890	8909	5
55	8928	8947	8966	8984	9003	9022	4
56	9041	9060	9079	9098	9117	9136	3
57	9154	9173	9192	9211	9230	9249	2
58	9268	9287	9306	9325	9344	9363	1
59	9382	9400	9419	9438	9457	9476	0
	60"	50"	40"	30"	20"	10"	
	Co-versed sine 67 degrees.						M

P. part { 1" 2" 3" 4" 5" 6" 7" 8" 9" / 2 4 6 7 9 11 13 15 17 }

M	Versed sine 23 degrees. 0"	10"	20"	30"	40"	50"	
0	079495	9514	9533	9552	9571	9590	59
1	9609	9628	9647	9666	9685	9704	58
2	9723	9742	9761	9780	9799	9817	57
3	9836	9855	9874	9893	9912	9931	56
4	9950	9969	9988	0007	0026	0045	55
5	080064	0083	0102	0121	0140	0159	54
6	0179	0198	0217	0236	0255	0274	53
7	0293	0312	0331	0350	0369	0388	52
8	0407	0426	0445	0464	0483	0502	51
9	0521	0540	0559	0578	0597	0617	50
10	080636	0655	0674	0693	0712	0731	49
11	0750	0769	0788	0807	0826	0846	48
12	0865	0884	0903	0922	0941	0960	47
13	0979	0998	1018	1037	1056	1075	46
14	1094	1113	1132	1151	1170	1190	45
15	1209	1228	1247	1266	1285	1305	44
16	1324	1343	1362	1381	1400	1419	43
17	1439	1458	1477	1496	1515	1534	42
18	1554	1573	1592	1611	1630	1650	41
19	1669	1688	1707	1726	1745	1765	40
20	081784	1803	1822	1842	1861	1880	39
21	1899	1918	1938	1957	1976	1995	38
22	2014	2034	2053	2072	2091	2111	37
23	2130	2149	2168	2188	2207	2226	36
24	2245	2265	2284	2303	2322	2342	35
25	2361	2380	2399	2419	2438	2457	34
26	2477	2496	2515	2534	2554	2573	33
27	2592	2612	2631	2650	2669	2689	32
28	2708	2727	2747	2766	2785	2805	31
29	2824	2843	2863	2882	2901	2921	30
30	082940	2959	2979	2998	3017	3037	29
31	3056	3075	3095	3114	3133	3153	28
32	3172	3191	3211	3230	3249	3269	27
33	3288	3308	3327	3346	3366	3385	26
34	3405	3424	3443	3463	3482	3502	25
35	3521	3540	3560	3579	3598	3618	24
36	3637	3657	3676	3696	3715	3734	23
37	3754	3773	3793	3812	3831	3851	22
38	3870	3890	3909	3929	3948	3968	21
39	3987	4006	4026	4045	4065	4084	20
40	084104	4123	4143	4162	4182	4201	19
41	4221	4240	4259	4279	4298	4318	18
42	4337	4357	4376	4396	4415	4435	17
43	4454	4474	4493	4513	4532	4552	16
44	4571	4591	4610	4630	4649	4669	15
45	4689	4708	4728	4747	4767	4786	14
46	4806	4825	4845	4864	4884	4903	13
47	4923	4943	4962	4982	5001	5021	12
48	5040	5060	5079	5099	5119	5138	11
49	5158	5177	5197	5216	5236	5256	10
50	085275	5295	5314	5334	5354	5373	9
51	5393	5412	5432	5452	5471	5491	8
52	5510	5530	5550	5569	5589	5609	7
53	5628	5648	5667	5687	5707	5726	6
54	5746	5766	5785	5805	5825	5844	5
55	5864	5884	5903	5923	5943	5962	4
56	5982	6002	6021	6041	6061	6080	3
57	6100	6120	6139	6159	6179	6198	2
58	6218	6238	6257	6277	6297	6317	1
59	6336	6356	6376	6395	6415	6435	0
	60"	50"	40"	30"	20"	10"	
	Co-versed sine 66 degrees.						M

P. part { 1" 2" 3" 4" 5" 6" 7" 8" 9" / 2 4 6 8 10 12 13 15 17 }

TABLE XLI. Natural Versed Sines.

M	Versed sine 24 degrees. 0"	10"	20"	30"	40"	50"	
0	086455	6474	6494	6514	6533	6553	59
1	6573	6593	6612	6632	6652	6672	58
2	6691	6711	6731	6751	6770	6790	57
3	6810	6830	6849	6869	6889	6909	56
4	6928	6948	6968	6988	7008	7027	55
5	7047	7067	7087	7106	7126	7146	54
6	7166	7186	7205	7225	7245	7265	53
7	7285	7304	7324	7344	7364	7384	52
8	7404	7423	7443	7463	7483	7503	51
9	7523	7542	7562	7582	7602	7622	50
10	087642	7661	7681	7701	7721	7741	49
11	7761	7781	7800	7820	7840	7860	48
12	7880	7920	7920	7940	7959	7979	47
13	7999	8019	8039	8059	8079	8099	46
14	8119	8138	8158	8178	8198	8218	45
15	8238	8258	8278	8298	8318	8338	44
16	8357	8377	8397	8417	8437	8457	43
17	8477	8497	8517	8537	8557	8577	42
18	8597	8617	8637	8657	8677	8697	41
19	8716	8736	8756	8776	8796	8816	40
20	088836	8856	8876	8896	8916	8936	39
21	8956	8976	8996	9016	9036	9056	38
22	9076	9096	9116	9136	9156	9176	37
23	9196	9216	9236	9256	9276	9296	36
24	9316	9336	9356	9376	9396	9417	35
25	9437	9457	9477	9497	9517	9537	34
26	9557	9577	9597	9617	9637	9657	33
27	9677	9697	9717	9737	9757	9778	32
28	9798	9818	9838	9858	9878	9898	31
29	9918	9938	9958	9978	9999	0019	30
30	090039	0059	0079	0099	0119	0139	29
31	0159	0180	0200	0220	0240	0260	28
32	0280	0300	0320	0341	0361	0381	27
33	0401	0421	0441	0461	0482	0502	26
34	0522	0542	0562	0582	0603	0623	25
35	0643	0663	0683	0703	0724	0744	24
36	0764	0784	0804	0824	0845	0865	23
37	0885	0905	0925	0946	0966	0986	22
38	1006	1026	1047	1067	1087	1107	21
39	1128	1148	1168	1188	1208	1229	20
40	091249	1269	1289	1310	1330	1350	19
41	1370	1391	1411	1431	1451	1472	18
42	1492	1512	1532	1553	1573	1593	17
43	1613	1634	1654	1674	1695	1715	16
44	1735	1755	1776	1796	1816	1837	15
45	1857	1877	1897	1918	1938	1958	14
46	1979	1999	2019	2040	2060	2080	13
47	2101	2121	2141	2162	2182	2202	12
48	2223	2243	2263	2284	2304	2324	11
49	2345	2365	2385	2406	2426	2446	10
50	092467	2487	2507	2528	2548	2569	9
51	2589	2609	2630	2650	2670	2691	8
52	2711	2732	2752	2772	2793	2813	7
53	2834	2854	2874	2895	2915	2936	6
54	2956	2976	2997	3017	3038	3058	5
55	3078	3099	3119	3140	3160	3181	4
56	3201	3222	3242	3262	3283	3303	3
57	3324	3344	3365	3385	3406	3426	2
58	3446	3467	3487	3508	3528	3549	1
59	3569	3590	3610	3631	3651	3672	0
	60"	50"	40"	30"	20"	10"	
	Co-versed sine 65 degrees.						M

P. part { 1" 2" 3" 4" 5" 6" 7" 8" 9" / 2 4 6 8 10 12 14 16 18

M	Versed sine 25 degrees. 0"	10"	20"	30"	40"	50"	
0	093692	3713	3733	3754	3774	3795	59
1	3815	3836	3856	3877	3897	3918	58
2	3938	3959	3979	4000	4020	4041	57
3	4061	4082	4102	4123	4143	4164	56
4	4185	4205	4226	4246	4267	4287	55
5	4308	4328	4349	4369	4390	4411	54
6	4431	4452	4472	4493	4513	4534	53
7	4555	4575	4596	4616	4637	4658	52
8	4678	4699	4719	4740	4760	4781	51
9	4802	4822	4843	4864	4884	4905	50
10	094925	4946	4967	4987	5008	5029	49
11	5049	5070	5090	5111	5132	5152	48
12	5173	5194	5214	5235	5256	5276	47
13	5297	5317	5338	5359	5379	5400	46
14	5421	5441	5462	5483	5503	5524	45
15	5545	5566	5586	5607	5628	5648	44
16	5669	5690	5710	5731	5752	5772	43
17	5793	5814	5835	5855	5876	5897	42
18	5917	5938	5959	5980	6000	6021	41
19	6042	6063	6083	6104	6125	6145	40
20	096166	6187	6208	6228	6249	6270	39
21	6291	6311	6332	6353	6374	6395	38
22	6415	6436	6457	6478	6498	6519	37
23	6540	6561	6582	6602	6623	6644	36
24	6665	6686	6706	6727	6748	6769	35
25	6790	6810	6831	6852	6873	6894	34
26	6914	6935	6956	6977	6998	7019	33
27	7039	7060	7081	7102	7123	7144	32
28	7164	7185	7206	7227	7248	7269	31
29	7290	7310	7331	7352	7373	7394	30
30	097415	7436	7456	7477	7498	7519	29
31	7540	7561	7582	7603	7624	7644	28
32	7665	7686	7707	7728	7749	7770	27
33	7791	7812	7833	7853	7874	7895	26
34	7916	7937	7958	7979	8000	8021	25
35	8042	8063	8084	8105	8126	8147	24
36	8167	8188	8209	8230	8251	8272	23
37	8293	8314	8335	8356	8377	8398	22
38	8419	8440	8461	8482	8503	8524	21
39	8545	8566	8587	8608	8629	8650	20
40	098671	8692	8713	8734	8755	8776	19
41	8797	8818	8839	8860	8881	8902	18
42	8923	8944	8965	8986	9007	9028	17
43	9049	9070	9091	9112	9133	9154	16
44	9175	9196	9218	9239	9260	9281	15
45	9302	9323	9344	9365	9386	9407	14
46	9428	9449	9470	9491	9512	9534	13
47	9555	9576	9597	9618	9639	9660	12
48	9681	9702	9723	9744	9766	9787	11
49	9808	9829	9850	9871	9892	9913	10
50	099935	9956	9977	9998	0019	0040	9
51	100061	0082	0104	0125	0146	0167	8
52	0188	0209	0231	0252	0273	0294	7
53	0315	0336	0358	0379	0400	0421	6
54	0442	0463	0485	0506	0527	0548	5
55	0569	0591	0612	0633	0654	0675	4
56	0696	0718	0739	0760	0781	0803	3
57	0824	0845	0866	0887	0909	0930	2
58	0951	0972	0994	1015	1036	1057	1
59	1078	1100	1121	1142	1163	1185	0
	60"	50"	40"	30"	20"	10"	
	Co-versed sine 64 degrees.						M

P. part { 1" 2" 3" 4" 5" 6" 7" 8" 9" / 2 4 6 8 10 13 15 17 19

TABLE XLI. Natural Versed Sines.

M	Versed sine 26 degrees. 0"	10"	20"	30"	40"	50"	
0	101206	1227	1248	1270	1291	1312	59
1	1334	1355	1376	1397	1419	1440	58
2	1461	1482	1504	1525	1546	1568	57
3	1589	1610	1631	1653	1674	1695	56
4	1717	1738	1759	1781	1802	1823	55
5	1844	1866	1887	1908	1930	1951	54
6	1972	1994	2015	2036	2058	2079	53
7	2100	2122	2143	2164	2186	2207	52
8	2229	2250	2271	2293	2314	2335	51
9	2357	2378	2399	2421	2442	2464	50
10	102485	2506	2528	2549	2570	2592	49
11	2613	2635	2656	2677	2699	2720	48
12	2742	2763	2784	2806	2827	2849	47
13	2870	2892	2913	2934	2956	2977	46
14	2999	3020	3042	3063	3084	3106	45
15	3127	3149	3170	3192	3213	3234	44
16	3256	3277	3299	3320	3342	3363	43
17	3385	3406	3428	3449	3471	3492	42
18	3514	3535	3557	3578	3600	3621	41
19	3642	3664	3685	3707	3728	3750	40
20	103771	3793	3815	3836	3858	3879	39
21	3901	3922	3944	3965	3987	4008	38
22	4030	4051	4073	4094	4116	4137	37
23	4159	4181	4202	4224	4245	4267	36
24	4288	4310	4331	4353	4374	4396	35
25	4418	4439	4461	4482	4504	4525	34
26	4547	4569	4590	4612	4633	4655	33
27	4677	4698	4720	4741	4763	4785	32
28	4806	4828	4849	4871	4893	4914	31
29	4936	4958	4979	5001	5022	5044	30
30	105066	5087	5109	5131	5152	5174	29
31	5195	5217	5239	5260	5282	5304	28
32	5325	5347	5369	5390	5412	5434	27
33	5455	5477	5499	5520	5542	5564	26
34	5585	5607	5629	5650	5672	5694	25
35	5716	5737	5759	5781	5802	5824	24
36	5846	5867	5889	5911	5933	5954	23
37	5976	5998	6019	6041	6063	6085	22
38	6106	6128	6150	6172	6193	6215	21
39	6237	6259	6280	6302	6324	6346	20
40	106367	6389	6411	6433	6454	6476	19
41	6498	6520	6541	6563	6585	6607	18
42	6629	6650	6672	6694	6716	6738	17
43	6759	6781	6803	6825	6847	6868	16
44	6890	6912	6934	6956	6977	6999	15
45	7021	7043	7065	7087	7108	7130	14
46	7152	7174	7196	7218	7239	7261	13
47	7283	7305	7327	7349	7370	7392	12
48	7414	7436	7458	7480	7502	7524	11
49	7545	7567	7589	7611	7633	7655	10
50	107677	7699	7720	7742	7764	7786	9
51	7808	7830	7852	7874	7896	7917	8
52	7939	7961	7983	8005	8027	8049	7
53	8071	8093	8115	8137	8159	8181	6
54	8202	8224	8246	8268	8290	8312	5
55	8334	8356	8378	8400	8422	8444	4
56	8466	8488	8510	8532	8554	8576	3
57	8598	8620	8642	8664	8686	8708	2
58	8730	8751	8773	8795	8817	8839	1
59	8861	8883	8905	8927	8949	8971	0
	60"	50"	40"	30"	20"	10"	
	Co-versed sine 63 degrees.						M

P. part { 1" 2" 3" 4" 5" 6" 7" 8" 9" / 2 4 6 9 11 13 15 17 19 }

M	Versed sine 27 degrees. 0"	10"	20"	30"	40"	50"	
0	108993	9015	9038	9060	9082	9104	59
1	9126	9148	9170	9192	9214	9236	58
2	9258	9280	9302	9324	9346	9368	57
3	9390	9412	9434	9456	9478	9500	56
4	9522	9544	9567	9589	9611	9633	55
5	9655	9677	9699	9721	9743	9765	54
6	9787	9809	9831	9853	9876	9898	53
7	9920	9942	9964	9986	0008	0030	52
8	110052	0074	0097	0119	0141	0163	51
9	0185	0207	0229	0251	0274	0296	50
10	110318	0340	0362	0384	0406	0429	49
11	0451	0473	0495	0517	0539	0561	48
12	0584	0606	0628	0650	0672	0694	47
13	0717	0739	0761	0783	0805	0828	46
14	0850	0872	0894	0916	0938	0961	45
15	0983	1005	1027	1049	1072	1094	44
16	1116	1138	1161	1183	1205	1227	43
17	1249	1272	1294	1316	1338	1361	42
18	1383	1405	1427	1449	1472	1494	41
19	1516	1538	1561	1583	1605	1627	40
20	111650	1672	1694	1717	1739	1761	39
21	1783	1806	1828	1850	1872	1895	38
22	1917	1939	1962	1984	2006	2028	37
23	2051	2073	2095	2118	2140	2162	36
24	2185	2207	2229	2252	2274	2296	35
25	2319	2341	2363	2385	2408	2430	34
26	2452	2475	2497	2520	2542	2564	33
27	2587	2609	2631	2654	2676	2698	32
28	2721	2743	2765	2788	2810	2833	31
29	2855	2877	2900	2922	2944	2967	30
30	112989	3012	3034	3056	3079	3101	29
31	3124	3146	3168	3191	3213	3236	28
32	3258	3280	3303	3325	3348	3370	27
33	3392	3415	3437	3460	3482	3505	26
34	3527	3549	3572	3594	3617	3639	25
35	3662	3684	3707	3729	3752	3774	24
36	3796	3819	3841	3864	3876	3909	23
37	3931	3954	3976	3999	4021	4044	22
38	4066	4089	4111	4134	4156	4179	21
39	4201	4224	4246	4269	4291	4314	20
40	114336	4359	4381	4404	4426	4449	19
41	4471	4494	4516	4539	4561	4584	18
42	4606	4629	4651	4674	4697	4719	17
43	4742	4764	4787	4809	4832	4854	16
44	4877	4900	4922	4945	4967	4990	15
45	5012	5035	5058	5080	5103	5125	14
46	5148	5170	5193	5216	5238	5261	13
47	5283	5306	5329	5351	5374	5396	12
48	5420	5442	5464	5487	5509	5532	11
49	5555	5577	5600	5623	5645	5668	10
50	115691	5713	5736	5758	5781	5804	9
51	5826	5849	5872	5894	5917	5940	8
52	5962	5985	6008	6030	6053	6076	7
53	6098	6121	6144	6166	6189	6212	6
54	6234	6257	6280	6302	6325	6348	5
55	6371	6393	6416	6439	6461	6484	4
56	6507	6529	6552	6575	6598	6620	3
57	6643	6666	6689	6711	6734	6757	2
58	6779	6802	6825	6848	6870	6893	1
59	6916	6939	6961	6984	7007	7030	0
	60"	50"	40"	30"	20"	10"	
	Co-versed sine 62 degrees.						M

P. part { 1" 2" 3" 4" 5" 6" 7" 8" 9" / 2 4 7 9 11 13 16 18 20 }

TABLE XLI. Natural Versed Sines.

M	Versed sine 28 degrees. 0"	10"	20"	30"	40"	50"	
0	117052	7075	7098	7121	7143	7166	59
1	7189	7212	7235	7257	7280	7303	58
2	7326	7348	7371	7394	7417	7440	57
3	7462	7485	7508	7531	7554	7576	56
4	7599	7622	7645	7668	7691	7713	55
5	7736	7759	7782	7805	7827	7850	54
6	7873	7896	7919	7942	7964	7987	53
7	8010	8033	8056	8079	8102	8124	52
8	8147	8170	8193	8216	8239	8262	51
9	8284	8307	8330	8353	8376	8399	50
10	118422	8445	8468	8490	8513	8536	49
11	8559	8582	8605	8628	8651	8674	48
12	8697	8719	8742	8765	8788	8811	47
13	8834	8857	8880	8903	8926	8949	46
14	8972	8995	9017	9040	9063	9086	45
15	9109	9132	9155	9178	9201	9224	44
16	9247	9270	9293	9316	9339	9362	43
17	9385	9408	9431	9454	9477	9500	42
18	9523	9546	9569	9592	9615	9638	41
19	9661	9684	9707	9730	9753	9776	40
20	119799	9822	9845	9868	9891	9914	39
21	9937	9960	9983	0006	0029	0052	38
22	120075	0098	0121	0144	0167	0190	37
23	0213	0236	0259	0282	0305	0328	36
24	0351	0374	0398	0421	0444	0467	35
25	0490	0513	0536	0559	0582	0605	34
26	0628	0651	0674	0698	0721	0744	33
27	0767	0790	0813	0836	0859	0882	32
28	0905	0929	0952	0975	0998	1021	31
29	1044	1067	1090	1113	1137	1160	30
30	121183	1206	1229	1252	1275	1299	29
31	1322	1345	1368	1391	1414	1437	28
32	1461	1484	1507	1530	1553	1576	27
33	1600	1623	1646	1669	1692	1715	26
34	1739	1762	1785	1808	1831	1855	25
35	1878	1901	1924	1947	1971	1994	24
36	2017	2040	2063	2087	2110	2133	23
37	2156	2180	2203	2226	2249	2272	22
38	2296	2319	2342	2365	2389	2412	21
39	2435	2458	2481	2505	2528	2551	20
40	122575	2598	2621	2644	2668	2691	19
41	2714	2737	2761	2784	2807	2831	18
42	2854	2877	2900	2924	2947	2970	17
43	2994	3017	3040	3063	3087	3110	16
44	3133	3157	3180	3203	3227	3250	15
45	3273	3297	3320	3343	3367	3390	14
46	3413	3437	3460	3483	3507	3530	13
47	3553	3577	3600	3623	3647	3670	12
48	3693	3717	3740	3763	3787	3810	11
49	3833	3857	3880	3904	3927	3950	10
50	123974	3997	4021	4044	4067	4091	9
51	4114	4137	4161	4184	4208	4231	8
52	4254	4278	4301	4325	4348	4372	7
53	4395	4418	4442	4465	4489	4512	6
54	4535	4559	4582	4606	4629	4653	5
55	4676	4700	4723	4746	4770	4793	4
56	4817	4840	4864	4887	4911	4934	3
57	4958	4981	5004	5028	5051	5075	2
58	5098	5122	5145	5169	5192	5216	1
59	5239	5263	5286	5310	5333	5357	0
	60"	50"	40"	30"	20"	10"	M
	Co-versed sine 61 degrees.						

P. part { 1" 2" 3" 4" 5" 6" 7" 8" 9" / 2 5 7 9 12 14 16 18 21

M	Versed sine 29 degrees. 0"	10"	20"	30"	40"	50"	
0	125380	5404	5427	5451	5474	5498	59
1	5521	5545	5568	5592	5615	5639	58
2	5662	5686	5710	5733	5757	5780	57
3	5804	5827	5851	5874	5898	5921	56
4	5945	5969	5992	6016	6039	6063	55
5	6086	6110	6133	6157	6181	6204	54
6	6228	6251	6275	6299	6322	6346	53
7	6369	6393	6416	6440	6464	6487	52
8	6511	6534	6558	6582	6605	6629	51
9	6653	6676	6700	6723	6747	6771	50
10	126794	6818	6842	6865	6889	6912	49
11	6936	6960	6983	7007	7031	7054	48
12	7078	7102	7125	7149	7173	7196	47
13	7220	7244	7267	7291	7315	7338	46
14	7362	7386	7409	7433	7457	7480	45
15	7504	7528	7551	7575	7599	7622	44
16	7646	7670	7694	7717	7741	7765	43
17	7788	7812	7836	7860	7883	7907	42
18	7931	7954	7978	8002	8026	8049	41
19	8073	8097	8121	8144	8168	8192	40
20	128216	8239	8263	8287	8311	8334	39
21	8358	8382	8406	8429	8453	8477	38
22	8501	8525	8548	8572	8596	8620	37
23	8643	8667	8691	8715	8739	8762	36
24	8786	8810	8834	8858	8881	8905	35
25	8929	8953	8977	9000	9024	9048	34
26	9072	9096	9120	[illegible]	[illegible]	9191	33
27	9215	9239	[illegible]	[illegible]	[illegible]	9334	32
28	9358	9382	[illegible]	[illegible]	[illegible]	9477	31
29	9501	9525	9549	9573	[illegible]	9620	30
30	129644	9668	9692	9716	9740	9764	29
31	9788	9811	9835	9859	9883	9907	28
32	9931	9955	9979	0003	0027	0050	27
33	130074	0098	0122	0146	0170	0194	26
34	0218	0242	0266	0290	0314	0337	25
35	0361	0385	0409	0433	0457	0481	24
36	0505	0529	0553	0577	0601	0625	23
37	0649	0673	0697	0721	0745	0769	22
38	0793	0817	0841	0865	0888	0912	21
39	0936	0960	0984	1008	1032	1056	20
40	131080	1104	1128	1152	1176	1200	19
41	1224	1248	1272	1296	1320	1344	18
42	1368	1392	1417	1441	1465	1489	17
43	1513	1537	1561	1585	1609	1633	16
44	1657	1681	1705	1729	1753	1777	15
45	1801	1825	1849	1873	1897	1921	14
46	1946	1970	1994	2018	2042	2066	13
47	2090	2114	2138	2162	2186	2210	12
48	2235	2259	2283	2307	2331	2355	11
49	2379	2403	2427	2451	2476	2500	10
50	132524	2548	2572	2596	2620	2644	9
51	2669	2693	2717	2741	2765	2789	8
52	2813	2838	2862	2886	2910	2934	7
53	2958	2982	3007	3031	3055	3079	6
54	3103	3127	3152	3176	3200	3224	5
55	3248	3272	3297	3321	3345	3369	4
56	3393	3418	3442	3466	3490	3514	3
57	3539	3563	3587	3611	3635	3660	2
58	3684	3708	3732	3756	3781	3805	1
59	3829	3853	3878	3902	3926	3950	0
	60"	50"	40"	30"	20"	10"	M
	Co-versed sine 60 degrees.						

P. part { 1" 2" 3" 4" 5" 6" 7" 8" 9" / 2 5 7 10 12 14 17 19 22

TABLE XLI. Natural Versed Sines.

Versed sine 30 degrees.

M	0″	10″	20″	30″	40″	50″	
0	133975	3999	4023	4047	4072	4096	59
1	4120	4144	4169	4193	4217	4241	58
2	4266	4290	4314	4338	4363	4387	57
3	4411	4436	4460	4484	4508	4533	56
4	4557	4581	4606	4630	4654	4678	55
5	4703	4727	4751	4776	4800	4824	54
6	4849	4873	4897	4922	4946	4970	53
7	4994	5019	5043	5067	5092	5116	52
8	5140	5165	5189	5213	5238	5262	51
9	5287	5311	5335	5360	5384	5408	50
10	135433	5457	5481	5506	5530	5555	49
11	5579	5603	5628	5652	5676	5701	48
12	5725	5750	5774	5798	5823	5847	47
13	5872	5896	5920	5945	5969	5994	46
14	6018	6042	6067	6091	6116	6140	45
15	6164	6189	6213	6238	6262	6287	44
16	6311	6336	6360	6384	6409	6433	43
17	6458	6482	6507	6531	6556	6580	42
18	6604	6629	6653	6678	6702	6727	41
19	6751	6776	6800	6825	6849	6874	40
20	136898	6923	6947	6972	6996	7021	39
21	7045	7070	7094	7119	7143	7168	38
22	7192	7217	7241	7266	7290	7315	37
23	[illegible]	7364	7388	7413	7437	7462	36
24	[illegible]	[illegible]	[illegible]	7560	7584	7609	35
25	[illegible]	[illegible]	[illegible]	7707	7732	7756	34
26	[illegible]	[illegible]	[illegible]	7855	7879	7904	33
27	[illegible]	[illegible]	[illegible]	8002	8027	8051	32
28	[illegible]	[illegible]	[illegible]	8149	8174	8199	31
29	8223	[illegible]	8272	8297	8322	8346	30
30	138372	8395	8420	8445	8469	8494	29
31	8519	8543	8568	8592	8617	8642	28
32	8666	8691	8716	8740	8765	8790	27
33	8814	8839	8863	8888	8913	8937	26
34	8962	8987	9011	9036	9061	9085	25
35	9110	9135	9159	9184	9209	9233	24
36	9258	9283	9307	9332	9357	9381	23
37	9406	9431	9455	9480	9505	9530	22
38	9554	9579	9604	9628	9653	9678	21
39	9703	9727	9752	9777	9801	9826	20
40	139851	9876	9900	9925	9950	9975	19
41	9999	0024	0049	0073	0098	0123	18
42	140148	0172	0197	0222	0247	0272	17
43	0296	0321	0346	0371	0395	0420	16
44	0445	0470	0494	0519	0544	0569	15
45	0594	0618	0643	0668	0692	0718	14
46	0742	0767	0792	0817	0842	0866	13
47	0891	0916	0941	0966	0991	1015	12
48	1040	1065	1090	1115	1140	1164	11
49	1189	1214	1239	1264	1288	1313	10
50	141338	1363	1388	1413	1438	1462	9
51	1487	1512	1537	1562	1587	1612	8
52	1636	1661	1686	1711	1736	1761	7
53	1786	1811	1836	1860	1885	1910	6
54	1935	1960	1985	2010	2035	2060	5
55	2085	2109	2134	2159	2184	2209	4
56	2234	2259	2284	2309	2334	2359	3
57	2384	2409	2434	2459	2483	2508	2
58	2533	2558	2583	2608	2633	2658	1
59	2683	2708	2733	2758	2783	2808	0
	60″	50″	40″	30″	20″	10″	
	Co-versed sine 59 degrees.						M

P. part	1″	2″	3″	4″	5″	6″	7″	8″	9″
	2	5	7	10	12	15	17	20	22

Versed sine 31 degrees.

M	0″	10″	20″	30″	40″	50″	
0	142833	2858	2883	2908	2933	2958	59
1	2983	3008	3033	3058	3082	3107	58
2	3132	3157	3182	3207	3232	3257	57
3	3282	3307	3332	3358	3383	3408	56
4	3433	3458	3483	3508	3533	3558	55
5	3583	3608	3633	3658	3683	3708	54
6	3733	3758	3783	3808	3833	3858	53
7	3883	3908	3933	3958	3983	4009	52
8	4034	4059	4084	4109	4134	4159	51
9	4184	4209	4234	4259	4284	4309	50
10	144335	4360	4385	4410	4435	4460	49
11	4485	4510	4535	4560	4586	4611	48
12	4636	4661	4686	4711	4736	4761	47
13	4786	4812	4837	4862	4887	4912	46
14	4937	4962	4988	5013	5038	5063	45
15	5088	5113	5138	5164	5189	5214	44
16	5239	5264	5289	5315	5340	5365	43
17	5390	5415	5440	5466	5491	5516	42
18	5541	5566	5592	5617	5642	5667	41
19	5692	5718	5743	5768	5793	5818	40
20	145844	5869	5894	5919	5944	5970	39
21	5995	6020	6045	6071	6096	6121	38
22	6146	6171	6197	6222	6247	6272	37
23	6298	6323	6348	6373	6399	6424	36
24	6449	6474	6500	6525	6550	6576	35
25	6601	6626	6651	6677	6702	6727	34
26	6752	6778	6803	6828	6854	6879	33
27	6904	6929	6955	6980	7005	7031	32
28	7056	7081	7107	7132	7157	7183	31
29	7208	7233	7259	7284	7309	7335	30
30	147360	7385	7411	7436	7461	7487	29
31	7512	7537	7563	7588	7613	7638	28
32	7664	7689	7715	7740	7765	7791	27
33	7816	7841	7867	7892	7918	7943	26
34	7968	7994	8019	8045	8070	8095	25
35	8121	8146	8171	8197	8222	8248	24
36	8273	8298	8324	8349	8375	8400	23
37	8425	8451	8476	8502	8527	8553	22
38	8578	8603	8629	8654	8680	8705	21
39	8731	8756	8782	8807	8832	8858	20
40	148883	8909	8934	8960	8985	9011	19
41	9036	9062	9087	9112	9138	9163	18
42	9189	9214	9240	9265	9291	9316	17
43	9342	9367	9393	9418	9444	9469	16
44	9495	9520	9546	9571	9597	9622	15
45	9648	9673	9699	9724	9750	9775	14
46	9801	9826	9852	9877	9903	9929	13
47	9954	9980	0005	0031	0056	0082	12
48	150107	0133	0158	0184	0210	0235	11
49	0261	0286	0312	0337	0363	0388	10
50	150414	0440	0465	0491	0516	0542	9
51	0567	0593	0619	0644	0670	0695	8
52	0721	0747	0772	0798	0823	0849	7
53	0875	0900	0926	0951	0977	1003	6
54	1028	1054	1080	1105	1131	1156	5
55	1182	1208	1233	1259	1285	1310	4
56	1336	1362	1387	1413	1438	1464	3
57	1490	1515	1541	1567	1592	1618	2
58	1644	1669	1695	1721	1746	1772	1
59	1798	1823	1849	1875	1900	1926	0
	60″	50″	40″	30″	20″	10″	
	Co-versed sine 58 degrees.						M

P. part	1″	2″	3″	4″	5″	6″	7″	8″	9″
	3	5	8	10	13	15	18	20	23

Versed sine 32 degrees.

M	0"	10"	20"	30"	40"	50"	
0	151952	1978	2003	2029	2055	2080	59
1	2106	2132	2157	2183	2209	2235	58
2	2260	2286	2312	2337	2363	2389	57
3	2415	2440	2466	2492	2518	2544	56
4	2569	2595	2621	2646	2672	2698	55
5	2742	2749	2775	2801	2826	2852	54
6	2878	2904	2930	2955	2981	3007	53
7	3033	3058	3084	3110	3136	3162	52
8	3187	3213	3239	3265	3291	3316	51
9	3342	3368	3394	3420	3445	3471	50
10	153497	3523	3549	3574	3600	3626	49
11	3652	3678	3704	3729	3755	3781	48
12	3807	3833	3859	3884	3910	3936	47
13	3962	3988	4014	4039	4065	4091	46
14	4117	4143	4169	4195	4220	4246	45
15	4272	4298	4324	4350	4376	4402	44
16	4427	4453	4479	4505	4531	4557	43
17	4583	4609	4635	4660	4686	4712	42
18	4738	4764	4790	4816	4842	4868	41
19	4894	4920	4945	4971	4997	5023	40
20	155049	5075	5101	5127	5153	5179	39
21	5205	5231	5257	5283	5309	5335	38
22	5360	5386	5412	5438	5464	5490	37
23	5516	5542	5568	5594	5620	5646	36
24	5672	5698	5724	5750	5776	5802	35
25	5828	5854	5880	5906	5932	5958	34
26	5984	6010	6036	6062	6088	6114	33
27	6140	6166	6192	6218	6244	6270	32
28	6296	6322	6348	6374	6400	6426	31
29	6452	6478	6504	6530	6556	6583	30
30	156609	6635	6661	6687	6713	6739	29
31	6765	6791	6817	6843	6869	6895	28
32	6921	6947	6973	7000	7026	7052	27
33	7078	7104	7130	7156	7182	7208	26
34	7234	7260	7286	7313	7339	7365	25
35	7391	7417	7443	7469	7495	7521	24
36	7547	7574	7600	7626	7652	7678	23
37	7704	7730	7757	7783	7809	7835	22
38	7861	7887	7913	7940	7966	7992	21
39	8018	8044	8070	8097	8123	8149	20
40	158175	8201	8227	8254	8280	8306	19
41	8332	8358	8384	8411	8437	8463	18
42	8489	8515	8542	8568	8594	8620	17
43	8646	8673	8699	8725	8751	8777	16
44	8804	8830	8856	8882	8909	8935	15
45	8961	8987	9013	9040	9066	9092	14
46	9118	9145	9171	9197	9223	9250	13
47	9276	9302	9328	9355	9381	9407	12
48	9433	9460	9486	9512	9538	9565	11
49	9591	9617	9644	9670	9696	9722	10
50	159749	9775	9801	9828	9854	9880	9
51	9906	9933	9959	9985	0012	0038	8
52	160064	0091	0117	0143	0170	0196	7
53	0222	0248	0275	0301	0327	0354	6
54	0380	0406	0433	0459	0485	0512	5
55	0538	0565	0591	0617	0644	0670	4
56	0696	0723	0749	0775	0802	0828	3
57	0854	0881	0907	0934	0960	0986	2
58	1013	1039	1065	1092	1118	1145	1
59	1171	1197	1224	1250	1277	1303	0
	60"	50"	40"	30"	20"	10"	M

Co-versed sine 57 degrees.

P. part	1"	2"	3"	4"	5"	6"	7"	8"	9"
	3	5	8	10	13	16	18	21	23

Versed sine 33 degrees.

M	0"	10"	20"	30"	40"	50"	
0	161329	1356	1382	1409	1435	1461	59
1	1488	1514	1541	1567	1594	1620	58
2	1646	1673	1699	1726	1752	1779	57
3	1805	1831	1858	1884	1911	1937	56
4	1964	1990	2017	2043	2070	2096	55
5	2122	2149	2175	2202	2228	2255	54
6	2281	2308	2334	2361	2387	2414	53
7	2440	2467	2493	2520	2546	2573	52
8	2599	2626	2652	2679	2705	2732	51
9	2757	2785	2811	2838	2864	2891	50
10	162918	2944	2970	2997	3023	3050	49
11	3076	3103	3130	3156	3183	3209	48
12	3236	3262	3289	3315	3342	3368	47
13	3395	3422	3448	3475	3501	3528	46
14	3554	3581	3608	3634	3661	3687	45
15	3714	3740	3767	3794	3820	3847	44
16	3873	3900	3927	3953	3980	4006	43
17	4033	4060	4086	4113	4139	4166	42
18	4193	4219	4246	4272	4299	4326	41
19	4352	4379	4406	4432	4459	4486	40
20	164512	4539	4565	4592	4619	4645	39
21	4672	4699	4725	4752	4779	4805	38
22	4832	4859	4885	4912	4939	4965	37
23	4992	5019	5045	5072	5098	5125	36
24	5152	5179	5206	5232	5259	5286	35
25	5312	5339	5366	5392	5419	5446	34
26	5473	5499	5526	5553	5579	5606	33
27	5633	5660	5686	5713	5740	5766	32
28	5793	5820	5847	5873	5900	5927	31
29	5954	5980	6007	6034	6061	6087	30
30	166114	6141	6168	6194	6221	6248	29
31	6275	6302	6328	6355	6382	6409	28
32	6435	6462	6489	6516	6543	6569	27
33	6596	6623	6650	6677	6703	6730	26
34	6757	6784	6811	6837	6864	6891	25
35	6918	6945	6971	6998	7025	7052	24
36	7079	7106	7132	7159	7186	7213	23
37	7240	7267	7293	7320	7347	7374	22
38	7401	7428	7455	7482	7508	7535	21
39	7562	7589	7616	7643	7669	7696	20
40	167723	7750	7777	7804	7831	7858	19
41	7885	7911	7938	7965	7992	8019	18
42	8046	8073	8100	8127	8153	8180	17
43	8207	8234	8261	8288	8315	8342	16
44	8369	8396	8423	8450	8477	8503	15
45	8530	8557	8584	8611	8638	8665	14
46	8692	8719	8746	8773	8800	8827	13
47	8854	8881	8908	8935	8962	8989	12
48	9016	9043	9069	9096	9123	9150	11
49	9177	9204	9231	9258	9285	9312	10
50	169339	9366	9393	9420	9447	9474	9
51	9501	9528	9555	9582	9609	9636	8
52	9663	9690	9717	9744	9771	9798	7
53	9826	9853	9880	9907	9934	9961	6
54	9988	0015	0042	0069	0096	0123	5
55	170150	0177	0204	0231	0258	0285	4
56	0312	0339	0366	0394	0421	0448	3
57	0475	0502	0529	0556	0583	0610	2
58	0637	0664	0691	0719	0746	0773	1
59	0800	0827	0854	0881	0908	0935	0
	60"	50"	40"	30"	20"	10"	M

Co-versed sine 56 degrees.

P. part	1"	2"	3"	4"	5"	6"	7"	8"	9"
	3	5	8	11	13	16	19	22	24

Versed sine 34 degrees.

M	0"	10"	20"	30"	40"	50"	
0	170962	0990	1017	1044	1071	1098	59
1	1125	1152	1179	1207	1234	1261	58
2	1288	1315	1342	1369	1396	1424	57
3	1451	1478	1505	1532	1559	1586	56
4	1614	1641	1668	1695	1722	1749	55
5	1777	1804	1831	1858	1885	1912	54
6	1940	1967	1994	2021	2048	2076	53
7	2103	2130	2157	2184	2212	2239	52
8	2266	2293	2320	2348	2375	2402	51
9	2429	2456	2484	2511	2538	2565	50
10	172593	2620	2647	2674	2701	2729	49
11	2756	2783	2810	2838	2865	2892	48
12	2920	2947	2974	3001	3028	3056	47
13	3083	3110	3137	3165	3192	3219	46
14	3247	3274	3301	3328	3356	3383	45
15	3410	3438	3465	3492	3519	3547	44
16	3574	3601	3629	3656	3683	3711	43
17	3738	3765	3793	3820	3847	3874	42
18	3902	3929	3956	3984	4011	4038	41
19	4066	4093	4120	4148	4175	4202	40
20	174230	4257	4284	4312	4339	4366	39
21	4394	4421	4449	4476	4503	4531	38
22	4558	4585	4613	4640	4667	4695	37
23	4722	4750	4777	4804	4832	4859	36
24	4887	4914	4941	4969	4996	5023	35
25	5051	5078	5106	5133	5161	5188	34
26	5215	5243	5270	5298	5325	5352	33
27	5380	5407	5435	5462	5490	5517	32
28	5544	5572	5599	5627	5654	5682	31
29	5709	5737	5764	5791	5819	5846	30
30	175874	5901	5929	5956	5984	6011	29
31	6039	6066	6094	6121	6149	6176	28
32	6203	6231	6258	6286	6313	6341	27
33	6368	6396	6423	6451	6478	6506	26
34	6533	6561	6588	6616	6643	6671	25
35	6698	6726	6754	6781	6809	6836	24
36	6864	6891	6919	6946	6974	7001	23
37	7029	7056	7084	7111	7139	7167	22
38	7194	7222	7249	7277	7304	7332	21
39	7359	7387	7415	7442	7470	7497	20
40	177525	7552	7580	7608	7635	7663	19
41	7690	7718	7746	7773	7801	7828	18
42	7856	7884	7911	7939	7966	7994	17
43	8022	8049	8077	8104	8132	8160	16
44	8187	8215	8243	8270	8298	8325	15
45	8353	8381	8408	8436	8464	8491	14
46	8519	8547	8574	8602	8630	8657	13
47	8685	8712	8740	8768	8795	8823	12
48	8851	8878	8906	8934	8961	8989	11
49	9017	9045	9072	9100	9128	9155	10
50	179183	9211	9238	9266	9294	9321	9
51	9349	9377	9405	9432	9460	9488	8
52	9515	9543	9571	9599	9626	9654	7
53	9682	9709	9737	9765	9793	9820	6
54	9848	9876	9904	9931	9959	9987	5
55	180015	0042	0070	0098	0126	0153	4
56	0181	0209	0237	0264	0292	0320	3
57	0348	0376	0403	0431	0459	0487	2
58	0514	0542	[illegible]	0598	0626	0653	1
59	0681	0709	0737	[illegible]	0792	0820	0
	60"	50"	40"	30"	20"	10"	M

Co-versed sine 55 degrees.

P. part: 1" 2" 3" 4" 5" 6" 7" 8" 9" — 3 5 8 11 14 16 19 22 25

Versed sine 35 degrees.

M	0"	10"	20"	30"	40"	50"	
0	180848	0876	0904	0931	0959	0987	59
1	1015	1043	1070	1098	1126	1154	58
2	1182	1210	1237	1265	1293	1321	57
3	1349	1377	1404	1432	1460	1488	56
4	1516	1544	1572	1599	1627	1655	55
5	1683	1711	1739	1767	1795	1822	54
6	1850	1878	1906	1934	1962	1990	53
7	2018	2045	2073	2101	2129	2157	52
8	2185	2213	2241	2269	2297	2324	51
9	2352	2380	2408	2436	2464	2492	50
10	182520	2548	2576	2604	2632	2660	49
11	2687	2715	2743	2771	2799	2827	48
12	2855	2883	2911	2939	2967	2995	47
13	3023	3051	3079	3107	3135	3163	46
14	3191	3219	3247	3275	3302	3330	45
15	3358	3386	3414	3442	3470	3498	44
16	3526	3554	3582	3610	3638	3666	43
17	3694	3722	3750	3778	3806	3834	42
18	3862	3890	3918	3946	3974	4003	41
19	4031	4059	4087	4115	4143	4171	40
20	184199	4227	4255	4283	4311	4339	39
21	4367	4395	4423	4451	4479	4507	38
22	4535	4563	4591	4620	4648	4676	37
23	4704	4732	4760	4788	4816	4844	36
24	4872	4900	4928	4956	4985	5013	35
25	5041	5069	5097	5125	5153	5181	34
26	5209	5237	5266	5294	5322	5350	33
27	5378	5406	5434	5462	5491	5519	32
28	5547	5575	5603	5631	5659	5687	31
29	5716	5744	5772	5800	5828	5856	30
30	185884	5912	5941	5969	5997	6025	29
31	6053	6082	6110	6138	6166	6194	28
32	6222	6251	6279	6307	6335	6363	27
33	6391	6420	6448	6476	6504	6533	26
34	6561	6589	6617	6645	6674	6702	25
35	6730	6758	6786	6815	6843	6871	24
36	6899	6927	6956	6984	7012	7040	23
37	7069	7097	7125	7153	7182	7210	22
38	7238	7266	7295	7323	7351	7379	21
39	7408	7436	7464	7493	7521	7549	20
40	187577	7605	7634	7662	7690	7718	19
41	7747	7775	7803	7832	7860	7888	18
42	7916	7945	7973	8001	8030	8058	17
43	8086	8115	8143	8171	8199	8228	16
44	8256	8284	8313	8341	8369	8398	15
45	8426	8454	8483	8511	8539	8568	14
46	8596	8624	8653	8681	8709	8738	13
47	8766	8794	8823	8851	8879	8908	12
48	8936	8965	8993	9021	9050	9078	11
49	9106	9135	9163	9191	9220	9248	10
50	189277	9305	9333	9362	9390	9419	9
51	9447	9475	9504	9532	9561	9589	8
52	9617	9646	9674	9703	9731	9759	7
53	9788	9816	9845	9873	9902	9930	6
54	9958	9987	0015	0044	0072	0101	5
55	190129	0157	0186	0214	0243	0271	4
56	0300	0328	0357	0385	0413	0442	3
57	0470	0499	0527	0556	0584	0613	2
58	0641	0670	0698	0727	0755	0784	1
59	0812	0841	0869	0898	0926	0955	0
	60"	50"	40"	30"	20"	10"	M

Co-versed sine 54 degrees.

P. part: 1" 2" 3" 4" 5" 6" 7" 8" 9" — 3 6 8 11 14 17 20 22 25

M	Versed sine 36 degrees. 0"	10"	20"	30"	40"	50"	
0	190983	1012	1040	1069	1097	1126	59
1	1154	1183	1211	1240	1268	1297	58
2	1325	1354	1382	1411	1439	1468	57
3	1496	1525	1553	1582	1610	1639	56
4	1667	1696	1725	1753	1782	1810	55
5	1839	1867	1896	1924	1953	1982	54
6	2010	2039	2067	2096	2124	2153	53
7	2182	2210	2239	2267	2296	2324	52
8	2353	2382	2410	2439	2467	2496	51
9	2525	2553	2582	2610	2639	2668	50
10	192696	2725	2753	2782	2811	2839	49
11	2868	2897	2925	2954	2982	3011	48
12	3039	3068	3097	3126	3154	3183	47
13	3211	3240	3269	3297	3326	3355	46
14	3383	3412	3441	3469	3498	3527	45
15	3555	3584	3613	3641	3670	3699	44
16	3727	3756	3785	3813	3842	3871	43
17	3900	3928	3957	3986	4014	4043	42
18	4072	4100	4129	4158	4187	4215	41
19	4244	4273	4301	4330	4359	4388	40
20	194416	4445	4474	4502	4531	4560	39
21	4589	4617	4646	4675	4704	4732	38
22	4771	4790	4819	4847	4876	4905	37
23	4934	4962	4991	5020	5049	5077	36
24	5106	5135	5164	5193	5221	5250	35
25	5279	5308	5336	5365	5394	5423	34
26	5452	5480	5509	5538	5567	5596	33
27	5624	5653	5682	5711	5740	5768	32
28	5797	5826	5855	5884	5912	5941	31
29	5970	5999	6028	6057	6085	6114	30
30	196143	6172	6201	6230	6259	6287	29
31	6316	6345	6374	6403	6432	6460	28
32	6489	6518	6547	6576	6605	6634	27
33	6663	6691	6720	6749	6778	6807	26
34	6836	6865	6894	6922	6951	6980	25
35	7009	7038	7067	7096	7125	7154	24
36	7183	7211	7240	7269	7298	7327	23
37	7356	7385	7414	7443	7472	7501	22
38	7530	7558	7587	7616	7645	7674	21
39	7703	7732	7761	7790	7819	7848	20
40	197877	7906	7935	7964	7993	8022	19
41	8051	8080	8108	8137	8166	8195	18
42	8224	8253	8282	8311	8340	8369	17
43	8398	8427	8456	8485	8514	8543	16
44	8572	8601	8630	8659	8688	8717	15
45	8746	8775	8804	8833	8862	8891	14
46	8920	8949	8978	9007	9036	9065	13
47	9094	9123	9152	9181	9211	9240	12
48	9269	9298	9327	9356	9385	9414	11
49	9443	9472	9501	9530	9559	9588	10
50	199617	9646	9675	9704	9734	9763	9
51	9792	9821	9850	9879	9908	9937	8
52	9966	9995	0024	0054	0083	0112	7
53	200141	0170	0199	0228	0257	0286	6
54	0315	0344	0374	0403	0432	0461	5
55	0490	0519	0548	0577	0607	0636	4
56	0665	0694	0723	0752	0781	0810	3
57	0840	0869	0898	0927	0956	0985	2
58	1015	1044	1073	1102	1131	1160	1
59	1189	1219	1248	1277	1306	1335	0
	60"	50"	40"	30"	20"	10"	
	Co-versed sine 53 degrees.						M

F. part { 1" 2" 3" 4" 5" 6" 7" 8" 9" / 3 6 9 12 14 17 20 23 26

M	Versed sine 37 degrees. 0"	10"	20"	30"	40"	50"	
0	201364	1394	1423	1452	1482	1510	59
1	1540	1569	1598	1627	1656	1686	58
2	1715	1744	1773	1802	1832	1861	57
3	1890	1919	1948	1978	2007	2036	56
4	2065	2094	2124	2153	2182	2211	55
5	2241	2270	2299	2328	2358	2387	54
6	2416	2445	2475	2504	2533	2562	53
7	2592	2621	2650	2679	2709	2738	52
8	2767	2796	2826	2855	2884	2913	51
9	2943	2972	3001	3031	3060	3089	50
10	203118	3148	3177	3206	3236	3265	49
11	3294	3324	3353	3382	3411	3441	48
12	3470	3499	3528	3558	3587	3617	47
13	3646	3675	3705	3734	3763	3793	46
14	3822	3851	3881	3910	3939	3969	45
15	3998	4027	4057	4086	4115	4145	44
16	4174	4203	4233	4262	4292	4321	43
17	4350	4380	4409	4438	4468	4497	42
18	4527	4556	4585	4615	4644	4673	41
19	4703	4732	4762	4791	4820	4850	40
20	204879	4909	4938	4967	4997	5026	39
21	5056	5085	5114	5144	5173	5203	38
22	5232	5261	5291	5320	5350	5379	37
23	5409	5438	5468	5497	5526	5556	36
24	5585	5615	5644	5674	5703	5733	35
25	5762	5792	5821	5850	5880	5909	34
26	5939	5968	5998	6027	6057	6086	33
27	6116	6145	6175	6204	6234	6263	32
28	6293	6322	6352	6381	6411	6440	31
29	6470	6499	6529	6558	6588	6617	30
30	206647	6676	6706	6735	6765	6794	29
31	6824	6853	6883	6912	6942	6971	28
32	7001	7030	7060	7090	7119	7149	27
33	7178	7208	7237	7267	7296	7326	26
34	7356	7385	7415	7444	7474	7503	25
35	7533	7562	7592	7622	7651	7681	24
36	7710	7740	7770	7799	7829	7858	23
37	7888	7917	7947	7977	8006	8036	22
38	8065	8095	8125	8154	8184	8213	21
39	8243	8273	8302	8332	8362	8391	20
40	208421	8450	8480	8510	8539	8569	19
41	8599	8628	8658	8688	8717	8747	18
42	8776	8806	8836	8865	8895	8925	17
43	8954	8984	9014	9043	9073	9103	16
44	9132	9162	9192	9221	9251	9281	15
45	9310	9340	9370	9399	9429	9459	14
46	9489	9518	9548	9578	9607	9637	13
47	9667	9696	9726	9756	9786	9815	12
48	9845	9875	9904	9934	9964	9994	11
49	210023	0053	0083	0112	0142	0172	10
50	210202	0231	0261	0291	0321	0350	9
51	0380	0410	0440	0469	0499	0529	8
52	0559	0588	0618	0648	0678	0707	7
53	0737	0767	0797	0827	0856	0886	6
54	0916	0946	0975	1005	1035	1065	5
55	1095	1124	1154	1184	1214	1244	4
56	1273	1303	1333	1363	1393	1422	3
57	1452	1482	1512	1542	1572	1601	2
58	1631	1661	1691	1721	1751	1780	1
59	1810	1840	1870	1900	1930	1959	0
	60"	50"	40"	30"	20"	10"	
	Co-versed sine 52 degrees.						M

F. part { 1" 2" 3" 4" 5" 6" 7" 8" 9" / 3 6 9 12 15 18 21 24 27

 TABLE XLI. Natural Versed Sines.

M	Versed sine 38 degrees						
M	0"	10"	20"	30"	40"	50"	
0	211989	2019	2049	2079	2109	2139	59
1	2168	2198	2228	2258	2288	2318	58
2	2348	2377	2407	2437	2467	2497	57
3	2527	2557	2587	2616	2646	2676	56
4	2706	2736	2766	2796	2826	2856	55
5	2886	2915	2945	2975	3005	3035	54
6	3065	3095	3125	3155	3185	3215	53
7	3244	3274	3304	3334	3364	3394	52
8	3424	3454	3484	3514	3544	3574	51
9	3604	3634	3664	3694	3724	3754	50
10	213783	3813	3843	3873	3903	3933	49
11	3963	3993	4023	4053	4083	4113	48
12	4143	4173	4203	4233	4263	4293	47
13	4323	4353	4383	4413	4443	4473	46
14	4503	4533	4563	4593	4623	4653	45
15	4683	4713	4743	4773	4803	4833	44
16	4863	4893	4923	4953	4983	5013	43
17	5043	5074	5103	5133	5164	5194	42
18	5224	5254	5284	5314	5344	5374	41
19	5404	5434	5464	5494	5524	5554	40
20	215584	5614	5644	5675	5705	5735	39
21	5765	5795	5825	5855	5885	5915	38
22	5945	5975	6005	6035	6066	6096	37
23	6126	6156	6186	6216	6246	6276	36
24	6307	6337	6367	6397	6427	6457	35
25	6487	6517	6548	6578	6608	6638	34
26	6668	6698	6728	6758	6789	6819	33
27	6849	6879	6909	6939	6969	7000	32
28	7030	7060	7090	7120	7150	7181	31
29	7211	7241	7271	7301	7331	7362	30
30	217392	7422	7452	7482	7513	7543	29
31	7573	7603	7633	7664	7694	7724	28
32	7754	7784	7815	7845	7875	7905	27
33	7935	7966	7996	8026	8056	8086	26
34	8117	8147	8177	8207	8238	8268	25
35	8298	8328	8359	8389	8419	8449	24
36	8480	8510	8540	8570	8601	8631	23
37	8661	8691	8722	8752	8782	8812	22
38	8843	8873	8903	8933	8964	8994	21
39	9024	9055	9085	9115	9145	9176	20
40	219206	9236	9267	9297	9327	9357	19
41	9388	9418	9448	9479	9509	9539	18
42	9570	9600	9630	9661	9691	9721	17
43	9751	9782	9812	9842	9873	9903	16
44	9933	9964	9994	0024	0055	0085	15
45	220116	0146	0176	0207	0237	0267	14
46	0298	0328	0358	0389	0419	0449	13
47	0480	0510	0541	0571	0601	0632	12
48	0662	0692	0723	0753	0784	0814	11
49	0844	0875	0905	0936	0966	0996	10
50	221027	1057	1088	1118	1148	1179	9
51	1209	1240	1270	1300	1331	1361	8
52	1392	1422	1452	1483	1513	1544	7
53	1574	1605	1635	1666	1696	1726	6
54	1757	1787	1818	1848	1879	1909	5
55	1940	1970	2000	2031	2061	2092	4
56	2122	2153	2183	2214	2244	2275	3
57	2305	2336	2366	2397	2427	2458	2
58	2488	2519	2549	2580	2610	2641	1
59	2671	2702	2732	2763	2793	2824	0
	60"	50"	40"	30"	20"	10"	
	Co-versed sine 51 degrees.						M

P. part	1"	2"	3"	4"	5"	6"	7"	8"	9"
	3	6	9	12	15	18	21	24	27

M	Versed sine 39 degrees						
M	0"	10"	20"	30"	40"	50"	
0	222854	2885	2915	2946	2976	3007	59
1	3037	3068	3098	3129	3159	3190	58
2	3220	3251	3281	3312	3342	3373	57
3	3404	3434	3465	3495	3526	3556	56
4	3587	3617	3648	3678	3709	3740	55
5	3770	3801	3831	3862	3892	3923	54
6	3954	3984	4015	4045	4076	4106	53
7	4137	4168	4198	4229	4259	4290	52
8	4321	4351	4382	4412	4443	4474	51
9	4504	4535	4565	4596	4627	4657	50
10	224688	4719	4749	4780	4810	4841	49
11	4872	4902	4933	4964	4994	5025	48
12	5056	5086	5117	5147	5178	5209	47
13	5239	5270	5301	5331	5362	5393	46
14	5423	5454	5485	5515	5546	5577	45
15	5607	5638	5669	5699	5730	5761	44
16	5791	5822	5853	5883	5914	5945	43
17	5976	6006	6037	6068	6098	6129	42
18	6160	6190	6221	6252	6283	6313	41
19	6344	6375	6406	6436	6467	6498	40
20	226528	6559	6590	6621	6651	6682	39
21	6713	6744	6774	6805	6836	6867	38
22	6897	6928	6959	6990	7020	7051	37
23	7082	7113	7143	7174	7205	7236	36
24	7266	7297	7328	7359	7390	7420	35
25	7451	7482	7513	7543	7574	7605	34
26	7636	7667	7697	7728	7759	7790	33
27	7821	7851	7882	7913	7944	7975	32
28	8005	8036	8067	8098	8129	8160	31
29	8190	8221	8252	8283	8314	8345	30
30	228375	8406	8437	8468	8499	8530	29
31	8560	8591	8622	8653	8684	8715	28
32	8746	8776	8807	8838	8869	8900	27
33	8931	8962	8993	9023	9054	9085	26
34	9116	9147	9178	9209	9240	9270	25
35	9301	9332	9363	9394	9425	9456	24
36	9487	9518	9549	9579	9610	9641	23
37	9672	9703	9734	9765	9796	9827	22
38	9858	9889	9920	9951	9981	0012	21
39	230043	0074	0105	0136	0167	0198	20
40	230229	0260	0291	0322	0353	0384	19
41	0415	0446	0477	0508	0539	0569	18
42	0600	0631	0662	0693	0724	0755	17
43	0786	0817	0848	0879	0910	0941	16
44	0972	1003	1034	1065	1096	1127	15
45	1158	1189	1220	1251	1282	1313	14
46	1344	1375	1406	1437	1468	1499	13
47	1530	1561	1592	1623	1654	1685	12
48	1716	1748	1779	1810	1841	1872	11
49	1903	1934	1965	1996	2027	2058	10
50	232089	2120	2151	2182	2213	2244	9
51	2275	2306	2338	2369	2400	2431	8
52	2462	2493	2524	2555	2586	2617	7
53	2648	2679	2710	2742	2773	2804	6
54	2835	2866	2897	2928	2959	2990	5
55	3021	3053	3084	3115	3146	3177	4
56	3208	3239	3270	3301	3333	3364	3
57	3395	3426	3457	3488	3519	3551	2
58	3582	3613	3644	3675	3706	3737	1
59	3769	3800	3831	3862	3893	3924	0
	60"	50"	40"	30"	20"	10"	
	Co-versed sine 50 degrees.						M

P. part	1"	2"	3"	4"	5"	6"	7"	8"	9"
	3	6	9	12	15	18	22	25	28

M	Versed sine 40 degrees. 0"	10"	20"	30"	40"	50"	
0	233956	3987	4018	4049	4080	4111	59
1	4143	4174	4205	4236	4267	4298	58
2	4330	4361	4392	4423	4454	4486	57
3	4517	4548	4579	4610	4642	4673	56
4	4704	4735	4766	4798	4829	4860	55
5	4891	4922	4954	4985	5016	5047	54
6	5079	5110	5141	5172	5204	5235	53
7	5266	5297	5328	5360	5391	5422	52
8	5453	5485	5516	5547	5578	5610	51
9	5641	5672	5704	5735	5766	5797	50
10	235829	5860	5891	5922	5954	5985	49
11	6016	6048	6079	6110	6141	6173	48
12	6204	6235	6267	6298	6329	6360	47
13	6392	6423	6454	6486	6517	6548	46
14	6580	6611	6642	6674	6705	6736	45
15	6768	6799	6830	6862	6893	6924	44
16	6956	6987	7018	7050	7081	7112	43
17	7144	7175	7206	7238	7269	7300	42
18	7332	7363	7394	7426	7457	7488	41
19	7519	7551	7583	7614	7645	7677	40
20	237708	7739	7771	7802	7834	7865	39
21	7896	7928	7959	7991	8022	8053	38
22	8085	8116	8148	8179	8210	8242	37
23	8273	8305	8336	8367	8399	8430	36
24	8462	8493	8525	8556	8587	8619	35
25	8650	8682	8713	8745	8776	8807	34
26	8839	8870	8902	8933	8965	8996	33
27	9028	9059	9090	9122	9153	9185	32
28	9216	9248	9279	9311	9342	9374	31
29	9405	9437	9468	9500	9531	9563	30
30	239594	9626	9657	9689	9720	9751	29
31	9783	9814	9846	9877	9909	9940	28
32	9972	0004	0035	0067	0098	0130	27
33	240161	0193	0224	0256	0287	0319	26
34	0350	0382	0413	0445	0476	0508	25
35	0539	0571	0603	0634	0666	0697	24
36	0729	0760	0792	0823	0855	0886	23
37	0918	0950	0981	1013	1044	1076	22
38	1107	1139	1171	1202	1234	1265	21
39	1297	1328	1360	1392	1423	1455	20
40	241486	1518	1550	1581	1613	1644	19
41	1676	1708	1739	1771	1802	1834	18
42	1866	1897	1929	1961	1992	2024	17
43	2055	2087	2119	2150	2182	2214	16
44	2245	2277	2308	2340	2372	2403	15
45	2435	2467	2498	2530	2562	2593	14
46	2625	2657	2688	2720	2752	2783	13
47	2815	2847	2878	2910	2942	2973	12
48	3005	3037	3068	3100	3132	3162	11
49	3195	3227	3258	3290	3322	3354	10
50	243385	3417	3449	3480	3512	3544	9
51	3575	3607	3639	3671	3702	3734	8
52	3766	3797	3829	3861	3893	3924	7
53	3956	3988	4020	4051	4083	4115	6
54	4147	4178	4210	4242	4274	4305	5
55	4337	4369	4401	4433	4464	4496	4
56	4528	4559	4591	4623	4655	4686	3
57	4718	4750	4782	4814	4845	4877	2
58	4909	4941	4972	5004	5036	5068	1
59	5100	5131	5163	5195	5227	5259	0
	60"	50"	40"	30"	20"	10"	
	Co-versed sine 49 degrees.						M

P. part	1"	2"	3"	4"	5"	6"	7"	8"	9"
	3	6	9	13	16	19	22	25	28

M	Versed sine 41 degrees. 0"	10"	20"	30"	40"	50"	
0	245290	5322	5354	5386	5418	5449	59
1	5481	5513	5545	5577	5609	5640	58
2	5672	5704	5736	5768	5800	5831	57
3	5863	5895	5927	5959	5991	6022	56
4	6054	6086	6118	6150	6182	6214	55
5	6245	6277	6309	6341	6373	6405	54
6	6437	6468	6500	6532	6564	6596	53
7	6628	6660	6692	6724	6755	6787	52
8	6819	6851	6883	6915	6947	6979	51
9	7011	7042	7074	7106	7138	7170	50
10	247202	7234	7266	7298	7330	7362	49
11	7394	7425	7457	7489	7521	6553	48
12	7585	7617	7649	7681	7713	7745	47
13	7777	7809	7841	7873	7905	7936	46
14	7968	8000	8032	8064	8096	8128	45
15	8160	8192	8224	8256	8288	8320	44
16	8352	8384	8416	8448	8480	8512	43
17	8544	8576	8608	8640	8672	8704	42
18	8736	8768	8800	8832	8864	8896	41
19	8928	8960	8992	9024	9056	9088	40
20	249120	9152	9184	9216	9248	9280	39
21	9312	9344	9376	9408	9440	9472	38
22	9504	9536	9568	9600	9632	9665	37
23	9697	9729	9761	9793	9825	9857	36
24	9889	9921	9953	9985	0017	0049	35
25	250081	0113	0145	0178	0210	0242	34
26	0274	0306	0338	0370	0402	0434	33
27	0466	0498	0531	0563	0595	0627	32
28	0659	0691	0723	0755	0787	0819	31
29	0852	0884	0916	0948	0980	1012	30
30	251044	1076	1109	1141	1173	1205	29
31	1237	1269	1301	1333	1366	1398	28
32	1430	1462	1494	1526	1559	1591	27
33	1623	1655	1687	1719	1752	1784	26
34	1816	1848	1880	1912	1944	1977	25
35	2009	2041	2073	2105	2138	2170	24
36	2202	2234	2266	2298	2331	2363	23
37	2395	2427	2459	2492	2524	2556	22
38	2588	2621	2653	2685	2717	2749	21
39	2782	2814	2846	2878	2910	2943	20
40	252975	3007	3039	3072	3104	3136	19
41	3168	3201	3233	3265	3297	3330	18
42	3362	3394	3426	3459	3491	3523	17
43	3555	3588	3620	3652	3684	3717	16
44	3749	3781	3814	3846	3878	3910	15
45	3943	3975	4007	4039	4072	4104	14
46	4136	4169	4201	4233	4266	4298	13
47	4330	4362	4395	4427	4459	4492	12
48	4524	4556	4589	4621	4653	4686	11
49	4718	4750	4783	4815	4847	4880	10
50	254912	4944	4977	5009	5041	5074	9
51	5106	5138	5171	5203	5235	5268	8
52	5300	5332	5365	5397	5429	5462	7
53	5494	5527	5559	5591	5624	5656	6
54	5688	5721	5753	5786	5818	5850	5
55	5883	5915	5948	5980	6012	6045	4
56	6077	6110	6142	6174	6207	6239	3
57	6272	6304	6336	6369	6401	6434	2
58	6466	6498	6531	6563	6596	6629	1
59	6661	6693	6725	6758	6790	6823	0
	60"	50"	40"	30"	20"	10"	
	Co-versed sine 48 degrees.						M

P. part	1"	2"	3"	4"	5"	6"	7"	8"	9"
	3	6	10	13	16	19	22	26	29

M	0"	10"	20"	30"	40"	50"	
	Versed sine 42 degrees.						
0	256855	6888	6920	6953	6985	7017	59
1	7050	7082	7115	7147	7180	7212	58
2	7245	7277	7310	7342	7374	7407	57
3	7439	7472	7504	7537	7569	7602	56
4	7634	7667	7699	7732	7764	7797	55
5	7829	7862	7894	7927	7959	7992	54
6	8024	8057	8089	8122	8154	8187	53
7	8219	8252	8284	8317	8349	8382	52
8	8414	8447	8479	8512	8544	8577	51
9	8609	8642	8675	8707	8740	8772	50
10	258805	8837	8870	8902	8935	8967	49
11	9000	9033	9065	9098	9130	9163	48
12	9195	9228	9261	9293	9326	9358	47
13	9391	9423	9456	9489	9521	9554	46
14	9586	9619	9652	9684	9717	9749	45
15	9782	9814	9847	9880	9912	9945	44
16	9977	0010	0043	0075	0108	0141	43
17	260173	0206	0239	0271	0304	0336	42
18	0369	0402	0434	0467	0499	0532	41
19	0565	0597	0630	0663	0695	0728	40
20	260761	0793	0826	0859	0891	0924	39
21	0957	0989	1022	1054	1087	1120	38
22	1152	1185	1218	1251	1283	1316	37
23	1349	1381	1414	1447	1479	1512	36
24	1545	1577	1610	1643	1675	1708	35
25	1741	1774	1806	1839	1872	1904	34
26	1937	1970	2003	2035	2068	2101	33
27	2133	2166	2199	2232	2264	2297	32
28	2330	2362	2395	2428	2461	2493	31
29	2526	2559	2592	2624	2657	2690	30
30	262723	2755	2788	2821	2854	2886	29
31	2919	2952	2985	3018	3050	3083	28
32	3116	3149	3181	3214	3247	3280	27
33	3313	3345	3378	3411	3444	3476	26
34	3509	3542	3575	3608	3640	3673	25
35	3706	3739	3772	3804	3837	3870	24
36	3903	3936	3969	4001	4034	4067	23
37	4100	4133	4166	4198	4231	4264	22
38	4297	4330	4363	4395	4428	4461	21
39	4494	4527	4560	4592	4625	4658	20
40	264691	4724	4757	4790	4822	4855	19
41	4888	4921	4954	4987	5020	5053	18
42	5085	5118	5151	5184	5217	5250	17
43	5283	5316	5348	5381	5414	5447	16
44	5480	5513	5546	5579	5612	5645	15
45	5677	5710	5743	5776	5809	5842	14
46	5875	5908	5941	5974	6007	6040	13
47	6073	6105	6138	6171	6204	6237	12
48	6270	6303	6336	6369	6402	6435	11
49	6468	6501	6534	6567	6600	6633	10
50	266666	6699	6731	6764	6798	6830	9
51	6863	6896	6929	6962	6995	7028	8
52	7061	7094	7127	7160	7193	7226	7
53	7259	7292	7325	7358	7391	7424	6
54	7457	7490	7523	7556	7589	7622	5
55	7655	7688	7721	7754	7787	7820	4
56	7853	7886	7919	7952	7985	8018	3
57	8051	8084	8117	8151	8184	8217	2
58	8250	8283	8316	8349	8382	8415	1
59	8448	8481	8514	8547	8580	8613	0
	60"	50"	40"	30"	20"	10"	
	Co-versed sine 47 degrees.						M

P. part { 1" 2" 3" 4" 5" 6" 7" 8" 9" / 3 7 10 13 16 20 23 26 29 }

M	0"	10"	20"	30"	40"	50"	
	Versed sine 43 degrees.						
0	268646	8679	8712	8745	8779	8812	59
1	8845	8878	8911	8944	8977	9010	58
2	9043	9076	9109	9142	9176	9209	57
3	9242	9275	9308	9341	9374	9407	56
4	9440	9473	9507	9540	9573	9606	55
5	9639	9672	9705	9738	9771	9805	54
6	9838	9871	9904	9937	9970	0003	53
7	270037	0070	0103	0136	0169	0202	52
8	0235	0269	0302	0335	0368	0401	51
9	0434	0467	0501	0534	0567	0600	50
10	270633	0666	0700	0733	0766	0799	49
11	0832	0865	0899	0932	0965	0998	48
12	1031	1065	1098	1131	1164	1197	47
13	1231	1264	1297	1330	1363	1397	46
14	1430	1463	1496	1529	1563	1596	45
15	1629	1662	1695	1729	1762	1795	44
16	1829	1862	1895	1928	1961	1995	43
17	2028	2061	2094	2128	2161	2194	42
18	2227	2260	2294	2327	2360	2394	41
19	2427	2460	2493	2527	2560	2593	40
20	272626	2660	2693	2726	2759	2793	39
21	2826	2859	2893	2926	2959	2992	38
22	3026	3059	3092	3126	3159	3192	37
23	3225	3259	3292	3325	3359	3392	36
24	3425	3459	3492	3525	3559	3592	35
25	3625	3659	3692	3725	3759	3792	34
26	3825	3859	3892	3925	3959	3992	33
27	4025	4059	4092	4125	4159	4192	32
28	4225	4259	4292	4325	4359	4392	31
29	4425	4459	4492	4526	4559	4592	30
30	274626	4659	4692	4726	4759	4793	29
31	4826	4859	4893	4926	4959	4993	28
32	5026	5060	5093	5126	5160	5193	27
33	5227	5260	5293	5327	5360	5394	26
34	5427	5460	5494	5527	5561	5594	25
35	5628	5661	5694	5728	5761	5795	24
36	5828	5862	5895	5928	5962	5995	23
37	6029	6062	6096	6129	6163	6196	22
38	6229	6263	6296	6330	6363	6397	21
39	6430	6464	6497	6531	6564	6598	20
40	276631	6665	6698	6731	6765	6798	19
41	6832	6865	6899	6932	6966	6999	18
42	7033	7066	7100	7133	7167	7200	17
43	7234	7267	7301	7334	7368	7401	16
44	7435	7468	7502	7535	7569	7603	15
45	7636	7670	7703	7736	7770	7804	14
46	7837	7871	7904	7938	7971	8005	13
47	8038	8072	8106	8139	8173	8206	12
48	8240	8273	8307	8340	8374	8408	11
49	8441	8475	8508	8542	8575	8609	10
50	278643	8676	8710	8743	8777	8810	9
51	8844	8878	8911	8945	8978	9012	8
52	9046	9079	9113	9146	9180	9214	7
53	9247	9281	9314	9348	9382	9415	6
54	9449	9483	9516	9550	9583	9617	5
55	9651	9684	9718	9752	9785	9819	4
56	9852	9886	9920	9953	9987	0021	3
57	280054	0088	0122	0155	0189	0223	2
58	0256	0290	0324	0357	0391	0425	1
59	0458	0492	0525	0559	0593	0627	0
	60"	50"	40"	30"	20"	10"	
	Co-versed sine 46 degrees.						M

P. part { 1" 2" 3" 4" 5" 6" 7" 8" 9" / 3 7 10 13 17 20 23 27 30 }

TABLE XLI. Natural Versed Sines.

Versed sine 44 degrees.

M	0"	10"	20"	30"	40"	50"	
0	280660	0694	0728	0761	0795	0829	59
1	0862	0896	0930	0963	0997	1031	58
2	1064	1098	1132	1166	1199	1233	57
3	1267	1300	1334	1368	1402	1435	56
4	1469	1503	1536	1570	1604	1638	55
5	1671	1705	1739	1772	1806	1840	54
6	1874	1907	1941	1975	2009	2042	53
7	2076	2110	2144	2177	2211	2245	52
8	2279	2312	2346	2380	2414	2448	51
9	2481	2515	2549	2583	2616	2650	50
10	282684	2718	2751	2785	2819	2853	49
11	2887	2920	2954	2988	3022	3056	48
12	3089	3123	3157	3191	3225	3258	47
13	3292	3326	3360	3394	3427	3461	46
14	3495	3529	3563	3597	3630	3664	45
15	3698	3732	3766	3800	3833	3867	44
16	3901	3934	3969	4003	4036	4070	43
17	4104	4138	4172	4206	4240	4273	42
18	4307	4341	4375	4409	4443	4477	41
19	4510	4544	4578	4612	4646	4680	40
20	284714	4748	4781	4815	4849	4883	39
21	4917	4951	4985	5019	5053	5086	38
22	5120	5154	5188	5222	5256	5290	37
23	5324	5358	5392	5426	5459	5493	36
24	5527	5561	5595	5629	5663	5697	35
25	5731	5765	5799	5833	5867	5901	34
26	5934	5968	6002	6036	6070	6104	33
27	6138	6172	6206	6240	6274	6308	32
28	6342	6376	6410	6444	6478	6512	31
29	6546	6580	6614	6648	6682	6716	30
30	286750	6784	6818	6852	6885	6919	29
31	6953	6987	7021	7055	7089	7123	28
32	7157	7191	7225	7259	7293	7327	27
33	7361	7395	7430	7464	7498	7532	26
34	7566	7600	7634	7668	7702	7736	25
35	7770	7804	7838	7872	7906	7940	24
36	7974	8008	8042	8076	8110	8144	23
37	8178	8212	8246	8280	8314	8349	22
38	8383	8417	8451	8485	8519	8553	21
39	8587	8621	8655	8689	8723	8757	20
40	288791	8826	8860	8894	8928	8962	19
41	8996	9030	9064	9098	9132	9166	18
42	9201	9235	9269	9303	9337	9371	17
43	9405	9439	9473	9508	9542	9576	16
44	9610	9644	9678	9712	9746	9780	15
45	9815	9849	9883	9917	9951	9985	14
46	290019	0054	0088	0122	0156	0190	13
47	0224	0258	0293	0327	0361	0395	12
48	0429	0463	0498	0532	0566	0600	11
49	0634	0668	0703	0737	0771	0805	10
50	290839	0874	0908	0942	0976	1010	9
51	1044	1079	1113	1147	1181	1215	8
52	1250	1284	1318	1352	1386	1421	7
53	1455	1489	1523	1558	1592	1626	6
54	1660	1694	1729	1763	1797	1831	5
55	1866	1900	1934	1968	2002	2037	4
56	2071	2105	2139	2174	2208	2242	3
57	2277	2311	2345	2379	2413	2448	2
58	2482	2516	2550	2585	2619	2653	1
59	2688	2722	2756	2790	2825	2859	0
	60"	50"	40"	30"	20"	10"	M

Co-versed sine 45 degrees.

P. part	1"	2"	3"	4"	5"	6"	7"	8"	9"
	3	7	10	14	17	20	24	27	31

Versed sine 45 degrees.

M	0"	10"	20"	30"	40"	50"	
0	292893	2928	2962	2996	3030	3065	59
1	3099	3133	3168	3202	3236	3270	58
2	3305	3339	3373	3408	3442	3476	57
3	3511	3545	3579	3613	3648	3682	56
4	3716	3751	3785	3819	3854	3888	55
5	3922	3957	3991	4025	4060	4094	54
6	4128	4163	4197	4231	4266	4300	53
7	4335	4369	4403	4438	4472	4506	52
8	4541	4575	4609	4644	4678	4712	51
9	4747	4781	4816	4850	4884	4919	50
10	294953	4987	5022	5056	5091	5125	49
11	5159	5194	5228	5263	5297	5331	48
12	5366	5400	5435	5469	5503	5538	47
13	5572	5607	5641	5675	5710	5744	46
14	5779	5813	5848	5882	5916	5951	45
15	5985	6020	6054	6089	6123	6157	44
16	6192	6226	6261	6295	6330	6364	43
17	6399	6433	6467	6502	6536	6571	42
18	6605	6640	6674	6709	6743	6778	41
19	6812	6847	6881	6916	6950	6984	40
20	297019	7053	7088	7122	7157	7191	39
21	7226	7260	7295	7329	7364	7398	38
22	7433	7467	7502	7536	7571	7605	37
23	7640	7674	7709	7743	7778	7812	36
24	7847	7881	7916	7951	7985	8020	35
25	8054	8089	8123	8158	8192	8227	34
26	8261	8296	8330	8365	8399	8434	33
27	8469	8503	8538	8572	8607	8641	32
28	8676	8710	8745	8780	8814	8849	31
29	8883	8918	8952	8987	9022	9056	30
30	299091	9125	9160	9194	9229	9264	29
31	9298	9333	9367	9402	9437	9471	28
32	9506	9540	9575	9610	9644	9679	27
33	9713	9748	9783	9817	9852	9886	26
34	9921	9956	9990	0025	0060	0094	25
35	300129	0163	0198	0233	0267	0302	24
36	0337	0371	0406	0441	0475	0510	23
37	0545	0579	0614	0648	0683	0718	22
38	0752	0787	0822	0856	0891	0926	21
39	0960	0995	1030	1064	1099	1134	20
40	301168	1203	1238	1273	1307	1342	19
41	1377	1411	1446	1481	1515	1550	18
42	1585	1619	1654	1689	1724	1758	17
43	1793	1828	1862	1897	1932	1966	16
44	2001	2036	2071	2105	2140	2175	15
45	2210	2244	2279	2314	2348	2383	14
46	2418	2453	2487	2522	2557	2592	13
47	2626	2661	2696	2731	2765	2800	12
48	2835	2870	2904	2939	2974	3009	11
49	3043	3078	3113	3148	3183	3217	10
50	303252	3287	3322	3356	3391	3426	9
51	3461	3496	3530	3565	3600	3635	8
52	3670	3704	3739	3774	3809	3844	7
53	3878	3913	3948	3983	4018	4052	6
54	4087	4122	4157	4192	4226	4261	5
55	4296	4331	4366	4401	4435	4470	4
56	4505	4540	4575	4610	4644	4679	3
57	4714	4749	4784	4819	4854	4888	2
58	4923	4958	4993	5028	5063	5098	1
59	5132	5167	5202	5237	5272	5307	0
	60"	50"	40"	30"	20"	10"	M

Co-versed sine 44 degrees.

P. part	1"	2"	3"	4"	5"	6"	7"	8"	9"
	3	7	10	14	17	21	24	28	31

M	Versed sine 46 degrees. 0″	10″	20″	30″	40″	50″	
0	305342	5377	5411	5446	5481	5516	59
1	5551	5586	5621	5656	5690	5725	58
2	5760	5795	5830	5865	5900	5935	57
3	5970	6005	6039	6074	6109	6144	56
4	6179	6214	6249	6284	6319	6354	55
5	6389	6424	6458	6493	6528	6563	54
6	6598	6633	6668	6703	6738	6773	53
7	6808	6843	6878	6913	6948	6983	52
8	7017	7052	7087	7122	7157	7192	51
9	7227	7262	7297	7332	7367	7402	50
10	307437	7472	7507	7542	7577	7612	49
11	7647	7682	7717	7752	7787	7822	48
12	7857	7892	7927	7962	7997	8032	47
13	8067	8102	8137	8172	8207	8242	46
14	8277	8312	8347	8382	8417	8452	45
15	8487	8522	8557	8592	8627	8662	44
16	8697	8732	8767	8802	8837	8872	43
17	8907	8942	8977	9012	9047	9083	42
18	9118	9153	9188	9223	9258	9293	41
19	9328	9363	9398	9433	9468	9503	40
20	309538	9573	9608	9644	9679	9714	39
21	9749	9784	9819	9854	9889	9924	38
22	9959	9994	0029	0064	0100	0135	37
23	310170	0205	0240	0275	0310	0345	36
24	0380	0416	0451	0486	0521	0556	35
25	0591	[illegible]	0661	0697	0732	0767	34
26	0802	0837	0872	0907	0942	0978	33
27	1013	1048	1083	1118	1153	1188	32
28	1224	1259	1294	1329	1364	1399	31
29	1434	1470	1505	1540	1575	1610	30
30	311645	1681	1716	1751	1786	1821	29
31	1856	1892	1927	1962	1997	2032	28
32	2068	2103	2138	2173	2208	2243	27
33	2279	2314	2349	2384	2419	2455	26
34	2490	2525	2560	2596	2631	2666	25
35	2701	2736	2772	2807	2842	2877	24
36	2912	2948	2983	3019	3053	3089	23
37	3124	3159	3194	3230	3265	3350	22
38	3335	3371	3406	3441	3476	3512	21
39	3547	3582	3617	3653	3688	3723	20
40	313758	3794	3829	3864	3899	3935	19
41	3970	4005	4041	4076	4111	4146	18
42	4182	4217	4252	4288	4323	4358	17
43	4393	4429	4464	4499	4535	4570	16
44	4605	4640	4676	4711	4746	4782	15
45	4817	4852	4888	4923	4958	4994	14
46	5029	5064	5100	5135	5170	5206	13
47	5241	5276	5312	5347	5382	5418	12
48	5453	5488	5524	5559	5594	5630	11
49	5665	5700	5736	5771	5806	5842	10
50	315877	5912	5948	5983	6019	6054	9
51	6089	6125	6160	6195	6231	6266	8
52	6302	6337	6372	6408	6443	6478	7
53	6514	6549	6585	6620	6655	6691	6

M.	Versed sine 47 degrees. 0″	10″	20″	30″	40″	50″	
0	318002	8037	8073	8108	8143	8179	59
1	8214	8250	8285	8321	8356	8392	58
2	8427	8463	8498	8534	8569	8605	57
3	8640	8676	8711	8747	8782	8818	56
4	8853	8889	8924	8960	8995	9031	55
5	9066	9102	9137	9173	9208	9244	54
6	9279	9315	9350	9386	9421	9457	53
7	9492	9528	9563	9599	9634	9670	52
8	9705	9741	9776	9812	9848	9883	51
9	9919	9954	9990	0025	0061	0096	50
10	320132	0177	0203	0239	0274	0310	49
11	0345	0381	0416	0452	0488	0523	48
12	0559	0594	0630	0665	0701	0737	47
13	0772	0808	0843	0879	0914	0950	46
14	0986	1021	1057	1092	1128	1164	45
15	1199	1235	1270	1306	1342	1377	44
16	1413	1448	1484	1520	1555	1591	43
17	1627	1662	1698	1733	1769	1805	42
18	1840	1876	1912	1947	1983	2018	41
19	2054	2090	2125	2161	2197	2232	40
20	322268	2304	2339	2375	2411	2446	39
21	2482	2518	2553	2589	2624	2660	38
22	2696	2731	2767	2803	2839	2874	37
23	2910	2946	2981	3017	3053	3088	36
24	3124	3160	3195	3231	3267	3302	35
25	3338	3374	3410	3445	3481	3517	34
26	3552	3588	3624	3660	3695	3731	33
27	3767	3802	3838	3874	3910	3945	32
28	3981	4017	4052	4088	4124	4160	31
29	4195	4231	4267	4313	4338	4374	30
30	324410	4446	4481	4517	4553	4589	29
31	4624	4660	4696	4732	4767	4803	28
32	4839	4875	4910	4946	4982	5018	27
33	5053	5089	5125	5161	5197	5232	26
34	5268	5304	5340	5375	5411	5447	25
35	5483	5519	5554	5590	5626	5662	24
36	5698	5733	5769	5805	5841	5877	23
37	5912	5948	5984	6020	6056	6092	22
38	6127	6163	6199	6235	6271	6306	21
39	6342	6378	6414	6450	6486	6521	20
40	326557	6593	6629	6645	6701	6737	19
41	6772	6808	6844	6880	6916	6952	18
42	6987	7023	7059	7095	7131	7167	17
43	7203	7239	7274	7310	7346	7382	16
44	7418	7414	7490	7526	7561	7597	15
45	7633	7669	7705	7741	7777	7813	14
46	7849	7884	7920	7956	7992	8028	13
47	8064	8100	8136	8172	8208	8243	12
48	8279	8315	8351	8387	8423	8459	11
49	8495	8531	8567	8603	8639	8675	10
50	328711	8746	8782	8818	8854	8890	9
51	8926	8962	8998	9034	9070	9106	8
52	9142	9178	9214	9250	9286	9322	7
53	9358	9394	9430	9465	9501	9537	6

M	Versed sine 48 degrees. 0″	10″	20″	30″	40″	50″	
0	330869	0905	0941	0977	1014	1050	59
1	1086	1122	1158	1194	1230	1266	58
2	1302	1338	1374	1410	1446	1482	57
3	1518	1554	1590	1626	1662	1698	56
4	1735	1771	1807	1843	1879	1915	55
5	1951	1987	2023	2059	2095	2131	54
6	2167	2204	2240	2276	2312	2348	53
7	2384	2420	2456	2492	2528	2564	52
8	2601	2637	2673	2709	2745	2781	51
9	2817	2853	2889	2926	2962	2998	50
10	333034	3070	3106	3142	3178	3215	49
11	3251	3287	3323	3359	3395	3431	48
12	3468	3504	3540	3576	3612	3648	47
13	3684	3721	3757	3793	3829	3865	46
14	3901	3938	3974	4010	4046	4082	45
15	4118	4155	4191	4227	4263	4299	44
16	4335	4372	4408	4444	4480	4516	43
17	4552	4589	4625	4661	4697	4733	42
18	4770	4806	4842	4878	4914	4951	41
19	4987	5023	5059	5095	5132	5168	40
20	335204	5240	5277	5313	5349	5385	39
21	5421	5458	5494	5530	5566	5603	38
22	5639	5675	5711	5748	5784	5820	37
23	5856	5893	5929	5965	6001	6038	36
24	6074	6110	6146	6183	6219	6255	35
25	6291	6328	6364	6400	6436	6473	34
26	6509	6545	6582	6618	6654	6690	33
27	6727	6763	6799	6835	6872	6908	32
28	6944	6981	7017	7053	7090	7126	31
29	7162	7198	7235	7271	7307	7344	30
30	337380	7416	7453	7489	7525	7562	29
31	7598	7634	7670	7707	7743	7779	28
32	7816	7852	7888	7925	7961	7997	27
33	8034	8070	8106	8143	8179	8216	26
34	8252	8288	8325	8361	8397	8434	25
35	8470	8506	8543	8579	8615	8652	24
36	8688	8725	8761	8797	8834	8870	23
37	8906	8943	8979	9015	9052	9088	22
38	9125	9161	9197	9234	9270	9307	21
39	9343	9379	9416	9452	9489	9525	20
40	339561	9598	9634	9671	9707	9743	19
41	9780	9816	9853	9889	9925	9962	18
42	340000	0035	0071	0108	0144	0180	17
43	0217	0253	0290	0326	0363	0399	16
44	0435	0472	0508	0545	0581	0618	15
45	0654	0691	0727	0764	0800	0836	14
46	0873	0909	0946	0982	1019	1055	13
47	1092	1128	1165	1201	1238	1274	12
48	1310	1347	1383	1420	1456	1493	11
49	1529	1566	1602	1639	1675	1712	10
50	341748	1785	1821	1858	1894	1931	9
51	1967	2004	2040	2077	2113	2150	8
52	2186	2223	2259	2296	2333	2369	7
53	2406	2442	2479	2515	2552	2588	6
54	2625	2661	2698	2734	2771	2807	5
55	2844	2881	2917	2954	2990	3027	4
56	3063	3100	3136	3173	3209	3246	3
57	3283	3319	3356	3392	3429	3465	2
58	3502	3539	3575	3612	3648	3685	1
59	3721	3758	3795	3831	3868	3904	0
	60″	50″	40″	30″	20″	10″	M
	Co-versed sine 41 degrees.						

P. part { 1″ 2″ 3″ 4″ 5″ 6″ 7″ 8″ 9″ / 4 7 11 15 18 22 25 29 33

M	Versed sine 49 degrees. 0″	10″	20″	30″	40″	50″	
0	343941	3978	4014	4051	4087	4124	59
1	4161	4197	4234	4270	4307	4344	58
2	4380	4417	4453	4490	4527	4563	57
3	4600	4636	4673	4710	4746	4783	56
4	4820	4856	4893	4929	4966	5003	55
5	5039	5076	5113	5149	5186	5223	54
6	5259	5296	5332	5369	5406	5442	53
7	5479	5516	5552	5589	5626	5662	52
8	5699	5736	5772	5809	5846	5882	51
9	5919	5956	5992	6029	6066	6102	50
10	346139	6176	6212	6249	6286	6323	49
11	6359	6396	6433	6469	6506	6543	48
12	6579	6616	6653	6690	6726	6763	47
13	6800	6836	6873	6910	6946	6983	46
14	7020	7057	7093	7130	7167	7204	45
15	7240	7277	7314	7350	7387	7424	44
16	7461	7497	7534	7571	7608	7644	43
17	7681	7718	7755	7791	7828	7865	42
18	7902	7938	7975	8012	8049	8085	41
19	8122	8159	8196	8232	8269	8306	40
20	348343	8380	8416	8453	8490	8527	39
21	8563	8600	8637	8674	8711	8747	38
22	8784	8821	8858	8895	8931	8968	37
23	9005	9042	9079	9115	9152	9189	36
24	9226	9263	9299	9336	9373	9410	35
25	9447	9483	9520	9557	9594	9631	34
26	9668	9704	9741	9778	9815	9852	33
27	9889	9925	9962	9999	0036	0073	32
28	350110	0147	0183	0220	0257	0294	31
29	0331	0368	0405	0441	0478	0515	30
30	350552	0589	0626	0663	0699	0736	29
31	0773	0810	0847	0884	0921	0958	28
32	0994	1031	1068	1105	1142	1179	27
33	1216	1253	1290	1326	1363	1400	26
34	1437	1474	1511	1548	1585	1622	25
35	1659	1696	1732	1769	1806	1843	24
36	1880	1917	1954	1991	2028	2065	23
37	2102	2139	2176	2212	2249	2286	22
38	2323	2360	2397	2434	2471	2508	21
39	2545	2582	2619	2656	2693	2730	20
40	352767	2804	2841	2878	2914	2951	19
41	2988	3025	3062	3099	3136	3173	18
42	3210	3247	3284	3321	3358	3395	17
43	3432	3469	3506	3543	3580	3617	16
44	3654	3691	3728	3765	3802	3839	15
45	3876	3913	3950	3987	4024	4061	14
46	4098	4135	4172	4209	4246	4283	13
47	4320	4357	4394	4431	4468	4505	12
48	4542	4579	4616	4653	4690	4727	11
49	4764	4802	4839	4876	4913	4950	10
50	354987	5024	5061	5098	5135	5172	9
51	5209	5246	5283	5320	5357	5394	8
52	5431	5469	5506	5543	5580	5617	7
53	5654	5691	5728	5765	5802	5839	6
54	5876	5913	5951	5988	6025	6062	5
55	6099	6136	6173	6210	6247	6284	4
56	6321	6359	6396	6433	6470	6507	3
57	6544	6581	6618	6655	6693	6730	2
58	6767	6804	6841	6878	6915	6952	1
59	6989	7027	7064	7101	7138	7175	0
	60″	50″	40″	30″	20″	10″	M
	Co-versed sine 40 degrees.						

P. part { 1″ 2″ 3″ 4″ 5″ 6″ 7″ 8″ 9″ / 4 7 11 15 18 22 26 29 33

Versed sine 30 degrees.

M	0″	10″	20″	30″	40″	50″	
0	357212	7250	7287	7324	7361	7398	59
1	7435	7472	7510	7547	7584	7621	58
2	7658	7695	7732	7770	7807	7844	57
3	7881	7918	7955	7993	8030	8067	56
4	8104	8141	8179	8216	8253	8290	55
5	8327	8364	8402	8439	8476	8513	54
6	8550	8588	8625	8662	8699	8736	53
7	8774	8811	8848	8885	8922	8960	52
8	8997	9034	9071	9108	9146	9183	51
9	9220	9257	9295	9332	9369	9406	50
10	359443	9481	9518	9555	9592	9630	49
11	9667	9704	9741	9779	9816	9853	48
12	9890	9928	9965	0002	0039	0077	47
13	360114	0151	0188	0226	0263	0300	46
14	0337	0375	0412	0449	0486	0524	45
15	0561	0598	0636	0673	0710	0747	44
16	0785	0822	0859	0897	0934	0971	43
17	1008	1046	1083	1120	1158	1195	42
18	1232	1269	1307	1344	1381	1419	41
19	1456	1493	1531	1568	1605	1643	40
20	361680	1717	1755	1792	1829	1867	39
21	1904	1941	1979	2016	2053	2091	38
22	2128	2165	2203	2240	2277	2315	37
23	2352	2389	2427	2464	2501	2539	36
24	2576	2613	2651	2688	2725	2763	35
25	2800	2838	2875	2912	2950	2987	34
26	3024	3062	3099	3137	3174	3211	33
27	3249	3286	3323	3361	3398	3436	32
28	3473	3510	3548	3585	3623	3660	31
29	3697	3735	3772	3810	3847	3884	30
30	363922	3959	3997	4034	4071	4109	29
31	4146	4184	4221	4259	4296	4333	28
32	4371	4408	4446	4483	4521	4558	27
33	4595	4633	4670	4708	4745	4783	26
34	4820	4857	4895	4932	4970	5007	25
35	5045	5082	5120	5157	5195	5232	24
36	5269	5307	5344	5382	5419	5457	23
37	5494	5532	5569	5607	5644	5682	22
38	5719	5757	5794	5832	5869	5907	21
39	5944	5982	6019	6057	6095	6132	20
40	366169	6207	6244	6282	6319	6357	19
41	6394	6432	6469	6507	6544	6582	18
42	6619	6657	6694	6732	6769	6807	17
43	6844	6882	6919	6957	6994	7032	16
44	7069	7107	7145	7182	7220	7257	15
45	7295	7332	7370	7407	7445	7482	14
46	7520	7558	7595	7633	7670	7708	13
47	7745	7783	7820	7858	7896	7933	12
48	7971	8008	8046	8084	8121	8159	11
49	8196	8234	8271	8309	8346	8384	10
50	368422	8459	8497	8534	8572	8610	9
51	8647	8685	8722	8760	8798	8835	8
52	8873	8910	8948	8986	9023	9061	7
53	9098	9136	9174	9211	9249	9287	6
54	9324	9362	9399	9437	9475	9512	5
55	9550	9588	9625	9663	9701	9738	4
56	9776	9813	9851	9889	9926	9964	3
57	370002	0039	0077	0115	0152	0190	2
58	0228	0265	0303	0341	0378	0416	1
59	0454	0491	0529	0567	0604	0642	0
	60″	50″	40″	30″	20″	10″	M

Co-versed sine 59 degrees.

P. part	1″	2″	3″	4″	5″	6″	7″	8″	9″
	4	7	11	15	19	22	26	30	34

Versed sine 31 degrees.

M	0″	10″	20″	30″	40″	50″	
0	370680	0718	0755	0793	0830	0868	59
1	0906	0944	0981	1019	1056	1094	58
2	1132	1170	1207	1245	1283	1320	57
3	1358	1396	1433	1471	1509	1547	56
4	1584	1622	1660	1697	1735	1773	55
5	1811	1848	1886	1924	1961	1999	54
6	2037	2075	2112	2150	2188	2226	53
7	2263	2301	2339	2377	2414	2452	52
8	2490	2528	2565	2603	2641	2679	51
9	2716	2754	2792	2830	2867	2905	50
10	372943	2981	3018	3056	3094	3132	49
11	3170	3207	3245	3283	3321	3358	48
12	3396	3434	3472	3510	3547	3585	47
13	3623	3661	3699	3736	3774	3812	46
14	3850	3887	3925	3963	4001	4039	45
15	4077	4114	4152	4190	4228	4266	44
16	4303	4341	4379	4417	4455	4493	43
17	4530	4568	4606	4644	4682	4720	42
18	4757	4795	4833	4871	4909	4947	41
19	4984	5022	5060	5098	5136	5174	40
20	375211	5249	5287	5325	5363	5401	39
21	5439	5476	5514	5552	5590	5628	38
22	5666	5704	5742	5779	5817	5855	37
23	5893	5931	5969	6007	6045	6083	36
24	6120	6158	6196	6234	6272	6310	35
25	6348	6386	6424	6461	6499	6537	34
26	6575	6613	6651	6689	6727	6765	33
27	6803	6841	6878	6916	6954	6992	32
28	7030	7068	7106	7144	7182	7220	31
29	7258	7296	7334	7372	7409	7447	30
30	377485	7523	7561	7599	7637	7675	29
31	7713	7751	7789	7827	7865	7903	28
32	7941	7979	8017	8055	8093	8131	27
33	8169	8207	8244	8282	8320	8358	26
34	8396	8434	8472	8510	8548	8586	25
35	8624	8662	8700	8738	8776	8814	24
36	8852	8890	8928	8966	9004	9042	23
37	9080	9118	9156	9194	9232	9270	22
38	9308	9346	9384	9422	9460	9498	21
39	9536	9574	9612	9650	9688	9726	20
40	379765	9803	9841	9879	9917	9955	19
41	9993	0031	0069	0107	0145	0183	18
42	380221	0259	0297	0335	0373	0411	17
43	0449	0487	0525	0563	0601	0639	16
44	0678	0716	0754	0792	0830	0868	15
45	0906	0944	0982	1021	1058	1096	14
46	1135	1173	1211	1249	1287	1325	13
47	1363	1401	1439	1477	1515	1554	12
48	1592	1630	1668	1706	1744	1782	11
49	1820	1858	1896	1935	1973	2011	10
50	382049	2087	2125	2163	2201	2240	9
51	2278	2316	2354	2392	2430	2468	8
52	2506	2545	2583	2621	2659	2697	7
53	2735	2773	2812	2850	2888	2926	6
54	2964	3002	3040	3079	3117	3155	5
55	3193	3231	3269	3308	3346	3384	4
56	3422	3460	3498	3537	3575	3613	3
57	3651	3689	3727	3766	3804	3842	2
58	3880	3918	3957	3995	4033	4071	1
59	4109	4148	4186	4224	4262	4300	0
	60″	50″	40″	30″	20″	10″	M

Co-versed sine 58 degrees.

P. part	1″	2″	3″	4″	5″	6″	7″	8″	9″
	4	8	11	15	19	23	27	30	34

M	Versed sine 52 degrees. 0"	10"	20"	30"	40"	50"	
0	384339	4377	4415	4453	4491	4530	59
1	4568	4606	4644	4682	4721	4759	58
2	4797	4835	4874	4912	4950	4988	57
3	5026	5065	5103	5141	5179	5218	56
4	5256	5294	5332	5371	5409	5447	55
5	5485	5524	5562	5600	5638	5677	54
6	5715	5753	5792	5830	5868	5906	53
7	5944	5983	6021	6059	6097	6136	52
8	6174	6212	6251	6289	6327	6365	51
9	6404	6442	6480	6518	6557	6595	50
10	386633	6672	6710	6748	6787	6825	49
11	6863	6901	6940	6978	7016	7055	48
12	7093	7131	7170	7208	7246	7285	47
13	7323	7361	7399	7438	7476	7514	46
14	7553	7591	7629	7668	7706	7744	45
15	7783	7821	7859	7898	7936	7974	44
16	8013	8051	8089	8128	8166	8204	43
17	8243	8281	8320	8358	8396	8435	42
18	8473	8511	8550	8588	8626	8665	41
19	8703	8742	8780	8818	8857	8895	40
20	388933	8972	9010	9049	9087	9125	39
21	9164	9202	9240	9279	9317	9356	38
22	9394	9432	9471	9509	9548	9586	37
23	9624	9663	9701	9740	9778	9816	36
24	9855	9893	9932	9970	0008	0047	35
25	390085	0124	0162	0201	0239	0277	34
26	0316	0354	0393	0431	0470	0508	33
27	0546	0585	0623	0662	0700	0739	32
28	0777	0816	0854	0892	0931	0969	31
29	1008	1046	1085	1123	1162	1200	30
30	391239	1277	1315	1354	1392	1431	29
31	1469	1508	1546	1585	1623	1662	28
32	1700	1739	1777	1816	1854	1923	27
33	1931	1970	2008	2047	2085	2124	26
34	2162	2201	2239	2278	2316	2355	25
35	2393	2432	2470	2509	2547	2586	24
36	2624	2663	2701	2740	2778	2817	23
37	2855	2894	2932	2971	3009	3048	22
38	3086	3125	3164	3202	3241	3279	21
39	3318	3356	3395	3433	3472	3510	20
40	393549	3587	3626	3665	3703	3742	19
41	3780	3819	3857	3896	3934	3973	18
42	4012	4050	4089	4127	4166	4204	17
43	4243	4282	4320	4359	4397	4436	16
44	4474	4513	4552	4590	4629	4668	15
45	4706	4745	4783	4822	4860	4899	14
46	4938	4976	5015	5053	5092	5131	13
47	5169	5208	5246	5285	5324	5362	12
48	5401	5440	5478	5517	5555	5594	11
49	5633	5672	5710	5748	5787	5826	10
50	395864	5903	5942	5980	6019	6058	9
51	6096	6135	6174	6212	6251	6289	8
52	6328	6367	6405	6444	6483	6521	7
53	6560	6599	6637	6676	6715	6753	6
54	6792	6831	6869	6908	6947	6985	5
55	7024	7063	7101	7140	7179	7217	4
56	7256	7295	7334	7372	7411	7450	3
57	7488	7527	7566	7604	7643	7682	2
58	7720	7759	7798	7837	7875	7914	1
59	7953	7992	8030	8069	8108	8146	0
	60"	50"	40"	30"	20"	10"	
	Co-versed sine 37 degrees.						M

P. part	1"	2"	3"	4"	5"	6"	7"	8"	9"
	4	8	12	15	19	23	27	31	35

M	Versed sine 53 degrees. 0"	10"	20"	30"	40"	50"	
0	398185	8224	8262	8301	8340	8379	59
1	8417	8456	8495	8534	8572	8611	58
2	8650	8688	8727	8766	8805	8843	57
3	8882	8921	8960	8998	9037	9076	56
4	9115	9153	9192	9231	9270	9308	55
5	9347	9386	9425	9463	9502	9541	54
6	9580	9619	9657	9696	9735	9774	53
7	9812	9851	9890	9929	9968	0006	52
8	400045	0084	0123	0161	0200	0239	51
9	0278	0317	0355	0394	0433	0472	50
10	400511	0549	0588	0627	0666	0705	49
11	0744	0782	0821	0860	0899	0938	48
12	0976	1015	1054	1093	1132	1171	47
13	1209	1248	1287	1326	1365	1404	46
14	1442	1481	1520	1559	1598	1637	45
15	1675	1714	1753	1792	1831	1870	44
16	1908	1947	1986	2025	2064	2103	43
17	2142	2181	2219	2258	2297	2336	42
18	2375	2414	2453	2491	2530	2569	41
19	2608	2647	2686	2725	2764	2803	40
20	402841	2880	2919	2958	2997	3036	39
21	3075	3114	3153	3191	3230	3269	38
22	3308	3347	3386	3425	3464	3503	37
23	3542	3581	3619	3658	3697	3736	36
24	3775	3814	3853	3892	3931	3970	35
25	4009	4048	4087	4125	4164	4203	34
26	4242	4281	4320	4359	4398	4437	33
27	4476	4515	4554	4593	4632	4671	32
28	4710	4749	4788	4827	4865	4904	31
29	4943	4982	5021	5060	5099	5138	30
30	405177	5216	5255	5294	5333	5372	29
31	5411	5450	5489	5528	5567	5606	28
32	5645	5684	5723	5762	5801	5840	27
33	5879	5918	5957	5996	6035	6074	26
34	6113	6152	6191	6230	6269	6308	25
35	6347	6386	6425	6464	6503	6542	24
36	6581	6620	6659	6698	6737	6776	23
37	6815	6854	6893	6932	6971	7010	22
38	7049	7089	7128	7167	7206	7245	21
39	7284	7323	7362	7401	7440	7479	20
40	407518	7557	7596	7635	7674	7713	19
41	7752	7791	7831	7870	7909	7948	18
42	7987	8026	8065	8104	8143	8182	17
43	8221	8260	8299	8339	8378	8417	16
44	8456	8495	8534	8573	8612	8651	15
45	8690	8729	8769	8808	8847	8886	14
46	8925	8964	9003	9042	9081	9121	13
47	9160	9199	9238	9277	9316	9355	12
48	9394	9433	9473	9512	9551	9590	11
49	9629	9668	9707	9746	9786	9825	10
50	409864	9903	9942	9981	0020	0060	9
51	410099	0138	0177	0216	0255	0295	8
52	0334	0373	0412	0451	0490	0529	7
53	0569	0608	0647	0686	0725	0764	6
54	0804	0843	0882	0921	0960	1000	5
55	1039	1078	1117	1156	1195	1235	4
56	1274	1313	1352	1391	1431	1470	3
57	1509	1548	1587	1627	1666	1705	2
58	1744	1783	1823	1862	1901	1940	1
59	1979	2019	2058	2097	2136	2176	0
	60"	50"	40"	30"	20"	10"	
	Co-versed sine 36 degrees.						M

P. part	1"	2"	3"	4"	5"	6"	7"	8"	9"
	4	8	12	16	19	23	27	31	35

	Versed sine 54 degrees.						
M	0"	10"	20"	30"	40"	50"	
0	412215	2254	2293	2332	2372	2411	59
1	2450	2489	2529	2568	2607	2646	58
2	2686	2725	2764	2803	2842	2882	57
3	2921	2960	2999	3039	3078	3117	56
4	3156	3196	3235	3274	3314	3353	55
5	3392	3431	3471	3510	3549	3588	54
6	3628	3667	3706	3745	3785	3824	53
7	3863	3903	3942	3981	4020	4060	52
8	4099	4138	4178	4217	4256	4295	51
9	4335	4374	4413	4453	4492	4531	50
10	414571	4610	4649	4688	4728	4767	49
11	4806	4845	4885	4924	4964	5003	48
12	5042	5082	5121	5160	5200	5239	47
13	5278	5318	5357	5396	5436	5475	46
14	5514	5554	5593	5632	5672	5711	45
15	5750	5790	5829	5868	5908	5947	44
16	5986	6026	6065	6105	6144	6183	43
17	6223	6262	6301	6341	6380	6419	42
18	6459	6498	6538	6577	6616	6656	41
19	6695	6734	6774	6813	6853	6892	40
20	416931	6971	7010	7050	7089	7128	39
21	7168	7207	7246	7286	7325	7365	38
22	7404	7443	7483	7522	7562	7601	37
23	7641	7680	7719	7759	7798	7838	36
24	7877	7916	7956	7995	8035	8074	35
25	8114	8153	8192	8232	8271	8311	34
26	8350	8390	8429	8468	8508	8547	33
27	8587	8626	8666	8705	8745	8784	32
28	8824	8863	8902	8942	8981	9021	31
29	9060	9100	9139	9179	9218	9258	30
30	419297	9337	9376	9415	9455	9494	29
31	9534	9573	9613	9652	9692	9731	28
32	9771	9810	9850	9889	9929	9968	27
33	420008	0047	0087	0126	0166	0205	26
34	0245	0284	0324	0363	0403	0442	25
35	0482	0521	0561	0600	0640	0679	24
36	0719	0758	0798	0837	0877	0916	23
37	0956	0995	1035	1075	1114	1154	22
38	1193	1233	1272	1312	1351	1391	21
39	1430	1470	1509	1549	1589	1628	20
40	421668	1707	1747	1786	1826	1865	19
41	1905	1945	1984	2024	2063	2103	18
42	2142	2182	2222	2261	2301	2340	17
43	2380	2419	2459	2499	2538	2578	16
44	2617	2657	2696	2736	2776	2815	15
45	2855	2894	2934	2974	3013	3053	14
46	3092	3132	3172	3211	3251	3290	13
47	3330	3370	3409	3449	3488	3528	12
48	3568	3607	3647	3687	3726	3766	11
49	3805	3845	3885	3924	3964	4004	10
50	424043	4083	4122	4162	4202	4241	9
51	4281	4321	4360	4400	4440	4479	8
52	4519	4559	4598	4638	4677	4717	7
53	4757	4796	4836	4876	4915	4955	6
54	4995	5034	5074	5114	5153	5193	5
55	5233	5272	5312	5352	5391	5431	4
56	5471	5511	5550	5590	5630	5669	3
57	5709	5749	5788	5828	5868	5907	2
58	5947	5987	6026	6066	6106	6146	1
59	6185	6225	6265	6304	6344	6384	0
	60"	50"	40"	30"	20"	10"	
	Co-versed sine 35 degrees.						M

P. part { 1" 2" 3" 4" 5" 6" 7" 8" 9" / 4 8 12 16 20 24 28 32 35 }

	Versed sine 55 degrees.						
M	0"	10"	20"	30"	40"	50"	
0	426424	6463	6503	6543	6582	6622	59
1	6662	6702	6741	6781	6821	6860	58
2	6900	6940	6980	7019	7059	7099	57
3	7139	7178	7218	7258	7298	7337	56
4	7377	7417	7457	7496	7536	7576	55
5	7616	7655	7695	7735	7775	7814	54
6	7854	7894	7934	7973	8013	8053	53
7	8093	8132	8172	8212	8252	8292	52
8	8331	8371	8411	8451	8490	8530	51
9	8570	8610	8650	8689	8729	8769	50
10	428809	8849	8888	8928	8968	9008	49
11	9048	9087	9127	9167	9207	9247	48
12	9286	9326	9366	9406	9446	9486	47
13	9525	9565	9605	9645	9685	9724	46
14	9764	9804	9844	9884	9924	9963	45
15	430003	0043	0083	0123	0163	0202	44
16	0242	0282	0322	0362	0402	0441	43
17	0481	0521	0561	0601	0641	0681	42
18	0720	0760	0800	0840	0880	0920	41
19	0960	1000	1039	1079	1119	1159	40
20	431199	1239	1279	1319	1358	1398	39
21	1438	1478	1518	1558	1598	1638	38
22	1677	1717	1757	1797	1837	1877	37
23	1917	1957	1997	2037	2076	2116	36
24	2156	2196	2236	2276	2316	2356	35
25	2396	2436	2476	2515	2555	2595	34
26	2635	2675	2715	2755	2795	2835	33
27	2875	2915	2955	2995	3035	3074	32
28	3114	3154	3194	3234	3274	3314	31
29	3354	3394	3434	3474	3514	3554	30
30	433594	3634	3674	3714	3754	3794	29
31	3834	3873	3913	3953	3993	4033	28
32	4073	4113	4153	4193	4233	4273	27
33	4313	4353	4393	4433	4473	4513	26
34	4553	4593	4633	4673	4713	4753	25
35	4793	4833	4873	4913	4953	4993	24
36	5033	5073	5113	5153	5193	5233	23
37	5273	5313	5353	5393	5433	5473	22
38	5513	5553	5593	5633	5673	5713	21
39	5753	5793	5833	5873	5913	5953	20
40	435993	6033	6074	6114	6154	6194	19
41	6234	6274	6314	6354	6394	6434	18
42	6474	6514	6554	6594	6634	6674	17
43	6714	6754	6794	6834	6875	6915	16
44	6955	6995	7035	7075	7115	7155	15
45	7195	7235	7275	7315	7355	7396	14
46	7436	7476	7516	7556	7596	7636	13
47	7676	7716	7756	7796	7836	7877	12
48	7917	7957	7997	8037	8077	8117	11
49	8157	8197	8237	8278	8318	8358	10
50	438398	8438	8478	8518	8558	8598	9
51	8639	8679	8719	8759	8799	8839	8
52	8879	8919	8960	9000	9040	9080	7
53	9120	9160	9200	9241	9281	9321	6
54	9361	9401	9441	9481	9522	9562	5
55	9602	9642	9682	9722	9763	9803	4
56	9843	9883	9923	9963	0004	0044	3
57	440084	0124	0164	0204	0245	0285	2
58	0325	0365	0405	0445	0486	0526	1
59	0566	0606	0646	0687	0727	0767	0
	60"	50"	40"	30"	20"	10"	
	Co-versed sine 34 degrees.						M

P. part { 1" 2" 3" 4" 5" 6" 7" 8" 9" / 4 8 12 16 20 24 8 32 36 }

M	Versed sine 56 degrees. 0″	10″	20″	30″	40″	50″	
0	440807	0847	0887	0928	0968	1008	59
1	1048	1088	1129	1169	1209	1249	58
2	1290	1330	1370	1410	1450	1491	57
3	1531	1571	1611	1651	1692	1732	56
4	1772	1812	1853	1893	1933	1973	55
5	2013	2054	2094	2134	2174	2215	54
6	2255	2295	2335	2376	2416	2456	53
7	2496	2537	2577	2617	2657	2698	52
8	2738	2778	2818	2859	2899	2939	51
9	2979	3020	3060	3100	3140	3181	50
10	443221	3261	3301	3342	3382	3422	49
11	3463	3503	3543	3584	3624	3664	48
12	3704	3745	3785	3825	3866	3906	47
13	3946	3986	4027	4067	4107	4148	46
14	4188	4228	4269	4309	4349	4389	45
15	4430	4470	4510	4551	4591	4631	44
16	4671	4712	4752	4793	4833	4873	43
17	4914	4954	4994	5035	5075	5115	42
18	5156	5196	5236	5277	5317	5357	41
19	5398	5438	5478	5519	5559	5599	40
20	445640	5680	5720	5761	5801	5841	39
21	5882	5922	5963	6003	6043	6084	38
22	6124	6164	6205	6245	6285	6326	37
23	6366	6407	6447	6487	6528	6568	36
24	6608	6649	6689	6730	6770	6810	35
25	6851	6891	6932	6972	7012	7053	34
26	7093	7134	7174	7214	7255	7295	33
27	7336	7376	7416	7457	7497	7538	32
28	7578	7618	7659	7699	7740	7780	31
29	7820	7861	7901	7942	7982	8023	30
30	448063	8103	8144	8184	8225	8265	29
31	8306	8346	8386	8427	8467	8508	28
32	8548	8589	8629	8670	8710	8750	27
33	8791	8831	8872	8912	8953	8993	26
34	9034	9074	9115	9155	9196	9236	25
35	9276	9317	9357	9398	9438	9479	24
36	9519	9560	9600	9641	9681	9722	23
37	9762	9803	9843	9884	9924	9965	22
38	450005	0046	0086	0127	0167	0208	21
39	0248	0289	0329	0370	0410	0451	20
40	450491	0532	0572	0613	0653	0694	19
41	0734	0775	0815	0856	0896	0937	18
42	0977	1018	1058	1099	1139	1180	17
43	1220	1261	1301	1342	1382	1423	16
44	1464	1504	1545	1585	1626	1666	15
45	1707	1747	1788	1828	1869	1910	14
46	1950	1991	2031	2072	2112	2153	13
47	2193	2234	2275	2315	2356	2396	12
48	2437	2477	2518	2559	2599	2640	11
49	2680	2721	2761	2802	2843	2883	10
50	452924	2964	3005	3045	3086	3127	9
51	3167	3208	3248	3289	3330	3370	8
52	3411	3451	3492	3533	3573	3614	7
53	3654	3695	3736	3776	3817	3857	6
54	3898	3939	3979	4020	4061	4101	5
55	4142	4182	4223	4264	4304	4345	4
56	4385	4426	4467	4507	4548	4589	3
57	4629	4670	4711	4751	4792	4832	2
58	4873	4914	4954	4995	5036	5076	1
59	5117	5158	5198	5239	5280	5320	0
	60″	50″	40″	30″	20″	10″	
	Co-versed sine 33 degrees.						M

P. part { 1″ 2″ 3″ 4″ 5″ 6″ 7″ 8″ 9″ / 4 8 12 16 20 24 28 32 36 }

M	Versed sine 57 degrees. 0″	10″	20″	30″	40″	50″	
0	455361	5402	5443	5483	5524	5564	59
1	5605	5646	5686	5727	5768	5808	58
2	5849	5890	5930	5971	6012	6052	57
3	6093	6134	6174	6215	6256	6296	56
4	6337	6378	6419	6459	6500	6541	55
5	6581	6622	6663	6703	6744	6785	54
6	6826	6866	6907	6948	6988	7029	53
7	7070	7111	7151	7192	7233	7273	52
8	7314	7355	7396	7436	7477	7518	51
9	7558	7599	7640	7681	7721	7762	50
10	457803	7844	7884	7925	7966	8007	49
11	8047	8088	8129	8170	8210	8251	48
12	8292	8333	8373	8414	8455	8496	47
13	8536	8577	8618	8659	8699	8740	46
14	8781	8822	8862	8903	8944	8985	45
15	9026	9066	9107	9148	9189	9229	44
16	9270	9311	9352	9393	9433	9474	43
17	9515	9556	9596	9637	9678	9719	42
18	9760	9800	9841	9882	9923	9964	41
19	460004	0045	0086	0127	0168	0209	40
20	460249	0290	0331	0372	0413	0453	39
21	0494	0535	0576	0617	0658	0698	38
22	0739	0780	0821	0862	0903	0943	37
23	0984	1025	1066	1107	1148	1188	36
24	1229	1270	1311	1352	1393	1433	35
25	1474	1515	1556	1597	1638	1679	34
26	1719	1760	1801	1842	1883	1924	33
27	1965	2005	2046	2087	2128	2169	32
28	2210	2251	2292	2332	2373	2414	31
29	2455	2496	2537	2578	2619	2660	30
30	462700	2741	2782	2823	2864	2905	29
31	2946	2987	3028	3068	3109	3150	28
32	3191	3232	3273	3314	3355	3396	27
33	3437	3478	3518	3559	3600	3641	26
34	3682	3723	3764	3805	3846	3887	25
35	3928	3969	4009	4050	4091	4132	24
36	4173	4214	4255	4296	4337	4378	23
37	4419	4460	4501	4542	4583	4624	22
38	4665	4705	4746	4787	4828	4869	21
39	4910	4951	4992	5033	5074	5115	20
40	465156	5197	5238	5279	5320	5361	19
41	5402	5443	5484	5525	5566	5607	18
42	5648	5689	5730	5771	5812	5853	17
43	5894	5935	5976	6017	6058	6099	16
44	6139	6180	6221	6262	6303	6344	15
45	6385	6426	6467	6508	6550	6591	14
46	6632	6673	6714	6755	6796	6837	13
47	6878	6919	6960	7001	7042	7083	12
48	7124	7165	7206	7247	7288	7329	11
49	7370	7411	7452	7493	7534	7575	10
50	467616	7657	7698	7739	7780	7821	9
51	7862	7903	7944	7986	8027	8068	8
52	8109	8150	8191	8232	8273	8314	7
53	8355	8396	8437	8478	8519	8560	6
54	8601	8642	8684	8725	8766	8807	5
55	8848	8889	8930	8971	9012	9053	4
56	9094	9135	9177	9218	9259	9300	3
57	9341	9382	9423	9464	9505	9546	2
58	9587	9629	9670	9711	9752	9793	1
59	9834	9875	9916	9957	9999	0040	0
	60″	50″	40″	30″	20″	10″	
	Co-versed sine 32 degrees.						M

P. part { 1″ 2″ 3″ 4″ 5″ 6″ 7″ 8″ 9″ / 4 8 12 16 20 25 29 33 37 }

 TABLE XLI. Natural Versed Sines.

M	Versed sine 58 degrees. 0″	10″	20″	30″	40″	50″	
0	470081	0122	0163	0204	0245	0286	59
1	0327	0369	0410	0451	0492	0533	58
2	0574	0615	0656	0698	0739	0780	57
3	0821	0862	0903	0944	0986	1027	56
4	1068	1109	1150	1191	1232	1274	55
5	1315	1356	1397	1438	1479	1521	54
6	1562	1603	1644	1685	1726	1767	53
7	1809	1850	1891	1932	1973	2014	52
8	2056	2097	2138	2179	2220	2262	51
9	2303	2344	2385	2426	2467	2509	50
10	472550	2591	2632	2673	2715	2756	49
11	2797	2838	2879	2921	2962	3003	48
12	3044	3085	3127	3168	3209	3250	47
13	3291	3333	3374	3415	3456	3498	46
14	3539	3580	3621	3662	3704	3745	45
15	3786	3827	3869	3910	3951	3992	44
16	4033	4075	4116	4157	4198	4240	43
17	4281	4322	4363	4405	4446	4487	42
18	4528	4570	4611	4652	4693	4735	41
19	4776	4817	4858	4900	4941	4982	40
20	475023	5065	5106	5147	5188	5230	39
21	5271	5312	5354	5395	5436	5477	38
22	5519	5560	5601	5643	5684	5725	37
23	5766	5808	5849	5890	5932	5973	36
24	6014	6055	6097	6138	6179	6221	35
25	6262	6303	6344	6386	6427	6468	34
26	6510	6551	6592	6634	6675	6716	33
27	6758	6799	6840	6882	6923	6964	32
28	7005	7047	7088	7129	7171	7212	31
29	7253	7295	7336	7377	7419	7460	30
30	477501	7543	7584	7625	7667	7708	29
31	7749	7791	7832	7874	7915	7956	28
32	7998	8039	8080	8122	8163	8204	27
33	8246	8287	8328	8370	8411	8453	26
34	8494	8535	8577	8618	8659	8701	25
35	8742	8783	8825	8866	8908	8949	24
36	8990	9032	9073	9115	9156	9197	23
37	9239	9280	9321	9363	9404	9446	22
38	9487	9528	9570	9611	9653	9694	21
39	9735	9777	9818	9860	9901	9942	20
40	479984	0025	0067	0108	0150	0191	19
41	480232	0274	0315	0357	0398	0439	18
42	0481	0522	0564	0605	0647	0688	17
43	0729	0771	0812	0854	0895	0937	16
44	0978	1020	1061	1102	1144	1185	15
45	1227	1268	1310	1351	1393	1434	14
46	1475	1517	1558	1600	1641	1683	13
47	1724	1766	1807	1849	1890	1932	12
48	1973	2014	2056	2097	2139	2180	11
49	2222	2263	2305	2346	2388	2429	10
50	482471	2512	2554	2595	2637	2678	9
51	2720	2761	2803	2844	2886	2927	8
52	2969	3010	3052	3093	3135	3176	7
53	3218	3259	3301	3342	3384	3425	6
54	3467	3508	3550	3591	3633	3674	5
55	3716	3757	3799	3840	3882	3923	4
56	3965	4006	4048	4090	4131	4173	3
57	4214	4256	4297	4339	4380	4422	2
58	4463	4505	4546	4588	4630	4671	1
59	4713	4754	4796	4837	4879	4920	0
	60″	50″	40″	30″	20″	10″	
	Co-versed sine 31 degrees.						M

P. part { 1″ 2″ 3″ 4″ 5″ 6″ 7″ 8″ 9″ / 4 8 12 17 21 25 29 33 37

M	Versed sine 59 degrees. 0″	10″	20″	30″	40″	50″	
0	484962	5003	5045	5087	5128	5170	59
1	5211	5253	5294	5336	5378	5419	58
2	5461	5502	5544	5585	5627	5669	57
3	5710	5752	5793	5835	5876	5918	56
4	5960	6001	6043	6084	6126	6168	55
5	6209	6251	6292	6334	6376	6417	54
6	6459	6500	6542	6584	6625	6667	53
7	6708	6750	6792	6833	6875	6916	52
8	6958	7000	7042	7083	7125	7166	51
9	7208	7249	7291	7333	7374	7416	50
10	487457	7499	7541	7582	7624	7666	49
11	7707	7749	7791	7832	7874	7915	48
12	7957	7999	8040	8082	8124	8165	47
13	8207	8249	8290	8332	8374	8415	46
14	8457	8499	8540	8582	8624	8665	45
15	8707	8749	8790	8832	8874	8915	44
16	8957	8999	9040	9082	9124	9165	43
17	9207	9249	9290	9332	9374	9415	42
18	9457	9499	9540	9582	9624	9666	41
19	9707	9749	9791	9832	9874	9916	40
20	489957	9999	0041	0083	0124	0166	39
21	490208	0249	0291	0333	0374	0416	38
22	0458	0500	0541	0583	0625	0667	37
23	0708	0750	0792	0833	0875	0917	36
24	0959	1000	1042	1084	1126	1167	35
25	1209	1251	1292	1334	1376	1418	34
26	1459	1501	1543	1585	1626	1668	33
27	1710	1752	1793	1835	1877	1919	32
28	1960	2002	2044	2086	2127	2169	31
29	2211	2253	2295	2336	2378	2420	30
30	492462	2503	2545	2587	2629	2671	29
31	2712	2754	2796	2838	2879	2921	28
32	2963	3005	3047	3088	3130	3172	27
33	3214	3256	3297	3339	3381	3423	26
34	3465	3506	3548	3590	3632	3674	25
35	3715	3757	3799	3841	3883	3924	24
36	3966	4008	4050	4092	4134	4175	23
37	4217	4259	4301	4343	4384	4426	22
38	4468	4510	4552	4594	4635	4677	21
39	4719	4761	4803	4845	4886	4928	20
40	494970	5012	5054	5096	5138	5179	19
41	5221	5263	5305	5347	5389	5431	18
42	5472	5514	5556	5598	5640	5682	17
43	5724	5765	5807	5849	5891	5933	16
44	5975	6017	6059	6100	6142	6184	15
45	6226	6268	6310	6352	6394	6435	14
46	6477	6519	6561	6603	6645	6687	13
47	6729	6771	6812	6854	6896	6938	12
48	6980	7022	7064	7106	7148	7190	11
49	7231	7273	7315	7357	7399	7441	10
50	497483	7525	7567	7609	7651	7693	9
51	7734	7776	7818	7860	7902	7944	8
52	7986	8028	8070	8112	8154	8196	7
53	8238	8280	8321	8363	8405	8447	6
54	8489	8531	8573	8615	8657	8699	5
55	8741	8783	8825	8867	8909	8951	4
56	8993	9035	9077	9119	9161	9202	3
57	9244	9286	9328	9370	9412	9454	2
58	9496	9538	9580	9622	9664	9706	1
59	9748	9790	9832	9874	9916	9958	0
	60″	50″	40″	30″	20″	10″	
	Co-versed sine 30 degrees.						M

P. part { 1″ 2″ 3″ 4″ 5″ 6″ 7″ 8″ 9″ / 4 8 13 17 21 25 29 33 38

Versed sine 60 degrees.

M	0"	10"	20"	30"	40"	50"	
0	500000	0042	0084	0126	0168	0210	59
1	0252	0294	0336	0378	0420	0462	58
2	0504	0546	0588	0630	0672	0714	57
3	0756	0798	0840	0882	0924	0966	56
4	1008	1050	1092	1134	1176	1218	55
5	1260	1302	1344	1386	1428	1470	54
6	1512	1554	1596	1638	1680	1722	53
7	1764	1806	1849	1891	1933	1975	52
8	2017	2059	2101	2143	2185	2227	51
9	2269	2311	2353	2395	2437	2479	50
10	502521	2563	2605	2647	2690	2732	49
11	2774	2816	2858	2900	2942	2984	48
12	3026	3068	3110	3152	3194	3236	47
13	3278	3321	3363	3405	3447	3489	46
14	3531	3573	3615	3657	3699	3741	45
15	3783	3826	3868	3910	3952	3994	44
16	4036	4078	4120	4162	4204	4247	43
17	4289	4331	4373	4415	4457	4499	42
18	4541	4583	4626	4668	4710	4752	41
19	4794	4836	4878	4920	4963	5005	40
20	505047	5089	5131	5173	5215	5257	39
21	5300	5342	5384	5426	5468	5510	38
22	5552	5595	5637	5679	5721	5763	37
23	5805	5847	5890	5932	5974	6016	36
24	6058	6100	6142	6185	6227	6269	35
25	6311	6353	6395	6438	6480	6522	34
26	6564	6606	6648	6691	6733	6775	33
27	6818	6859	6901	6944	6986	7028	32
28	7070	7112	7155	7197	7239	7281	31
29	7323	7365	7408	7450	7492	7534	30
30	507576	7619	7661	7703	7745	7787	29
31	7830	7872	7914	7956	7998	8041	28
32	8083	8125	8167	8210	8252	8294	27
33	8336	8378	8421	8463	8505	8547	26
34	8589	8632	8674	8716	8758	8801	25
35	8843	8885	8927	8970	9012	9054	24
36	9096	9138	9181	9223	9265	9307	23
37	9350	9392	9434	9476	9519	9561	22
38	9603	9645	9688	9730	9772	9814	21
39	9857	9899	9941	9983	0026	0068	20
40	510110	0153	0195	0237	0279	0322	19
41	0364	0406	0448	0491	0533	0575	18
42	0618	0660	0702	0744	0787	0829	17
43	0871	0914	0956	0998	1040	1083	16
44	1125	1167	1210	1252	1294	1337	15
45	1379	1421	1463	1506	1548	1590	14
46	1633	1675	1717	1760	1802	1844	13
47	1886	1929	1971	2013	2056	2098	12
48	2140	2183	2225	2267	2310	2351	11
49	2394	2437	2479	2521	2564	2606	10
50	512648	2691	2733	2775	2818	2860	9
51	2902	2945	2987	3029	3072	3114	8
52	3156	3199	3241	3283	3326	3368	7
53	3410	3453	3495	3538	3580	3622	6
54	3665	3707	3749	3792	3834	3876	5
55	3919	3961	4004	4046	4088	4131	4
56	4173	4215	4258	4300	4343	4385	3
57	4427	4470	4512	4554	4597	4639	2
58	4682	4724	4766	4809	4851	4894	1
59	4936	4978	5021	5063	5106	5148	0
	60"	50"	40"	30"	20"	10"	M

Co-versed sine 29 degrees.

P. part	1"	2"	3"	4"	5"	6"	7"	8"	9"
	4	8	13	17	21	25	30	34	38

Versed sine 61 degrees.

M	0"	10"	20"	30"	40"	50"	
0	515190	5233	5275	5318	5360	5402	59
1	5445	5487	5530	5572	5614	5657	58
2	5699	5742	5784	5827	5869	5911	57
3	5954	5996	6039	6081	6124	6166	56
4	6208	6251	6293	6336	6378	6421	55
5	6463	6505	6548	6590	6633	6675	54
6	6718	6760	6803	6845	6887	6930	53
7	6972	7015	7057	7100	7142	7185	52
8	7227	7269	7312	7354	7397	7439	51
9	7482	7524	7567	7609	7652	7694	50
10	517737	7779	7822	7864	7906	7949	49
11	7991	8034	8076	8119	8161	8204	48
12	8246	8289	8331	8374	8416	8459	47
13	8501	8544	8586	8629	8671	8714	46
14	8756	8799	8841	8884	8926	8969	45
15	9011	9054	9096	9139	9181	9224	44
16	9266	9309	9351	9394	9436	9479	43
17	9521	9564	9606	9649	9691	9734	42
18	9777	9819	9862	9904	9947	9989	41
19	520032	0074	0117	0159	0202	0244	40
20	520287	0329	0372	0415	0457	0500	39
21	0542	0585	0627	0670	0712	0755	38
22	0797	0840	0883	0925	0968	1010	37
23	1053	1095	1138	1180	1223	1266	36
24	1308	1351	1393	1436	1478	1521	35
25	1564	1606	1649	1691	1734	1776	34
26	1819	1862	1904	1947	1989	2032	33
27	2075	2117	2160	2202	2245	2287	32
28	2330	2373	2415	2458	2500	2543	31
29	2586	2628	2671	2713	2756	2799	30
30	522841	2884	2926	2969	3012	3054	29
31	3097	3140	3182	3225	3267	3310	28
32	3353	3395	3438	3480	3523	3566	27
33	3608	3651	3694	3736	3779	3821	26
34	3864	3907	3949	3992	4035	4077	25
35	4120	4163	4205	4248	4291	4333	24
36	4376	4418	4461	4504	4546	4589	23
37	4632	4674	4717	4760	4802	4845	22
38	4888	4930	4973	5016	5058	5101	21
39	5144	5186	5229	5272	5314	5357	20
40	525400	5442	5485	5528	5570	5613	19
41	5656	5698	5741	5784	5826	5869	18
42	5912	5954	5997	6040	6083	6125	17
43	6168	6211	6253	6296	6339	6381	16
44	6424	6467	6510	6552	6595	6638	15
45	6680	6723	6766	6808	6851	6894	14
46	6937	6979	7022	7065	7107	7150	13
47	7193	7236	7278	7321	7364	7407	12
48	7449	7492	7535	7577	7620	7663	11
49	7706	7748	7791	7834	7877	7919	10
50	527962	8005	8048	8091	8133	8176	9
51	8218	8261	8304	8347	8389	8432	8
52	8475	8518	8561	8603	8646	8689	7
53	8732	8774	8817	8860	8903	8945	6
54	8988	9031	9074	9116	9159	9202	5
55	9245	9288	9330	9373	9416	9459	4
56	9502	9544	9587	9630	9673	9715	3
57	9758	9801	9844	9886	9929	9972	2
58	530015	0058	0100	0143	0186	0229	1
59	0271	0314	0357	0400	0443	0486	0
	60"	50"	40"	30"	20"	10"	M

Co-versed sine 28 degrees.

P. part	1"	2"	3"	4"	5"	6"	7"	8"	9"
	4	9	13	17	21	26	30	34	38

TABLE XLI. Natural Versed Sines.

M	0"	10"	20"	30"	40"	50"	
	Versed sine 62 degrees.						
0	530528	0571	0614	0657	0700	0742	59
1	0785	0828	0871	0914	0957	0999	58
2	1042	1085	1128	1171	1213	1256	57
3	1299	1342	1385	1428	1470	1513	56
4	1556	1599	1642	1685	1727	1770	55
5	1813	1856	1899	1942	1984	2027	54
6	2070	2113	2156	2199	2242	2284	53
7	2327	2370	2413	2456	2499	2542	52
8	2584	2627	2670	2713	2756	2799	51
9	2842	2884	2927	2970	3013	3056	50
10	533099	3142	3185	3227	3270	3313	49
11	3356	3399	3442	3485	3528	3570	48
12	3613	3656	3699	3742	3785	3828	47
13	3871	3914	3956	3999	4042	4085	46
14	4128	4171	4214	4257	4300	4343	45
15	4385	4428	4471	4514	4557	4600	44
16	4643	4686	4729	4772	4815	4858	43
17	4900	4943	4986	5029	5072	5115	42
18	5158	5201	5244	5287	5330	5373	41
19	5416	5458	5501	5544	5587	5630	40
20	535673	5716	5759	5802	5845	5888	39
21	5931	5974	6017	6060	6103	6156	38
22	6188	6231	6274	6317	6360	6403	37
23	6446	6489	6532	6575	6618	6661	36
24	6704	6747	6790	6833	6876	6919	35
25	6962	7005	7048	7091	7134	7177	34
26	7220	7263	7306	7349	7392	7435	33
27	7477	7520	7563	7606	7649	7692	32
28	7735	7778	7821	7864	7907	7950	31
29	7993	8036	8079	8122	8165	8208	30
30	538251	8294	8337	8380	8423	8466	29
31	8509	8552	8595	8638	8681	8724	28
32	8768	8811	8854	8897	8940	8983	27
33	9026	9069	9112	9155	9198	9241	26
34	9284	9327	9370	9413	9456	9499	25
35	9542	9585	9628	9671	9714	9757	24
36	9800	9843	9886	9929	9972	0015	23
37	540058	0102	0145	0188	0231	0274	22
38	0317	0360	0403	0446	0489	0532	21
39	0575	0618	0661	0704	0747	0790	20
40	540834	0877	0920	0963	1006	1049	19
41	1092	1135	1178	1221	1264	1307	18
42	1350	1394	1437	1480	1523	1566	17
43	1609	1652	1695	1738	1781	1824	16
44	1867	1911	1954	1997	2040	2083	15
45	2126	2169	2212	2255	2298	2342	14
46	2385	2428	2471	2514	2557	2600	13
47	2643	2686	2730	2773	2816	2859	12
48	2902	2945	2988	3031	3075	3118	11
49	3161	3204	3247	3290	3333	3376	10
50	543420	3473	3506	3549	3592	3635	9
51	3678	3722	3765	3808	3851	3894	8
52	3937	3980	4024	4067	4110	4153	7
53	4196	4239	4282	4326	4369	4412	6
54	4455	4498	4541	4585	4628	4671	5
55	4714	4757	4800	4844	4887	4930	4
56	4973	5016	5059	5103	5146	5189	3
57	5232	5275	5318	5362	5405	5448	2
58	5491	5534	5578	5621	5664	5707	1
59	5750	5794	5837	5880	5923	5966	0
	60"	50"	40"	30"	20"	10"	
	Co-versed sine 27 degrees.						M

P. part { 1" 2" 3" 4" 5" 6" 7" 8" 9" / 4 9 13 17 21 26 30 34 39 }

M	0"	10"	20"	30"	40"	50"	
	Versed sine 63 degrees.						
0	546010	6053	6096	6139	6182	6225	59
1	6269	6312	6355	6398	6442	6485	58
2	6528	6571	6614	6658	6701	6744	57
3	6787	6830	6874	6917	6960	7003	56
4	7047	7090	7133	7176	7219	7263	55
5	7306	7349	7392	7436	7479	7522	54
6	7565	7609	7652	7695	7738	7781	53
7	7825	7868	7911	7954	7998	8041	52
8	8084	8127	8171	8214	8257	8300	51
9	8344	8387	8430	8473	8517	8560	50
10	548603	8647	8690	8733	8776	8820	49
11	8863	8906	8949	8993	9036	9079	48
12	9122	9166	9209	9252	9296	9339	47
13	9382	9425	9469	9512	9555	9599	46
14	9642	9685	9728	9772	9815	9858	45
15	9902	9945	9988	0031	0075	0118	44
16	550161	0205	0248	0291	0335	0378	43
17	0421	0464	0508	0551	0594	0638	42
18	0681	0724	0768	0811	0854	0898	41
19	0941	0984	1028	1071	1114	1157	40
20	551201	1244	1287	1331	1374	1417	39
21	1461	1504	1547	1591	1634	1677	38
22	1721	1764	1807	1851	1894	1937	37
23	1981	2024	2068	2111	2154	2198	36
24	2241	2284	2328	2371	2414	2458	35
25	2501	2544	2588	2631	2674	2718	34
26	2761	2805	2848	2891	2935	2978	33
27	3021	3065	3108	3151	3195	3238	32
28	3282	3325	3368	3412	3455	3498	31
29	3542	3585	3629	3672	3715	3759	30
30	553802	3846	3889	3932	3976	4019	29
31	4063	4106	4149	4193	4236	4280	28
32	4323	4366	4410	4453	4497	4540	27
33	4583	4627	4670	4714	4757	4800	26
34	4844	4887	4931	4974	5017	5061	25
35	5104	5148	5191	5235	5278	5321	24
36	5365	5408	5452	5495	5539	5582	23
37	5625	5669	5712	5756	5799	5843	22
38	5856	5929	5973	6016	6060	6103	21
39	6147	6190	6234	6277	6320	6364	20
40	556407	6451	6494	6538	6581	6625	19
41	6668	6712	6755	6798	6842	6885	18
42	6929	6972	7016	7059	7103	7146	17
43	7190	7233	7277	7320	7363	7407	16
44	7450	7494	7537	7581	7624	7668	15
45	7711	7755	7798	7842	7885	7929	14
46	7972	8016	8059	8103	8146	8190	13
47	8233	8277	8320	8364	8407	8451	12
48	8494	8538	8581	8625	8668	8712	11
49	8755	8799	8842	8886	8929	8973	10
50	559016	9060	9103	9147	9190	9234	9
51	9277	9321	9364	9408	9451	9495	8
52	9538	9582	9626	9669	9713	9756	7
53	9800	9843	9887	9930	9974	0017	6
54	560061	0104	0148	0181	0235	0279	5
55	0322	0366	0409	0453	0496	0540	4
56	0583	0627	0670	0714	0758	0801	3
57	0845	0888	0932	0975	1019	1062	2
58	1106	1150	1193	1237	1280	1324	1
59	1367	1411	1455	1498	1542	1585	0
	60"	50"	40"	30"	20"	10"	
	Co-versed sine 26 degrees.						M

P. part { 1" 2" 3" 4" 5" 6" 7" 8" 9" / 4 9 13 17 22 26 30 35 39 }

M	Versed sine 64 degrees. 0"	10"	20"	30"	40"	50"	
0	561629	1673	1716	1760	1803	1847	59
1	1890	1934	1977	2021	2065	2108	58
2	2152	2195	2239	2283	2326	2370	57
3	2413	2457	2501	2544	2588	2631	56
4	2675	2719	2762	2806	2849	2893	55
5	2937	2980	3024	3067	3111	3155	54
6	3198	3242	3285	3329	3373	3416	53
7	3460	3504	3547	3591	3634	3678	52
8	3722	3765	3809	3853	3896	3940	51
9	3983	4027	4071	4114	4158	4202	50
10	564245	4289	4332	4376	4420	4463	49
11	4507	4551	4594	4638	4682	4725	48
12	4769	4813	4856	4900	4944	4987	47
13	5031	5074	5118	5162	5205	5249	46
14	5293	5336	5380	5424	5467	5511	45
15	5555	5598	5642	5686	5729	5773	44
16	5817	5860	5904	5948	5991	6035	43
17	6079	6123	6166	6210	6254	6297	42
18	6341	6385	6428	6472	6516	6559	41
19	6603	6647	6690	6734	6778	6822	40
20	566865	6909	6953	6996	7040	7084	39
21	7127	7171	7215	7259	7302	7346	38
22	7390	7433	7477	7521	7565	7608	37
23	7652	7696	7739	7783	7827	7871	36
24	7914	7958	8002	8045	8089	8133	35
25	8177	8220	8264	8308	8352	8395	34
26	8439	8483	8526	8570	8614	8658	33
27	8701	8745	8789	8833	8876	8920	32
28	8964	9008	9051	9095	9139	9183	31
29	9226	9270	9314	9358	9401	9445	30
30	569489	9533	9576	9620	9664	9708	29
31	9751	9795	9839	9883	9927	9970	28
32	570014	0058	0102	0145	0189	0233	27
33	0277	0320	0364	0408	0452	0496	26
34	0539	0583	0627	0671	0715	0758	25
35	0802	0846	0890	0933	0977	1021	24
36	1065	1109	1152	1196	1240	1284	23
37	1328	1371	1415	1459	1503	1547	22
38	1590	1634	1678	1722	1766	1810	21
39	1853	1897	1941	1985	2029	2072	20
40	572116	2160	2204	2248	2292	2335	19
41	2379	2423	2467	2511	2554	2598	18
42	2642	2686	2730	2774	2817	2861	17
43	2905	2949	2993	3037	3080	3124	16
44	3168	3212	3256	3300	3344	3387	15
45	3431	3475	3519	3563	3607	3651	14
46	3694	3738	3782	3826	3870	3914	13
47	3958	4001	4045	4089	4133	4177	12
48	4221	4265	4308	4352	4396	4440	11
49	4484	4528	4572	4616	4659	4703	10
50	574747	4791	4835	4879	4923	4967	9
51	5010	5054	5098	5142	5186	5230	8
52	5274	5318	5362	5405	5449	5493	7
53	5537	5581	5625	5669	5713	5757	6
54	5801	5844	5888	5932	5976	6020	5
55	6064	6108	6152	6196	6240	6284	4
56	6327	6371	6415	6459	6503	6547	3
57	6591	6635	6679	6723	6767	6811	2
58	6855	6898	6942	6986	7030	7074	1
59	7118	7162	7206	7250	7294	7338	0
	60"	50"	40"	30"	20"	10"	
	Co-versed sine 25 degrees.						M

P. part { 1" 2" 3" 4" 5" 6" 7" 8" 9" / 4 9 13 17 22 26 31 35 39

M	Versed sine 65 degrees. 0"	10"	20"	30"	40"	50"	
0	577382	7426	7470	7514	7558	7601	59
1	7645	7689	7733	7777	7821	7865	58
2	7909	7953	7997	8041	8085	8129	57
3	8173	8217	8261	8305	8349	8393	56
4	8437	8481	8524	8568	8612	8656	55
5	8700	8744	8788	8832	8876	8920	54
6	8964	9008	9052	9096	9140	9184	53
7	9228	9272	9316	9360	9404	9448	52
8	9492	9536	9580	9624	9668	9712	51
9	9756	9800	9844	9888	9932	9976	50
10	580020	0064	0108	0152	0196	0240	49
11	0284	0328	0372	0416	0460	0504	48
12	0548	0592	0636	0680	0724	0768	47
13	0812	0856	0900	0944	0988	1032	46
14	1076	1120	1164	1208	1252	1296	45
15	1340	1384	1428	1472	1516	1560	44
16	1604	1648	1693	1737	1781	1825	43
17	1869	1913	1957	2001	2045	2089	42
18	2133	2177	2221	2265	2309	2353	41
19	2397	2441	2485	2529	2573	2617	40
20	582662	2706	2750	2794	2838	2882	39
21	2926	2970	3014	3058	3102	3146	38
22	3190	3234	3278	3323	3367	3411	37
23	3455	3499	3543	3587	3631	3675	36
24	3719	3763	3807	3851	3896	3940	35
25	3984	4028	4072	4116	4160	4204	34
26	4248	4292	4336	4381	4425	4469	33
27	4513	4557	4601	4645	4689	4733	32
28	4777	4822	4866	4910	4954	4998	31
29	5042	5086	5130	5174	5219	5263	30
30	585307	5351	5395	5439	5483	5527	29
31	5571	5616	5660	5704	5748	5792	28
32	5836	5880	5924	5969	6013	6057	27
33	6101	6145	6189	6233	6278	6322	26
34	6366	6410	6454	6498	6542	6586	25
35	6631	6675	6719	6763	6807	6851	24
36	6896	6940	6984	7028	7072	7116	23
37	7160	7205	7249	7293	7337	7381	22
38	7425	7470	7514	7558	7602	7646	21
39	7690	7735	7779	7823	7867	7911	20
40	587955	8000	8044	8088	8132	8176	19
41	8221	8265	8309	8353	8397	8441	18
42	8486	8530	8574	8618	8662	8707	17
43	8751	8795	8839	8883	8928	8972	16
44	9016	9060	9104	9149	9193	9237	15
45	9281	9325	9370	9414	9458	9502	14
46	9546	9591	9635	9679	9723	9767	13
47	9812	9856	9900	9944	9989	0033	12
48	590077	0121	0165	0210	0254	0298	11
49	0342	0387	0431	0475	0519	0563	10
50	590608	0652	0696	0740	0785	0829	9
51	0873	0917	0962	1006	1050	1094	8
52	1139	1183	1227	1271	1316	1360	7
53	1404	1449	1493	1537	1581	1635	6
54	1670	1714	1758	1802	1847	1891	5
55	1935	1979	2024	2068	2112	2156	4
56	2201	2245	2289	2333	2378	2422	3
57	2466	2511	2555	2599	2643	2688	2
58	2732	2776	2821	2865	2909	2953	1
59	2998	3042	3086	3130	3175	3219	0
	60"	50"	40"	30"	20"	10"	
	Co-versed sine 24 degrees.						M

P. part { 1" 2" 3" 4" 5" 6" 7" 8" 9" / 4 9 13 18 22 26 31 35 40

TABLE XLI Natural Versed Sines.

M	Versed sine 66 degrees.						
M	0"	10"	20"	30"	40"	50"	
0	593263	3308	3352	3396	3441	3485	59
1	3529	3573	3618	3662	3706	3751	58
2	3795	3839	3884	3928	3972	4016	57
3	4061	4105	4149	4194	4238	4282	56
4	4327	4371	4415	4460	4504	4548	55
5	4592	4637	4681	4725	4770	4814	54
6	4858	4903	4947	4991	5036	5080	53
7	5124	5169	5213	5257	5302	5346	52
8	5390	5435	5479	5523	5568	5612	51
9	5656	5701	5745	5789	5834	5878	50
10	595922	5967	6011	6056	6100	6144	49
11	6189	6233	6277	6322	6366	6410	48
12	6455	6499	6543	6588	6632	6677	47
13	6721	6765	6810	6854	6898	6943	46
14	6987	7031	7076	7121	7165	7209	45
15	7253	7298	7342	7386	7431	7475	44
16	7520	7564	7608	7653	7697	7741	43
17	7786	7830	7875	7919	7963	8008	42
18	8052	8097	8141	8185	8230	8274	41
19	8319	8363	8407	8452	8496	8541	40
20	598585	8629	8674	8718	8763	8807	39
21	8851	8896	8940	8985	9029	9074	38
22	9118	9162	9207	9251	9296	9340	37
23	9384	9429	9473	9518	9562	9607	36
24	9651	9695	9740	9784	9829	9873	35
25	9918	9962	0006	0051	0095	0140	34
26	600184	0229	0273	0317	0362	0406	33
27	0451	0495	0540	0583	0629	0673	32
28	0717	0762	0806	0851	0895	0940	31
29	0983	1029	1073	1118	1162	1206	30
30	601251	1295	1340	1384	1429	1473	29
31	1518	1562	1607	1651	1696	1740	28
32	1785	1829	1873	1918	1962	2007	27
33	2051	2096	2140	2185	2229	2274	26
34	2318	2363	2407	2452	2496	2541	25
35	2585	2630	2674	2719	2763	2808	24
36	2852	2897	2941	2986	3030	3075	23
37	3119	3164	3208	3253	3297	3342	22
38	3386	3431	3475	3520	3564	3609	21
39	3653	3698	3742	3787	3831	3876	20
40	603920	3965	4009	4054	4098	4143	19
41	4187	4232	4276	4321	4365	4410	18
42	4454	4499	4544	4588	4633	4677	17
43	4722	4766	4811	4855	4900	4944	16
44	4989	5033	5078	5123	5167	5212	15
45	5256	5301	5345	5390	5434	5479	14
46	5523	5568	5613	5657	5702	5746	13
47	5791	5835	5880	5924	5969	6014	12
48	6058	6103	6147	6192	6236	6281	11
49	6325	6370	6415	6459	6504	6548	10
50	606593	6637	6682	6727	6771	6816	9
51	6860	6905	6949	6994	7039	7083	8
52	7128	7172	7217	7262	7306	7351	7
53	7395	7440	7485	7529	7574	7618	6
54	7663	7707	7752	7797	7841	7886	5
55	7930	7975	8020	8064	8109	8153	4
56	8198	8243	8287	8332	8377	8421	3
57	8466	8510	8555	8600	8644	8689	2
58	8733	8778	8823	8867	8912	8957	1
59	9001	9046	9090	9135	9180	9224	0
	60"	50"	40"	30"	20"	10"	
	Co-versed sine 23 degrees.						M

P. part { 1" 2" 3" 4" 5" 6" 7" 8" 9" / 4 9 13 18 22 27 31 36 40

M	Versed sine 67 degrees.						
M	0"	10"	20"	30"	40"	50"	
0	609269	9313	9358	9403	9447	9492	59
1	9537	9581	9626	9671	9715	9760	58
2	9804	9849	9894	9938	9983	0028	57
3	610072	0117	0161	0206	0251	0296	56
4	0340	0385	0429	0474	0519	0563	55
5	0608	0653	0697	0742	0787	0831	54
6	0876	0921	0965	1010	1055	1099	53
7	1144	1189	1233	1278	1323	1367	52
8	1412	1457	1501	1546	1591	1635	51
9	1680	1725	1769	1814	1859	1903	50
10	611948	1993	2038	2082	2127	2172	49
11	2216	2261	2306	2350	2395	2440	48
12	2484	2529	2574	2618	2663	2708	47
13	2753	2797	2842	2887	2931	2976	46
14	3021	3066	3110	3155	3200	3244	45
15	3289	3334	3378	3423	3468	3513	44
16	3557	3602	3647	3691	3736	3781	43
17	3826	3870	3915	3960	4005	4049	42
18	4094	4139	4183	4228	4273	4318	41
19	4362	4407	4452	4497	4541	4586	40
20	614631	4675	4720	4765	4810	4854	39
21	4899	4944	4989	5033	5078	5123	38
22	5168	5212	5257	5302	5347	5391	37
23	5436	5481	5526	5570	5615	5660	36
24	5705	5749	5794	5839	5884	5928	35
25	5973	6018	6063	6108	6152	6197	34
26	6242	6287	6331	6376	6421	6466	33
27	6510	6555	6600	6645	6690	6734	32
28	6779	6824	6869	6913	6958	7003	31
29	7048	7093	7137	7182	7227	7272	30
30	617317	7361	7406	7451	7496	7541	29
31	7585	7630	7675	7720	7765	7809	28
32	7854	7899	7944	7989	8033	8078	27
33	8123	8168	8213	8257	8302	8347	26
34	8392	8437	8482	8526	8571	8616	25
35	8661	8706	8750	8795	8840	8885	24
36	8930	8974	9019	9064	9109	9154	23
37	9199	9243	9288	9333	9378	9423	22
38	9468	9512	9557	9602	9647	9692	21
39	9737	9781	9826	9871	9916	9961	20
40	620006	0050	0095	0140	0185	0230	19
41	0275	0320	0364	0409	0454	0499	18
42	0544	0589	0634	0678	0723	0768	17
43	0813	0858	0903	0948	0992	1037	16
44	1082	1127	1172	1217	1262	1307	15
45	1351	1396	1441	1486	1531	1576	14
46	1621	1666	1710	1755	1800	1845	13
47	1890	1935	1980	2025	2069	2114	12
48	2159	2204	2249	2294	2339	2384	11
49	2429	2473	2518	2563	2608	2653	10
50	622698	2743	2788	2833	2878	2922	9
51	2968	3012	3057	3102	3147	3192	8
52	3237	3282	3327	3371	3416	3461	7
53	3506	3551	3596	3641	3686	3731	6
54	3776	3821	3866	3910	3955	4000	5
55	4045	4090	4135	4180	4225	4270	4
56	4315	4360	4405	4450	4495	4539	3
57	4584	4629	4674	4719	4764	4809	2
58	4854	4899	4944	4989	5034	5079	1
59	5124	5169	5214	5259	5304	5348	0
	60"	50"	40"	30"	20"	10"	
	Co-versed sine 22 degrees.						M

P. part { 1" 2" 3" 4" 5" 6" 7" 8" 9" / 4 9 13 18 22 27 31 36 40

TABLE XLI. Natural Versed Sines.

M	Versed sine 68 degrees. 0″	10″	20″	30″	40″	50″	
0	625393	5438	5483	5528	5573	5618	59
1	5663	5708	5753	5798	5843	5888	58
2	5933	5978	6023	6068	6113	6158	57
3	6203	6248	6293	6338	6383	6428	56
4	6472	6517	6562	6607	6652	6697	55
5	6742	6787	6832	6877	6922	6967	54
6	7012	7057	7102	7147	7192	7237	53
7	7282	7327	7372	7417	7462	7507	52
8	7552	7597	7642	7687	7732	7777	51
9	7822	7867	7912	7957	8002	8047	50
10	628092	8137	8182	8227	8272	8317	49
11	8362	8407	8452	8497	8542	8587	48
12	8632	8677	8722	8767	8812	8857	47
13	8902	8947	8992	9037	9082	9127	46
14	9172	9217	9262	9307	9352	9398	45
15	9443	9488	9533	9578	9623	9668	44
16	9713	9758	9803	9848	9893	9938	43
17	9983	0028	0073	0118	0163	0208	42
18	630253	0298	0343	0388	0433	0478	41
19	0524	0569	0614	0659	0704	0749	40
20	630794	0839	0884	0929	0974	1019	39
21	1064	1109	1154	1199	1244	1290	38
22	1335	1380	1425	1470	1515	1560	37
23	1605	1650	1695	1740	1785	1830	36
24	1875	1921	1966	2011	2056	2101	35
25	2146	2191	2236	2281	2326	2371	34
26	2416	2462	2507	2552	2597	2642	33
27	2687	2732	2777	2822	2867	2912	32
28	2958	3003	3048	3093	3138	3183	31
29	3228	3273	3318	3363	3409	3454	30
30	633499	3544	3589	3634	3679	3724	29
31	3769	3815	3860	3905	3950	3995	28
32	4040	4085	4130	4175	4221	4266	27
33	4311	4356	4401	4446	4491	4536	26
34	4582	4627	4672	4717	4762	4807	25
35	4852	4898	4943	4988	5033	5078	24
36	5123	5168	5213	5259	5304	5349	23
37	5394	5439	5484	5530	5575	5620	22
38	5665	5710	5755	5800	5846	5891	21
39	5936	5981	6026	6071	6116	6162	20
40	636207	6252	6297	6342	6387	6433	19
41	6478	6523	6568	6613	6658	6704	18
42	6749	6794	6839	6884	6929	6975	17
43	7020	7065	7110	7155	7201	7246	16
44	7291	7336	7381	7426	7472	7517	15
45	7562	7607	7652	7698	7743	7788	14
46	7833	7878	7923	7969	8014	8059	13
47	8104	8149	8195	8240	8285	8330	12
48	8375	8421	8466	8511	8556	8601	11
49	8647	8692	8737	8782	8827	8873	10
50	638918	8963	9008	9054	9099	9144	9
51	9189	9234	9280	9325	9370	9415	8
52	9460	9506	9551	9596	9641	9687	7
53	9732	9777	9822	9868	9913	9958	6
54	640003	0048	0094	0139	0184	0229	5
55	0275	0320	0365	0410	0456	0501	4
56	0546	0591	0637	0682	0727	0772	3
57	0817	0863	0908	0953	0998	1044	2
58	1089	1134	1179	1225	1270	1315	1
59	1360	1406	1451	1496	1542	1587	0
	60″	50″	40″	30″	20″	10″	
	Co-versed sine 21 degrees.						M

P. part	1″	2″	3″	4″	5″	6″	7″	8″	9″
	5	9	14	18	23	27	32	36	41

M	Versed sine 69 degrees. 0″	10″	20″	30″	40″	50″	
0	641632	1677	1723	1778	1813	1858	59
1	1904	1949	1994	2039	2085	2130	58
2	2175	2221	2266	2311	2356	2402	57
3	2447	2492	2537	2583	2628	2673	56
4	2719	2764	2809	2854	2900	2945	55
5	2990	3036	3081	3126	3171	3217	54
6	3262	3307	3353	3398	3443	3488	53
7	3534	3579	3624	3670	3715	3760	52
8	3806	3851	3896	3941	3987	4032	51
9	4077	4122	4168	4213	4259	4304	50
10	644349	4395	4440	4485	4530	4576	49
11	4621	4666	4712	4757	4802	4848	48
12	4893	4938	4984	5029	5074	5120	47
13	5165	5210	5256	5301	5346	5392	46
14	5437	5482	5528	5573	5618	5664	45
15	5709	5754	5800	5845	5890	5936	44
16	5981	6026	6072	6117	6162	6208	43
17	6253	6298	6344	6389	6434	6480	42
18	6525	6571	6616	6661	6707	6752	41
19	6797	6843	6888	6933	6979	7024	40
20	647069	7115	7160	7206	7251	7296	39
21	7342	7387	7433	7478	7523	7569	38
22	7614	7659	7705	7750	7795	7841	37
23	7886	7931	7977	8022	8068	8113	36
24	8159	8204	8249	8295	8340	8385	35
25	8431	8476	8522	8567	8612	8658	34
26	8703	8748	8794	8839	8885	8930	33
27	8975	9021	9066	9112	9157	9202	32
28	9248	9293	9339	9384	9429	9475	31
29	9520	9566	9611	9656	9701	9747	30
30	649793	9838	9883	9929	9974	0020	29
31	650065	0111	0156	0201	0247	0292	28
32	0338	0383	0428	0474	0519	0565	27
33	0610	0656	0701	0746	0792	0837	26
34	0883	0928	0974	1019	1064	1110	25
35	1155	1201	1246	1292	1337	1383	24
36	1428	1473	1519	1564	1610	1655	23
37	1701	1746	1792	1837	1882	1928	22
38	1973	2019	2064	2110	2155	2201	21
39	2246	2292	2337	2383	2428	2473	20
40	652519	2564	2610	2655	2701	2746	19
41	2792	2837	2882	2928	2973	3019	18
42	3064	3110	3155	3201	3246	3291	17
43	3337	3382	3428	3474	3519	3565	16
44	3610	3656	3701	3746	3792	3837	15
45	3883	3928	3974	4019	4065	4110	14
46	4156	4201	4247	4292	4338	4383	13
47	4429	4474	4520	4565	4611	4656	12
48	4702	4747	4793	4838	4884	4929	11
49	4975	5020	5066	5111	5157	5202	10
50	655248	5293	5339	5384	5430	5475	9
51	5521	5566	5612	5657	5703	5749	8
52	5794	5840	5885	5931	5976	6022	7
53	6067	6113	6158	6204	6249	6295	6
54	6340	6386	6431	6477	6522	6568	5
55	6613	6659	6705	6750	6796	6841	4
56	6887	6933	6978	7024	7069	[illegible]	3
57	7160	7205	7251	7297	7342	7388	2
58	7433	7479	7524	7570	7615	7661	1
59	7707	7752	7798	7843	7889	7934	0
	60″	50″	40″	30″	20″	10″	
	Co-versed sine 20 degrees.						M

P. part	1″	2″	3″	4″	5″	6″	7″	8″	9″
	5	9	14	18	23	27	32	36	41

TABLE XLI. Natural Versed Sines.

Versed sine 70 degrees.

M	0"	10"	20"	30"	40"	50"	
0	657980	8025	8071	8117	8162	8208	59
1	8253	8299	8344	8390	8435	8481	58
2	8527	8572	8618	8663	8709	8754	57
3	8800	8846	8891	8937	8982	9028	56
4	9073	9119	9165	9210	9256	9301	55
5	9347	9393	9438	9484	9529	9575	54
6	9620	9666	9712	9757	9803	9848	53
7	9894	9940	9985	0031	0076	0122	52
8	660168	0213	0259	0304	0350	0396	51
9	0441	0487	0532	0578	0624	0669	50
10	660715	0760	0806	0852	0897	0943	49
11	0988	1034	1080	1125	1171	1216	48
12	1262	1308	1353	1399	1445	1490	47
13	1536	1581	1627	1673	1718	1764	46
14	1810	1855	1901	1946	1992	2038	45
15	2083	2129	2175	2220	2266	2311	44
16	2357	2403	2448	2494	2540	2585	43
17	2631	2677	2722	2768	2813	2859	42
18	2905	2950	2996	3042	3087	3133	41
19	3179	3224	3270	3316	3361	3407	40
20	663453	3498	3544	3589	3635	3681	39
21	3726	3772	3818	3863	3909	3955	38
22	4000	4046	4092	4137	4183	4229	37
23	4274	4320	4366	4411	4457	4503	36
24	4548	4594	4640	4685	4731	4777	35
25	4822	4868	4914	4960	5005	5051	34
26	5097	5142	5188	5234	5279	5325	33
27	5371	5416	5462	5508	5553	5599	32
28	5645	5690	5736	5782	5828	5873	31
29	5919	5965	6010	6056	6102	6147	30
30	666193	6239	6285	6330	6376	6422	29
31	6467	6513	6559	6604	6650	6696	28
32	6742	6787	6833	6879	6924	6970	27
33	7016	7061	7107	7153	7199	7244	26
34	7290	7336	7381	7427	7473	7519	25
35	7564	7610	7656	7702	7747	7793	24
36	7839	7885	7930	7976	8022	8068	23
37	8113	8159	8205	8250	8296	8342	22
38	8388	8433	8479	8525	8571	8616	21
39	8662	8708	8754	8799	8845	8891	20
40	668937	8982	9028	9074	9120	9165	19
41	9211	9257	9303	9348	9394	9440	18
42	9486	9531	9577	9623	9669	9714	17
43	9760	9806	9852	9897	9943	9989	16
44	670035	0081	0126	0172	0218	0264	15
45	0309	0355	0401	0447	0492	0538	14
46	0584	0630	0676	0721	0767	0813	13
47	0859	0904	0950	0996	1042	1088	12
48	1133	1179	1225	1271	1317	1362	11
49	1408	1454	1500	1545	1591	1637	10
50	671683	1729	1774	1820	1866	1912	9
51	1958	2003	2049	2095	2141	2187	8
52	2232	2278	2324	2370	2416	2461	7
53	2507	2553	2599	2645	2690	2736	6
54	2782	2828	2874	2920	2965	3011	5
55	3057	3103	3149	3194	3240	3286	4
56	[illegible]	3378	3423	3469	3515	3561	3
57	[illegible]	3653	3698	3744	3790	3836	2
58	3882	3928	3973	4019	4065	4111	1
59	4157	4203	4248	4294	4340	4386	0
	60"	50"	40"	30"	20"	10"	M

Co-versed sine 19 degrees.

P. part	1"	2"	3"	4"	5"	6"	7"	8"	9"
	5	9	14	18	23	27	32	37	41

Versed sine 71 degrees.

M	0"	10"	20"	30"	40"	50"	
0	674432	4478	4524	4569	4615	4661	59
1	4707	4753	4799	4844	4890	4936	58
2	4982	5028	5074	5120	5165	5211	57
3	5257	5303	5349	5395	5441	5486	56
4	5532	5578	5624	5670	5716	5762	55
5	5807	5853	5899	5945	5991	6037	54
6	6083	6128	6174	6220	6266	6312	53
7	6358	6404	6450	6495	6541	6587	52
8	6633	6679	6725	6771	6817	6862	51
9	6908	6954	7000	7046	7092	7138	50
10	677184	7230	7275	7321	7367	7413	49
11	7459	7505	7551	7597	7643	7688	48
12	7734	7780	7826	7872	7918	7964	47
13	8010	8056	8101	8147	8193	8239	46
14	8285	8331	8377	8423	8469	8515	45
15	8561	8606	8652	8698	8744	8790	44
16	8836	8882	8928	8974	9020	9066	43
17	9111	9157	9203	9249	9295	9341	42
18	9387	9433	9479	9525	9571	9617	41
19	9663	9708	9754	9800	9846	9892	40
20	679938	9984	0030	0076	0122	0168	39
21	680214	0260	0306	0352	0397	0443	38
22	0489	0535	0581	0627	0673	0719	37
23	0765	0811	0857	0903	0949	0995	36
24	1041	1087	1133	1179	1224	1270	35
25	1316	1362	1408	1454	1500	1546	34
26	1592	1638	1684	1730	1776	1822	33
27	1868	1914	1960	2006	2052	2098	32
28	2144	2190	2236	2282	2328	2374	31
29	2420	2465	2511	2557	2603	2649	30
30	682695	2741	2787	2833	2879	2925	29
31	2971	3017	3063	3109	3155	3201	28
32	3247	3293	3339	3385	3431	3477	27
33	3523	3569	3615	3661	3707	3753	26
34	3799	3845	3891	3937	3983	4029	25
35	4075	4121	4167	4213	4259	4305	24
36	4351	4397	4443	4489	4535	4581	23
37	4627	4673	4719	4765	4811	4857	22
38	4903	4949	4995	5041	5087	5133	21
39	5179	5225	5271	5317	5363	5409	20
40	685455	5501	5547	5593	5639	5685	19
41	5731	5777	5823	5869	5915	5962	18
42	6008	6054	6100	6146	6192	6238	17
43	6284	6330	6376	6422	6468	6514	16
44	6560	6606	6652	6698	6744	6790	15
45	6836	6882	6928	6974	7020	7066	14
46	7112	7159	7205	7251	7297	7343	13
47	7389	7435	7481	7527	7573	7619	12
48	7665	7711	7757	7803	7849	7895	11
49	7941	7987	8034	8080	8126	8172	10
50	688218	8264	8310	8356	8402	8448	9
51	8494	8540	8586	8632	8678	8725	8
52	8771	8817	8863	8909	8955	9001	7
53	9047	9093	9139	9185	9231	9277	6
54	9324	9370	9416	9462	9508	9554	5
55	9600	9646	9692	9738	9784	9831	4
56	9877	9923	9969	0015	0061	0107	3
57	690153	0199	0245	0291	0338	0384	2
58	0430	0476	0522	0568	0614	0660	1
59	0706	0752	0799	0845	0891	0937	0
	60"	50"	40"	30"	20"	10"	M

Co-versed sine 18 degrees.

P. part	1"	2"	3"	4"	5"	6"	7"	8"	9"
	5	9	14	18	23	28	32	37	41

Verfed fine 72 degrees.

M	0"	10"	20"	30"	40"	50"	
0	690983	1029	1075	1121	1167	1214	59
1	1260	1306	1352	1398	1444	1490	58
2	1536	1582	1629	1675	1721	1767	57
3	1813	1859	1905	1951	1998	2044	56
4	2090	2136	2182	2228	2274	2320	55
5	2367	2413	2459	2505	2551	2597	54
6	2643	2690	2736	2782	2828	2874	53
7	2920	2966	3012	3059	3105	3151	52
8	3197	3243	3289	3335	3382	3428	51
9	3474	3520	3566	3612	3659	3705	50
10	693751	3797	3843	3889	3935	3982	49
11	4028	4074	4120	4166	4212	4259	48
12	4305	4351	4397	4443	4489	4536	47
13	4582	4628	4674	4720	4766	4813	46
14	4859	4905	4951	4997	5043	5090	45
15	5136	5182	5228	5274	5320	5367	44
16	5413	5459	5505	5551	5597	5644	43
17	5690	5736	5782	5828	5875	5921	42
18	5967	6013	6059	6106	6152	6198	41
19	6244	6290	6336	6383	6429	6475	40
20	696521	6567	6614	6660	6706	6752	39
21	6798	6845	6891	6937	6983	7029	38
22	7076	7122	7168	7214	7260	7307	37
23	7353	7399	7445	7491	7538	7584	36
24	7630	7676	7723	7769	7815	7861	35
25	7907	7954	8000	8046	8092	8138	34
26	8185	8231	8277	8323	8370	8416	33
27	8462	8508	8554	8601	8647	8693	32
28	8739	8786	8832	8878	8924	8971	31
29	9017	9063	9109	9155	9202	9248	30
30	699294	9340	9387	9433	9479	9525	29
31	9572	9618	9664	9710	9757	9803	28
32	9849	9895	9942	9988	0034	0080	27
33	700127	0173	0219	0265	0312	0358	26
34	0404	0450	0497	0543	0589	0635	25
35	0682	0728	0774	0820	0867	0913	24
36	0959	1005	1052	1098	1144	1191	23
37	1237	1283	1329	1376	1422	1468	22
38	1514	1561	1607	1653	1700	1746	21
39	1792	1838	1885	1931	1977	2023	20
40	702070	2116	2162	2209	2255	2301	19
41	2347	2394	2440	2486	2533	2579	18
42	2625	2671	2718	2764	2810	2857	17
43	2903	2949	2995	3042	3088	3134	16
44	3181	3227	3273	3320	3366	3412	15
45	3458	3505	3551	3597	3644	3690	14
46	3736	3783	3829	3875	3921	3968	13
47	4014	4060	4107	4153	4199	4246	12
48	4292	4338	4385	4431	4477	4524	11
49	4570	4616	4662	4709	4755	4801	10
50	704848	4894	4940	4987	5033	5079	9
51	5126	5172	5218	5265	5311	5357	8
52	5404	5450	5496	5543	5589	5635	7
53	5682	5728	5774	5821	5867	5913	6
54	5960	6006	6052	6099	6145	6191	5
55	6238	6284	6330	6377	6423	6469	4
56	6516	6562	6608	6655	6701	6748	3
57	6794	6840	6887	6933	6979	7026	2
58	7072	7118	7165	7211	7257	7304	1
59	7350	7396	7443	7489	7536	7582	0
	60"	50"	40"	30"	20"	10"	M

Co-verfed fine 17 degrees.

P. part	1"	2"	3"	4"	5"	6"	7"	8"	9"
	5	9	14	18	23	28	32	37	42

Verfed fine 73 degrees.

M	0"	10"	20"	30"	40"	50"	
0	707628	7675	7721	7767	7814	7860	59
1	7906	7953	7999	8046	8092	8138	58
2	8185	8231	8277	8324	8370	8417	57
3	8463	8509	8556	8602	8648	8695	56
4	8741	8788	8834	8880	8927	8973	55
5	9019	9066	9112	9159	9205	9251	54
6	9298	9344	9390	9437	9483	9530	53
7	9576	9623	9669	9715	9762	9808	52
8	9855	9901	9947	9994	0040	0086	51
9	710133	0179	0226	0272	0318	0365	50
10	710411	0458	0504	0551	0597	0643	49
11	0690	0736	0783	0829	0875	0922	48
12	0968	1015	1061	1107	1154	1200	47
13	1247	1293	1340	1386	1432	1479	46
14	1525	1572	1618	1664	1711	1757	45
15	1804	1850	1897	1943	1989	2036	44
16	2082	2129	2175	2222	2268	2314	43
17	2361	2407	2454	2500	2547	2593	42
18	2639	2686	2732	2778	2825	2872	41
19	2918	2965	3011	3057	3104	3150	40
20	713197	3243	3290	3336	3383	3429	39
21	3475	3522	3568	3615	3661	3708	38
22	3754	3801	3847	3894	3940	3986	37
23	4033	4079	4126	4172	4219	4265	36
24	4312	4358	4405	4451	4497	4544	35
25	4590	4637	4683	4730	4776	4823	34
26	4869	4916	4962	5009	5055	5102	33
27	5148	5195	5241	5287	5334	5380	32
28	5427	5473	5520	5566	5613	5659	31
29	5706	5752	5799	5845	5892	5938	30
30	715985	6031	6078	6124	6171	6217	29
31	6264	6310	6357	6403	6450	6496	28
32	6543	6589	6636	6682	6728	6775	27
33	6821	6868	6914	6961	7007	7054	26
34	7100	7147	7193	7240	7286	7333	25
35	7380	7426	7473	7519	7566	7612	24
36	7659	7705	7752	7798	7845	7891	23
37	7938	7984	8031	8077	8124	8170	22
38	8217	8263	8310	8356	8403	8449	21
39	8496	8542	8589	8635	8682	8728	20
40	718775	8821	8868	8915	8961	9008	19
41	9054	9101	9147	9194	9240	9287	18
42	9333	9380	9426	9473	9519	9566	17
43	9612	9659	9706	9752	9799	9845	16
44	9892	9938	9985	0031	0078	0124	15
45	720171	0218	0264	0311	0357	0404	14
46	0450	0497	0543	0590	0636	0683	13
47	0730	0776	0823	0869	0916	0962	12
48	1009	1055	1102	1149	1195	1242	11
49	1288	1335	1381	1428	1474	1521	10
50	721568	1614	1661	1707	1754	1800	9
51	1847	1894	1940	1987	2033	2080	8
52	2126	2173	2220	2266	2313	2359	7
53	2406	2452	2499	2546	2592	2639	6
54	2685	2732	2779	2825	2872	2918	5
55	2965	3011	3058	3105	3151	3198	4
56	3244	3291	3338	3384	3431	3477	3
57	3524	3570	3617	3664	3710	3757	2
58	3803	3850	3897	3943	3990	4036	1
59	4083	4130	4176	4223	4269	4316	0
	60"	50"	40"	30"	20"	10"	M

Co-verfed fine 16 degrees.

P. part	1"	2"	3"	4"	5"	6"	7"	8"	9"
	5	9	14	19	23	28	33	37	42

M	0"	10"	20"	30"	40"	50"	
	Versed sine 74 degrees.						
0	724363	4410	4456	4502	4549	4596	59
1	4642	4689	4735	4782	4829	4875	58
2	4922	4969	5015	5062	5108	5155	57
3	5202	5248	5295	5341	5388	5435	56
4	5481	5528	5575	5621	5668	5714	55
5	5761	5808	5854	5901	5948	5994	54
6	6041	6087	6134	6181	6227	6274	53
7	6321	6367	6414	6460	6507	6554	52
8	6600	6647	6694	6740	6787	6834	51
9	6880	6927	6973	7020	7067	7113	50
10	727160	7207	7253	7300	7347	7393	49
11	7440	7487	7533	7580	7626	7673	48
12	7720	7766	7813	7860	7906	7953	47
13	8000	8046	8093	8140	8186	8133	46
14	8280	8326	8373	8420	8466	8513	45
15	8560	8606	8653	8700	8746	8793	44
16	8840	8886	8933	8980	9026	9073	43
17	9120	9166	9213	9260	9306	9353	42
18	9400	9446	9493	9540	9586	9633	41
19	9680	9726	9773	9820	9866	9913	40
20	729960	0006	0053	0100	0146	0193	39
21	730240	0286	0333	0380	0427	0473	38
22	0520	0567	0613	0660	0707	0753	37
23	0800	0847	0893	0940	0987	1033	36
24	1080	1127	1174	1220	1267	1314	35
25	1360	1407	1454	1500	1547	1594	34
26	1641	1687	1734	1781	1827	1874	33
27	1921	1968	2014	2061	2108	2154	32
28	2201	2248	2294	2341	2388	2435	31
29	2481	2528	2575	2621	2668	2715	30
30	732762	2808	2855	2902	2949	2995	29
31	3042	3089	3135	3182	3229	3276	28
32	3322	3369	3416	3462	3509	3556	27
33	3603	3649	3696	3743	3790	3836	26
34	3883	3930	3977	4023	4070	4117	25
35	4163	4210	4257	4304	4350	4397	24
36	4444	4491	4537	4584	4631	4678	23
37	4724	4771	4818	4865	4911	4958	22
38	5005	5052	5098	5145	5192	5239	21
39	5285	5332	5379	5426	5472	5519	20
40	735566	5613	5659	5706	5753	5800	19
41	5846	5893	5940	5987	6033	6080	18
42	6127	6174	6220	6267	6314	6361	17
43	6407	6454	6501	6548	6595	6641	16
44	6688	6735	6782	6828	6875	6922	15
45	6969	7016	7062	7109	7156	7203	14
46	7249	7296	7343	7390	7437	7483	13
47	7530	7577	7624	7670	7717	7764	12
48	7811	7858	7904	7951	7998	8045	11
49	8092	8138	8185	8232	8279	8325	10
50	738372	8419	8466	8513	8559	8606	9
51	8653	8700	8747	8793	8840	8887	8
52	8934	8981	9027	9074	9121	9168	7
53	9215	9261	9308	9355	9402	9449	6
54	9495	9542	9589	9636	9683	9730	5
55	9776	9823	9870	9917	9964	0010	4
56	740057	0104	0151	0198	0244	0291	3
57	0338	0385	0432	0479	0525	0572	2
58	0619	0666	0713	0760	0806	0853	1
59	0900	0947	0994	1040	1087	1134	0
	60"	50"	40"	30"	20"	10"	
	Co-versed sine 15 degrees.						M

P. part { 1" 2" 3" 4" 5" 6" 7" 8" 9" / 5 9 14 19 23 28 33 37 42

M	0"	10"	20"	30"	40"	50"	
	Versed sine 75 degrees.						
0	741181	1228	1275	1321	1368	1415	59
1	1462	1509	1556	1602	1649	1696	58
2	1743	1790	1837	1883	1930	1977	57
3	2024	2071	2118	2165	2211	2258	56
4	2305	2352	2399	2446	2492	2539	55
5	2586	2633	2680	2727	2774	2820	54
6	2867	2914	2961	3008	3055	3101	53
7	3148	3195	3242	3289	3336	3383	52
8	3429	3466	3523	3570	3617	3664	51
9	3711	3757	3804	3851	3898	3945	50
10	743992	4039	4086	4132	4179	4226	49
11	4273	4320	4367	4414	4460	4507	48
12	4554	4601	4648	4695	4742	4789	47
13	4835	4882	4929	4976	5023	5070	46
14	5117	5164	5211	5257	5304	5351	45
15	5398	5445	5492	5539	5586	5632	44
16	5679	5726	5773	5820	5867	5914	43
17	5961	6008	6054	6101	6148	6195	42
18	6242	6289	6336	6383	6430	6477	41
19	6523	6570	6617	6664	6711	6758	40
20	746805	6852	6899	6946	6992	7039	39
21	7086	7133	7180	7227	7274	7321	38
22	7368	7414	7461	7508	7555	7602	37
23	7649	7696	7743	7790	7837	7884	36
24	7931	7978	8024	8071	8118	8165	35
25	8212	8259	8306	8353	8400	8447	34
26	8494	8541	8588	8634	8681	8728	33
27	8775	8822	8869	8916	8963	9010	32
28	9057	9104	9151	9198	9245	9291	31
29	9338	9385	9432	9479	9526	9573	30
30	749620	9667	9714	9761	9808	9855	29
31	9902	9949	9996	0042	0089	0136	28
32	750183	0230	0277	0324	0371	0418	27
33	0466	0512	0559	0606	0653	0700	26
34	0747	0794	0841	0888	0934	0981	25
35	1028	1075	1122	1169	1216	1263	24
36	1310	1357	1404	1451	1498	1545	23
37	1592	1639	1686	1733	1780	1827	22
38	1874	1921	1968	2015	2062	2108	21
39	2155	2202	2249	2296	2343	2390	20
40	752437	2484	2531	2578	2625	2672	19
41	2719	2766	2813	2860	2907	2954	18
42	3001	3048	3095	3142	3189	3236	17
43	3283	3330	3377	3424	3471	3518	16
44	3565	3612	3659	3706	3753	3800	15
45	3847	3894	3941	3988	4035	4082	14
46	4129	4176	4223	4270	4317	4364	13
47	4411	4458	4505	4552	4599	4646	12
48	4693	4740	4787	4834	4881	4928	11
49	4975	5022	5069	5116	5163	5210	10
50	755257	5304	5351	5398	5445	5492	9
51	5539	5586	5633	5680	5727	5774	8
52	5821	5868	5915	5962	6009	6056	7
53	6103	6150	6197	6244	6291	6338	6
54	6385	6432	6479	6526	6573	6620	5
55	6667	6714	6761	6808	6855	6902	4
56	6949	6996	7043	7090	7137	7184	3
57	7231	7278	7326	7373	7420	7467	2
58	7514	7561	7608	7655	7702	7749	1
59	7796	7843	7890	7937	7984	8031	0
	60"	50"	40"	30"	20"	10"	
	Co-versed sine 14 degrees.						M

P. part { 1" 2" 3" 4" 5" 6" 7" 8" 9" / 5 9 14 19 23 28 33 37 42

M	Versed sine 76 degrees. 0"	10"	20"	30"	40"	50"	
0	758078	8125	8172	8219	8266	8313	59
1	8360	8407	8454	8501	8549	8596	58
2	8643	8690	8737	8784	8831	8878	57
3	8925	8972	9019	9066	9113	9160	56
4	9207	9254	9301	9348	9395	9443	55
5	9490	9537	9584	9631	9678	9725	54
6	9772	9819	9866	9913	9960	0007	53
7	760054	0101	0148	0196	0243	0290	52
8	0337	0384	0431	0478	0525	0572	51
9	0619	0666	0713	0760	0807	0855	50
10	760902	0949	0996	1043	1090	1137	49
11	1184	1231	1278	1325	1372	1419	48
12	1467	1514	1561	1608	1655	1702	47
13	1749	1796	1843	1890	1937	1984	46
14	2032	2079	2126	2173	2220	2267	45
15	2314	2361	2408	2455	2502	2550	44
16	2597	2644	2691	2738	2785	2832	43
17	2880	2926	2973	3021	3068	3115	42
18	3162	3209	3256	3303	3350	3397	41
19	3444	3492	3539	3586	3633	3680	40
20	763727	3774	3821	3868	3916	3963	39
21	4010	4057	4104	4151	4198	4245	38
22	4292	4340	4387	4434	4481	4528	37
23	4575	4622	4669	4717	4764	4811	36
24	4858	4905	4952	4999	5046	5094	35
25	5141	5188	5235	5282	5329	5376	34
26	5423	5471	5518	5565	5612	5659	33
27	5706	5753	5800	5848	5895	5942	32
28	5989	6036	6083	6130	6178	6225	31
29	6272	6319	6366	6413	6460	6507	30
30	766555	6602	6649	6696	6743	6790	29
31	6837	6885	6932	6979	7026	7073	28
32	7120	7168	7215	7262	7309	7356	27
33	7403	7450	7498	7545	7592	7639	26
34	7686	7733	7781	7828	7875	7922	25
35	7969	8016	8063	8111	8158	8205	24
36	8252	8299	8346	8394	8441	8488	23
37	8535	8582	8629	8677	8724	8771	22
38	8818	8865	8912	8960	9007	9054	21
39	9101	9148	9195	9243	9290	9337	20
40	769384	9431	9478	9526	9573	9620	19
41	9667	9714	9762	9809	9856	9903	18
42	9950	9997	0045	0092	0139	0186	17
43	770233	0281	0328	0375	0422	0469	16
44	0516	0564	0611	0658	0705	0752	15
45	0800	0847	0894	0941	0988	1036	14
46	1083	1130	1177	1224	1272	1319	13
47	1366	1413	1460	1508	1555	1602	12
48	1649	1696	1744	1791	1838	1885	11
49	1932	1980	2027	2074	2121	2168	10
50	772216	2263	2310	2357	2404	2452	9
51	2499	2546	2593	2640	2688	2735	8
52	2782	2829	2877	2924	2971	3018	7
53	3065	3113	3160	3207	3254	3301	6
54	3349	3396	3443	3490	3538	3585	5
55	3632	3679	3726	3774	3821	3868	4
56	3915	3963	4010	4057	4104	4152	3
57	4199	4246	4293	4340	4388	4435	2
58	4482	4529	4577	4624	4671	4718	1
59	4766	4813	4860	4907	4954	5002	0
	60"	50"	40"	30"	20"	10"	
	Co-versed sine 13 degrees.						M

P. part	1"	2"	3"	4"	5"	6"	7"	8"	9"
	5	9	14	19	24	28	33	38	43

M	Versed sine 77 degrees. 0"	10"	20"	30"	40"	50"	
0	775049	5096	5143	5191	5238	5285	59
1	5332	5380	5427	5474	5521	5568	58
2	5616	5663	5710	5758	5805	5852	57
3	5899	5947	5994	6041	6088	6136	56
4	6183	6230	6277	6325	6372	6419	55
5	6466	6514	6561	6608	6655	6703	54
6	6750	6797	6844	6892	6939	6986	53
7	7033	7081	7128	7175	7222	7270	52
8	7317	7364	7412	7459	7506	7553	51
9	7601	7648	7695	7743	7790	7837	50
10	777884	7931	7979	8026	8073	8121	49
11	8168	8215	8262	8310	8357	8404	48
12	8452	8499	8546	8593	8641	8688	47
13	8735	8782	8830	8877	8924	8972	46
14	9019	9066	9113	9161	9208	9255	45
15	9303	9350	9397	9444	9492	9539	44
16	9587	9634	9681	9728	9775	9823	43
17	9870	9917	9965	0012	0059	0107	42
18	780154	0201	0248	0296	0343	0390	41
19	0438	0485	0532	0579	0627	0674	40
20	780721	0769	0816	0863	0911	0958	39
21	1005	1052	1100	1147	1194	1242	38
22	1289	1336	1384	1431	1478	1526	37
23	1573	1620	1668	1715	1762	1809	36
24	1857	1904	1951	1999	2046	2093	35
25	2141	2188	2235	2283	2330	2377	34
26	2425	2472	2519	2567	2614	2661	33
27	2708	2756	2803	2850	2898	2945	32
28	2992	3040	3087	3134	3182	3229	31
29	3276	3324	3371	3418	3466	3513	30
30	783560	3608	3655	3702	3749	3797	29
31	3844	3892	3939	3986	4034	4081	28
32	4128	4176	4223	4270	4318	4365	27
33	4412	4460	4507	4554	4602	4649	26
34	4697	4744	4791	4839	4886	4933	25
35	4981	5028	5075	5123	5170	5217	24
36	5265	5312	5359	5407	5454	5501	23
37	5549	5596	5643	5691	5738	5786	22
38	5833	5880	5928	5975	6022	6070	21
39	6117	6164	6212	6259	6307	6354	20
40	786401	6449	6496	6543	6591	6638	19
41	6685	6733	6780	6828	6875	6922	18
42	6970	7017	7064	7112	7159	7206	17
43	7254	7301	7349	7396	7443	7491	16
44	7538	7585	7633	7680	7728	7775	15
45	7822	7870	7917	7964	8012	8059	14
46	8107	8154	8201	8249	8296	8344	13
47	8391	8438	8486	8533	8580	8628	12
48	8675	8723	8770	8817	8865	8912	11
49	8960	9007	9054	9102	9149	9196	10
50	789244	9291	9339	9386	9433	9481	9
51	9528	9576	9623	9670	9718	9765	8
52	9813	9860	9907	9955	0002	0050	7
53	790097	0144	0192	0239	0287	0334	6
54	0381	0429	0476	0524	0571	0618	5
55	0666	0713	0761	0808	0856	0903	4
56	0950	0998	1045	1093	1140	1187	3
57	1235	1282	1330	1377	1424	1472	2
58	1518	1567	1614	1662	1709	1756	1
59	1804	1851	1899	1946	1993	2041	0
	60"	50"	40"	30"	20"	10"	
	Co-versed sine 12 degrees.						M

P. part	1"	2"	3"	4"	5"	6"	7"	8"	9"
	5	9	14	19	24	28	33	38	43

TABLE XLI. Natural Versed Sines.

M	0"	10"	20"	30"	40"	50"		M	0"	10"	20"	30"	40"	50"	
	Versed sine 78 degrees.								Versed sine 79 degrees.						
0	792088	2136	2183	2231	2278	2325	59	0	809191	9239	9286	9334	9381	9429	59
1	2373	2420	2468	2515	2563	2610	58	1	9477	9524	9571	9619	9667	9715	58
2	2657	2705	2752	2800	2847	2895	57	2	9762	9810	9857	9905	9953	0000	57
3	2942	2989	3037	3084	3132	3179	56	3	810048	0095	0143	0191	0238	0286	56
4	3227	3274	3321	3369	3416	3464	55	4	0333	0381	0429	0476	0524	0571	55
5	3511	3559	3606	3653	3701	3748	54	5	0619	0667	0714	0762	0809	0857	54
6	3796	3843	3891	3938	3986	4033	53	6	0905	0952	1000	1047	1095	1143	53
7	4080	4128	4175	4223	4270	4318	52	7	1190	1238	1285	1333	1381	1428	52
8	4365	4413	4460	4507	4555	4602	51	8	1476	1523	1571	1619	1666	1714	51
9	4650	4697	4745	4792	4840	4887	50	9	1762	1809	1857	1904	1952	2000	50
10	794935	4982	5029	5077	5124	5172	49	10	812047	2095	2142	2190	2238	2285	49
11	5219	5267	5314	5362	5409	5456	48	11	2333	2381	2428	2476	2523	2571	48
12	5504	5551	5599	5646	5694	5741	47	12	2619	2666	2714	2762	2809	2857	47
13	5789	5836	5884	5931	5979	6026	46	13	2904	2952	3000	3047	3095	3143	46
14	6073	6121	6168	6216	6263	6311	45	14	3190	3238	3285	3333	3381	3428	45
15	6358	6406	6453	6501	6548	6596	44	15	3476	3524	3571	3619	3666	3714	44
16	6643	6691	6738	6785	6833	6880	43	16	3762	3809	3857	3905	3952	4000	43
17	6928	6975	7023	7070	7118	7165	42	17	4048	4095	4143	4190	4238	4286	42
18	7213	7260	7308	7355	7403	7450	41	18	4333	4381	4429	4476	4524	4572	41
19	7498	7545	7593	7640	7687	7735	40	19	4619	4667	4715	4762	4810	4857	40
20	797782	7830	7877	7925	7972	8020	39	20	814905	4953	5000	5048	5096	5143	39
21	8067	8115	8162	8210	8257	8305	38	21	5191	5239	5286	5334	5382	5429	38
22	8352	8400	8447	8495	8542	8590	37	22	5477	5524	5572	5620	5667	5715	37
23	8637	8685	8732	8780	8827	8875	36	23	5763	5810	5858	5906	5953	6001	36
24	8922	8970	9017	9065	9112	9160	35	24	6049	6096	6144	6192	6239	6287	35
25	9207	9255	9302	9350	9397	9445	34	25	6335	6382	6430	6478	6525	6573	34
26	9492	9540	9587	9634	9682	9730	33	26	6621	6668	6716	6764	6811	6859	33
27	9777	9824	9872	9919	9967	0015	32	27	6906	6954	7002	7049	7097	7145	32
28	800062	0110	0157	0205	0252	0300	31	28	7192	7240	7288	7335	7383	7431	31
29	0347	0395	0442	0490	0537	0585	30	29	7478	7526	7574	7621	7669	7717	30
30	800632	0680	0727	0775	0822	0870	29	30	817764	7812	7860	7907	7955	8003	29
31	0917	0965	1012	1060	1107	1155	28	31	8050	8098	8146	8194	8241	8289	28
32	1202	1250	1297	1345	1392	1440	27	32	8337	8384	8432	8480	8527	8575	27
33	1487	1535	1582	1630	1677	1725	26	33	8623	8670	8718	8766	8813	8861	26
34	1772	1820	1867	1915	1962	2010	25	34	8909	8956	9004	9052	9099	9147	25
35	2058	2105	2153	2200	2248	2295	24	35	9195	9242	9290	9338	9385	9433	24
36	2343	2390	2438	2485	2533	2580	23	36	9481	9529	9576	9624	9672	9719	23
37	2628	2675	2723	2770	2818	2865	22	37	9767	9815	9862	9910	9958	0005	22
38	2913	2961	3008	3056	3103	3151	21	38	820053	0101	0148	0196	0244	0292	21
39	3198	3246	3293	3341	3388	3436	20	39	0339	0387	0435	0482	0530	0578	20
40	803483	3531	3578	3626	3674	3721	19	40	820625	0673	0721	0769	0816	0864	19
41	3769	3816	3864	3911	3959	4006	18	41	0912	0959	1007	1055	1102	1150	18
42	4054	4101	4149	4196	4244	4292	17	42	1198	1245	1293	1341	1389	1436	17
43	4339	4387	4434	4482	4529	4577	16	43	1484	1532	1579	1627	1675	1723	16
44	4624	4672	4719	4767	4815	4862	15	44	1770	1818	1866	1913	1961	2009	15
45	4910	4957	5005	5052	5100	5147	14	45	2056	2104	2152	2200	2247	2295	14
46	5195	5243	5290	5338	5385	5433	13	46	2343	2390	2438	2486	2534	2581	13
47	5480	5528	5575	5623	5671	5718	12	47	2629	2677	2724	2772	2820	2868	12
48	5766	5813	5861	5908	5956	6003	11	48	2915	2963	3011	3058	3106	3154	11
49	6051	6099	6146	6194	6241	6289	10	49	3201	3259	3297	3345	3392	3440	10
50	806336	6384	6432	6479	6527	6574	9	50	823488	3536	3583	3631	3679	3726	9
51	6622	6669	6717	6764	6812	6860	8	51	3774	3822	3870	3917	3965	4013	8
52	6907	6955	7002	7050	7097	7165	7	52	4061	4108	4156	4204	4251	4299	7
53	7193	7240	7288	7335	7383	7430	6	53	4347	4395	4442	4490	4538	4586	6
54	7478	7526	7573	7621	7668	7716	5	54	4633	4681	4729	4776	4824	4872	5
55	7763	7811	7859	7906	7954	8001	4	55	4920	4967	5015	5063	5111	5158	4
56	8049	8097	8144	8192	8239	8287	3	56	5206	5254	5302	5349	5397	5445	3
57	8334	8382	8430	8477	8525	8572	2	57	5492	5540	5588	5636	5683	5731	2
58	8620	8668	8715	8763	8810	8858	1	58	5779	5827	5874	5922	5970	6018	1
59	8905	8953	9001	9048	9096	9143	0	59	6065	6113	6161	6209	6256	6304	0
	60"	50"	40"	30"	20"	10"			60"	50"	40"	30"	20"	10"	
	Coversed sine 11 degrees.						M		Coversed sine 10 degrees.						M

P. part	1"	2"	3"	4"	5"	6"	7"	8"	9"
(78°)	5	9	14	19	24	28	33	38	43
(79°)	5	10	14	19	24	29	33	38	43

Verſed ſine 80 degrees.

M	0"	10"	20"	30"	40"	50"	
0	826352	6400	6447	6495	6543	6591	59
1	6638	6686	6734	6782	6829	6877	58
2	6925	6973	7020	7068	7116	7164	57
3	7211	7259	7307	7355	7402	7450	56
4	7498	7546	7593	7641	7689	7737	55
5	7784	7832	7880	7928	7975	8023	54
6	8071	8119	8166	8214	8262	8310	53
7	8357	8405	8453	8501	8549	8596	52
8	8644	8692	8740	8787	8835	8883	51
9	8931	8978	9026	9074	9122	9169	50
10	829217	9265	9313	9361	9408	9456	49
11	9504	9552	9599	9647	9695	9743	48
12	9791	9838	9886	9934	9982	0029	47
13	830077	0125	0173	0220	0268	0316	46
14	0364	0412	0459	0507	0555	0603	45
15	0650	0698	0746	0794	0841	0889	44
16	0937	0985	1033	1081	1128	1176	43
17	1224	1272	1319	1367	1415	1463	42
18	1511	1558	1606	1654	1702	1750	41
19	1797	1845	1893	1941	1989	2036	40
20	832084	2132	2180	2227	2275	2323	39
21	2371	2419	2466	2514	2562	2610	38
22	2658	2705	2753	2801	2849	2897	37
23	2944	2992	3040	3088	3136	3183	36
24	3231	3279	3327	3375	3422	3470	35
25	3518	3566	3614	3661	3709	3757	34
26	3805	3853	3901	3948	3996	4044	33
27	4092	4140	4187	4235	4283	4331	32
28	4379	4426	4474	4522	4570	4618	31
29	4666	4713	4761	4809	4857	4905	30
30	834952	5000	5048	5096	5144	5191	29
31	5239	5287	5335	5383	5431	5478	28
32	5526	5574	5622	5670	5718	5765	27
33	5813	5861	5909	5957	6004	6052	26
34	6100	6148	6196	6244	6291	6339	25
35	6387	6435	6483	6531	6578	6626	24
36	6674	6722	6770	6818	6865	6913	23
37	6961	7009	7057	7105	7152	7200	22
38	7248	7296	7344	7392	7439	7487	21
39	7535	7583	7631	7679	7726	7774	20
40	837822	7870	7918	7966	8013	8061	19
41	8109	8157	8205	8253	8300	8348	18
42	8396	8444	8492	8540	8588	8635	17
43	8683	8731	8779	8827	8875	8922	16
44	8970	9018	9066	9114	9162	9210	15
45	9257	9305	9353	9401	9449	9497	14
46	9545	9592	9640	9688	9736	9784	13
47	9832	9880	9927	9975	0023	0071	12
48	840119	0167	0215	0262	0310	0358	11
49	0406	0454	0502	0550	0597	0645	10
50	840693	0741	0789	0837	0885	0932	9
51	0980	1028	1076	1124	1172	1220	8
52	1268	1315	1363	1411	1459	1507	7
53	1555	1603	1650	1698	1746	1794	6
54	1842	1890	1938	1986	2033	2081	5
55	2129	2177	2225	2273	2321	2369	4
56	2416	2464	2512	2560	2608	2656	3
57	2704	2752	2799	2847	2895	2943	2
58	2991	3039	3087	3135	3182	3230	1
59	3278	3326	3374	3422	3470	3518	0
	60"	50"	40"	30"	20"	10"	M

Co-verſed ſine 9 degrees.

P. part { 1" 2" 3" 4" 5" 6" 7" 8" 9" / 5 10 14 19 24 29 33 38 43

Verſed ſine 81 degrees.

M	0"	10"	20"	30"	40"	50"	
0	843566	3613	3661	3709	3757	3805	59
1	3853	3901	3949	3997	4044	4092	58
2	4140	4188	4236	4284	4332	4380	57
3	4428	4475	4523	4571	4619	4667	56
4	4715	4763	4811	4859	4906	4954	55
5	5002	5050	5098	5146	5194	5242	54
6	5290	5338	5385	5433	5481	5529	53
7	5577	5625	5673	5721	5769	5817	52
8	5864	5912	5960	6008	6056	6104	51
9	6152	6200	6248	6296	6343	6391	50
10	846439	6487	6535	6583	6631	6679	49
11	6727	6775	6823	6870	6918	6966	48
12	7014	7062	7110	7158	7206	7254	47
13	7302	7350	7397	7445	7493	7541	46
14	7589	7637	7685	7733	7781	7829	45
15	7877	7925	7972	8020	8068	8116	44
16	8164	8212	8260	8308	8356	8404	43
17	8452	8500	8547	8595	8643	8691	42
18	8739	8787	8835	8883	8931	8979	41
19	9027	9075	9123	9171	9218	9266	40
20	849314	9362	9410	9458	9506	9554	39
21	9602	9650	9698	9746	9794	9842	38
22	9889	9937	9985	0033	0081	0129	37
23	850177	0225	0273	0321	0369	0417	36
24	0465	0513	0561	0608	0656	0704	35
25	0752	0800	0848	0896	0944	0992	34
26	1040	1088	1136	1184	1232	1280	33
27	1328	1376	1423	1471	1519	1567	32
28	1615	1663	1711	1759	1807	1855	31
29	1903	1951	1999	2047	2095	2143	30
30	852191	2239	2286	2334	2382	2430	29
31	2478	2526	2574	2622	2670	2718	28
32	2766	2814	2862	2910	2958	3006	27
33	3054	3102	3150	3198	3246	3294	26
34	3341	3389	3437	3485	3533	3581	25
35	3629	3677	3725	3773	3821	3869	24
36	3917	3965	4013	4061	4109	4157	23
37	4205	4253	4301	4349	4397	4445	22
38	4493	4540	4588	4636	4684	4732	21
39	4780	4828	4876	4924	4972	5020	20
40	855068	5116	5164	5212	5260	5308	19
41	5356	5404	5452	5500	5548	5596	18
42	5644	5692	5740	5788	5836	5884	17
43	5932	5980	6028	6076	6124	6172	16
44	6220	6267	6315	6363	6411	6459	15
45	6507	6555	6603	6651	6699	6747	14
46	6795	6843	6891	6939	6987	7035	13
47	7083	7131	7179	7227	7275	7323	12
48	7371	7419	7467	7515	7563	7611	11
49	7659	7707	7755	7803	7851	7899	10
50	857947	7995	8043	8091	8139	8187	9
51	8235	8283	8331	8379	8427	8475	8
52	8523	8571	8619	8667	8715	8763	7
53	8811	8859	8907	8955	9003	9051	6
54	9099	9147	9195	9243	9291	9339	5
55	9387	9435	9483	9531	9579	9627	4
56	9675	9723	9771	9819	9867	9915	3
57	9963	0011	0059	0107	0155	0203	2
58	860251	0299	0347	0395	0443	0491	1
59	0539	0587	0635	0683	0731	0779	0
	60"	50"	40"	30"	20"	10"	M

Co-verſed ſine 8 degrees.

P. part { 1" 2" 3" 4" 5" 6" 7" 8" 9" / 5 10 14 19 24 29 34 38 43

TABLE XII. Natural Versed Sines.

M	Versed sine 82 degrees 0"	10"	20"	30"	40"	50"	
0	860827	0875	0923	0971	1019	1067	59
1	1115	1163	1211	1259	1307	1355	58
2	1403	1451	1499	1547	1595	1643	57
3	1691	1739	1787	1835	1883	1931	56
4	1979	2027	2075	2123	2171	2219	55
5	2267	2315	2363	2411	2459	2507	54
6	2555	2603	2651	2700	2748	2796	53
7	2844	2892	2940	2988	3036	3084	52
8	3132	3180	3228	3276	3324	3372	51
9	3420	3468	3516	3564	3612	3660	50
10	863708	3756	3804	3852	3900	3948	49
11	3996	4044	4092	4140	4188	4236	48
12	4284	4332	4380	4429	4477	4525	47
13	4573	4621	4669	4717	4765	4813	46
14	4861	4909	4957	5005	5053	5101	45
15	5149	5197	5245	5293	5341	5389	44
16	5437	5485	5533	5581	5629	5678	43
17	5726	5774	5822	5870	5918	5966	42
18	6014	6062	6110	6158	6206	6254	41
19	6302	6350	6398	6446	6494	6542	40
20	866590	6638	6686	6735	6783	6831	39
21	6879	6927	6975	7023	7071	7119	38
22	7167	7215	7263	7311	7359	7407	37
23	7455	7503	7551	7599	7647	7696	36
24	7744	7792	7840	7888	7936	7984	35
25	8032	8080	8128	8176	8224	8272	34
26	8320	8368	8416	8464	8513	8561	33
27	8609	8657	8705	8753	8801	8849	32
28	8897	8945	8993	9041	9089	9137	31
29	9185	9233	9282	9330	9378	9426	30
30	869474	9522	9570	9618	9666	9714	29
31	9762	9810	9858	9906	9954	0003	28
32	870051	0099	0147	0195	0243	0291	27
33	0339	0387	0435	0483	0531	0579	26
34	0627	0676	0724	0772	0820	0868	25
35	0916	0964	1012	1060	1108	1156	24
36	1204	1252	1301	1349	1397	1445	23
37	1493	1541	1589	1637	1685	1733	22
38	1781	1829	1878	1926	1974	2022	21
39	2070	2118	2166	2214	2262	2310	20
40	872358	2406	2455	2503	2551	2599	19
41	2647	2695	2743	2791	2839	2887	18
42	2935	2983	3032	3080	3128	3176	17
43	3224	3272	3320	3368	3416	3464	16
44	3512	3561	3609	3657	3705	3753	15
45	3801	3849	3897	3945	3993	4042	14
46	4090	4138	4186	4234	4282	4330	13
47	4378	4426	4474	4522	4571	4619	12
48	4667	4715	4763	4811	4859	4907	11
49	4955	5003	5052	5100	5148	5196	10
50	875244	5292	5340	5388	5436	5484	9
51	5533	5581	5629	5677	5725	5773	8
52	5821	5869	5917	5966	6014	6062	7
53	6110	6158	6206	6254	6302	6350	6
54	6399	6447	6495	6543	6591	6639	5
55	6687	6735	6783	6832	6880	6928	4
56	6976	7024	7072	7120	7168	7216	3
57	7265	7313	7361	7409	7457	7505	2
58	7553	7601	7649	7698	7746	7794	1
59	7842	7890	7938	7986	8034	8083	0
	60"	50"	40"	30"	20"	10"	
	Co-versed sine 7 degrees.						M

P. part { 1" 2" 3" 4" 5" 6" 7" 8" 9" / 5 10 14 19 24 29 34 38 43 }

M	Versed sine 83 degrees 0"	10"	20"	30"	40"	50"	
0	878131	8179	8227	8275	8323	8371	59
1	8419	8468	8516	8564	8612	8660	58
2	8708	8756	8804	8852	8901	8949	57
3	8997	9045	9093	9141	9189	9237	56
4	9286	9334	9382	9430	9478	9526	55
5	9574	9623	9671	9719	9767	9815	54
6	9863	9911	9959	0008	0056	0104	53
7	880152	0200	0248	0296	0344	0393	52
8	0441	0489	0537	0585	0633	0682	51
9	0730	0778	0826	0874	0922	0970	50
10	881018	1067	1115	1163	1211	1259	49
11	1307	1355	1403	1452	1500	1548	48
12	1596	1644	1692	1740	1789	1837	47
13	1885	1933	1981	2029	2077	2126	46
14	2174	2222	2270	2318	2366	2414	45
15	2463	2511	2559	2607	2655	2703	44
16	2751	2800	2848	2896	2944	2992	43
17	3040	3089	3137	3185	3233	3281	42
18	3329	3377	3426	3474	3522	3570	41
19	3618	3666	3714	3763	3811	3859	40
20	883907	3955	4003	4052	4100	4148	39
21	4196	4244	4292	4340	4389	4437	38
22	4485	4533	4581	4629	4678	4726	37
23	4774	4822	4870	4918	4967	5015	36
24	5063	5111	5159	5207	5255	5304	35
25	5352	5400	5448	5496	5544	5593	34
26	5641	5689	5737	5785	5833	5882	33
27	5930	5978	6026	6074	6122	6171	32
28	6219	6267	6315	6363	6411	6460	31
29	6508	6556	6604	6652	6700	6749	30
30	886797	6845	6893	6941	6989	7038	29
31	7086	7134	7182	7230	7278	7327	28
32	7375	7423	7471	7519	7568	7616	27
33	7664	7712	7760	7808	7857	7905	26
34	7953	8001	8049	8097	8146	8194	25
35	8242	8290	8338	8387	8435	8483	24
36	8531	8579	8627	8676	8724	8772	23
37	8820	8868	8917	8965	9013	9061	22
38	9109	9157	9206	9254	9302	9350	21
39	9398	9447	9495	9543	9591	9639	20
40	889687	9736	9784	9832	9880	9928	19
41	9977	0025	0073	0121	0169	0218	18
42	890266	0314	0362	0410	0458	0507	17
43	0555	0603	0651	0699	0748	0796	16
44	0844	0892	0940	0989	1037	1085	15
45	1133	1181	1230	1278	1326	1374	14
46	1422	1470	1519	1567	1615	1663	13
47	1711	1760	1808	1856	1904	1952	12
48	2000	2049	2097	2145	2193	2242	11
49	2290	2338	2386	2434	2483	2531	10
50	892579	2627	2675	2724	2772	2820	9
51	2868	2916	2965	3013	3061	3109	8
52	3157	3206	3254	3302	3350	3398	7
53	3447	3495	3543	3591	3640	3688	6
54	3736	3784	3832	3881	3929	3977	5
55	4025	4073	4122	4170	4218	4266	4
56	4314	4363	4411	4459	4507	4555	3
57	4604	4652	4700	4748	4797	4845	2
58	4893	4941	4989	5038	5086	5134	1
59	5182	5230	5279	5327	5375	5423	0
	60"	50"	40"	30"	20"	10"	
	Co-versed sine 6 degrees.						M

P. part { 1" 2" 3" 4" 5" 6" 7" 8" 9" / 5 10 14 19 24 29 34 39 43 }

M	Versed sine 84 degrees. 0"	10"	20"	30"	40"	50"	
0	895472	5520	5568	5616	5664	5713	59
1	5761	5809	5857	5905	5954	6002	58
2	6050	6098	6147	6195	6243	6291	57
3	6339	6388	6436	6484	6532	6581	56
4	6629	6677	6725	6773	6822	6870	55
5	6918	6966	7015	7063	7111	7159	54
6	7207	7256	7304	7352	7400	7449	53
7	7497	7545	7593	7641	7690	7738	52
8	7786	7834	7883	7931	7979	8027	51
9	8076	8124	8172	8220	8268	8317	50
10	898365	8413	8461	8510	8558	8606	49
11	8654	8703	8751	8799	8847	8895	48
12	8944	8992	9040	9088	9137	9185	47
13	9233	9281	9330	9378	9426	9474	46
14	9523	9571	9619	9667	9715	9764	45
15	9812	9860	9908	9957	0005	0053	44
16	900101	0150	0198	0246	0294	0343	43
17	0391	0439	0487	0536	0584	0632	42
18	0680	0728	0777	0825	0873	0921	41
19	0970	1018	1066	1114	1163	1211	40
20	901259	1307	1356	1404	1452	1500	39
21	1549	1597	1645	1693	1742	1790	38
22	1838	1886	1935	1983	2031	2079	37
23	2128	2176	2224	2272	2321	2369	36
24	2417	2465	2514	2562	2610	2658	35
25	2707	2755	2803	2851	2900	2948	34
26	2996	3044	3093	3141	3189	3237	33
27	3286	3334	3382	3430	3479	3527	32
28	3575	3623	3672	3720	3768	3816	31
29	3865	3913	3961	4009	4058	4106	30
30	904154	4203	4251	4299	4347	4396	29
31	4444	4492	4540	4589	4637	4685	28
32	4733	4782	4830	4878	4926	4975	27
33	5023	5071	5119	5168	5216	5264	26
34	5313	5361	5409	5457	5506	5554	25
35	5602	5650	5698	5747	5795	5843	24
36	5892	5940	5988	6036	6085	6133	23
37	6181	6230	6278	6326	6374	6423	22
38	6471	6519	6567	6616	6664	6712	21
39	6761	6809	6857	6905	6954	7002	20
40	907050	7098	7147	7195	7243	7291	19
41	7340	7388	7436	7485	7533	7581	18
42	7629	7678	7726	7774	7823	7871	17
43	7919	7967	8016	8064	8112	8160	16
44	8209	8257	8305	8354	8402	8450	15
45	8498	8547	8595	8643	8692	8740	14
46	8788	8836	8885	8933	8981	9029	13
47	9078	9126	9174	9223	9271	9319	12
48	9367	9416	9464	9512	9561	9609	11
49	9657	9705	9754	9802	9850	9899	10
50	909947	9995	0043	0092	0140	0188	9
51	910237	0285	0333	0381	0430	0478	8
52	0526	0575	0623	0671	0719	0768	7
53	0816	0864	0913	0961	1009	1057	6
54	1106	1154	1202	1251	1299	1347	5
55	1395	1444	1492	1540	1589	1637	4
56	1685	1733	1782	1830	1878	1927	3
57	1975	2023	2072	2120	2168	2216	2
58	2265	2313	2361	2410	2458	2506	1
59	2554	2603	2651	2699	2748	2796	0
	60"	50"	40"	30"	20"	10"	
	Co-versed sine 5 degrees.						M

P. part { 1" 2" 3" 4" 5" 6" 7" 8" 9" / 5 10 14 19 24 29 34 39 43

M	Versed sine 85 degrees. 0"	10"	20"	30"	40"	50"	
0	912844	2893	2941	2989	3037	3086	59
1	3134	3182	3231	3279	3327	3376	58
2	3424	3472	3520	3569	3617	3665	57
3	3714	3762	3810	3859	3907	3955	56
4	4003	4052	4100	4148	4197	4245	55
5	4293	4342	4390	4438	4486	4535	54
6	4583	4631	4680	4728	4776	4825	53
7	4873	4921	4970	5018	5066	5114	52
8	5163	5211	5259	5308	5356	5404	51
9	5453	5501	5549	5598	5646	5694	50
10	915742	5791	5839	5887	5936	5984	49
11	6032	6081	6129	6177	6226	6274	48
12	6322	6370	6419	6467	6515	6564	47
13	6612	6660	6709	6757	6805	6854	46
14	6902	6950	6999	7047	7095	7143	45
15	7192	7240	7288	7337	7385	7433	44
16	7482	7530	7578	7627	7675	7723	43
17	7772	7820	7868	7917	7965	8013	42
18	8061	8110	8158	8206	8255	8303	41
19	8351	8400	8448	8496	8545	8593	40
20	918641	8690	8738	8786	8835	8883	39
21	8931	8980	9028	9076	9125	9173	38
22	9221	9270	9318	9366	9415	9463	37
23	9511	9559	9608	9656	9704	9753	36
24	9801	9849	9898	9946	9994	0043	35
25	920091	0139	0188	0236	0284	0333	34
26	0381	0429	0478	0526	0574	0623	33
27	0671	0719	0768	0816	0864	0913	32
28	0961	1009	1058	1106	1154	1203	31
29	1251	1299	1348	1396	1444	1493	30
30	921541	1589	1638	1686	1734	1783	29
31	1831	1879	1928	1976	2024	2073	28
32	2121	2169	2218	2266	2314	2363	27
33	2411	2459	2508	2556	2604	2653	26
34	2701	2749	2798	2846	2894	2943	25
35	2991	3039	3088	3136	3184	3233	24
36	3281	3329	3378	3426	3474	3523	23
37	3571	3619	3668	3716	3764	3813	22
38	3861	3909	3958	4006	4054	4103	21
39	4151	4199	4248	4296	4344	4393	20
40	924441	4489	4538	4586	4635	4683	19
41	4731	4780	4828	4876	4925	4973	18
42	5021	5070	5118	5166	5215	5263	17
43	5311	5360	5408	5456	5505	5553	16
44	5601	5650	5698	5746	5795	5843	15
45	5892	5940	5988	6037	6085	6133	14
46	6182	6230	6278	6327	6375	6423	13
47	6472	6520	6568	6617	6665	6713	12
48	6762	6810	6859	6907	6955	7004	11
49	7052	7100	7149	7197	7245	7294	10
50	927342	7390	7439	7487	7535	7584	9
51	7632	7681	7729	7777	7826	7874	8
52	7922	7971	8019	8067	8116	8164	7
53	8212	8261	8309	8357	8406	8454	6
54	8503	8551	8599	8648	8696	8744	5
55	8793	8841	8889	8938	8986	9034	4
56	9083	9131	9180	9228	9276	9325	3
57	9373	9421	9470	9518	9566	9615	2
58	9663	9712	9760	9808	9857	9905	1
59	9953	0002	0050	0098	0147	0195	0
	60"	50"	40"	30"	20"	10"	
	Co-versed sine 4 degrees.						M

P. part { 1" 2" 3" 4" 5" 6" 7" 8" 9" / 5 10 15 19 24 29 34 39 43

M	Versed sine 86 degrees. 0"	10"	20"	30"	40"	50"	
0	930244	0292	0340	0389	0437	0485	59
1	0534	0582	0630	0679	0727	0776	58
2	0824	0872	0921	0969	1017	1066	57
3	1114	1162	1211	1259	1308	1356	56
4	1404	1453	1501	1549	1598	1646	55
5	1694	1743	1791	1840	1888	1936	54
6	1985	2033	2081	2130	2178	2227	53
7	2275	2323	2372	2420	2468	2517	52
8	2565	2614	2662	2710	2759	2807	51
9	2855	2904	2952	3000	3049	3097	50
10	933146	3194	3242	3291	3339	3387	49
11	3436	3484	3533	3581	3629	3678	48
12	3726	3774	3823	3871	3920	3968	47
13	4016	4065	4113	4161	4210	4258	46
14	4307	4355	4403	4452	4500	4548	45
15	4597	4645	4694	4742	4790	4839	44
16	4887	4936	4984	5032	5081	5129	43
17	5177	5226	5274	5322	5371	5419	42
18	5468	5516	5564	5613	5661	5710	41
19	5758	5806	5855	5903	5952	6000	40
20	936048	6097	6145	6193	6242	6290	39
21	6339	6387	6435	6484	6532	6580	38
22	6629	6677	6726	6774	6822	6871	37
23	6919	6968	7016	7064	7113	7161	36
24	7209	7258	7306	7355	7403	7451	35
25	7500	7548	7597	7645	7693	7742	34
26	7790	7839	7887	7935	7984	8032	33
27	8080	8129	8177	8226	8274	8322	32
28	8371	8419	8468	8516	8564	8613	31
29	8661	8710	8758	8806	8855	8903	30
30	938951	9000	9048	9097	9145	9193	29
31	9242	9290	9339	9387	9435	9484	28
32	9532	9581	9629	9677	9726	9774	27
33	9823	9871	9919	9968	0016	0064	26
34	940113	0161	0210	0258	0306	0355	25
35	0403	0452	0500	0548	0597	0645	24
36	0694	0742	0790	0839	0887	0936	23
37	0984	1032	1081	1129	1178	1226	22
38	1274	1323	1371	1420	1468	1516	21
39	1565	1613	1662	1710	1758	1807	20
40	941855	1904	1952	2000	2049	2097	19
41	2146	2194	2242	2291	2339	2388	18
42	2436	2484	2533	2581	2630	2678	17
43	2726	2775	2823	2872	2920	2968	16
44	3017	3065	3114	3162	3210	3259	15
45	3307	3356	3404	3452	3501	3549	14
46	3598	3646	3694	3743	3791	3840	13
47	3888	3936	3985	4033	4082	4130	12
48	4178	4227	4275	4324	4372	4421	11
49	4469	4517	4566	4614	4663	4711	10
50	944759	4808	4856	4905	4953	5001	9
51	5050	5098	5147	5195	5243	5292	8
52	5340	5389	5437	5485	5534	5582	7
53	5631	5679	5728	5776	5824	5873	6
54	5921	5970	6018	6066	6115	6163	5
55	6212	6260	6308	6357	6405	6454	4
56	6502	6551	6599	6647	6696	6744	3
57	6793	6841	6889	6938	6986	7035	2
58	7083	7131	7180	7228	7277	7325	1
59	7374	7422	7470	7519	7567	7616	0
	60"	50"	40"	30"	20"	10"	
	Co-versed sine 3 degrees.						M

P. part { 1" 2" 3" 4" 5" 6" 7" 8" 9" / 5 10 15 19 24 29 34 39 44 }

M	Versed sine 87 degrees. 0"	10"	20"	30"	40"	50"	
0	947664	7712	7761	7809	7858	7906	59
1	7955	8003	8051	8100	8148	8197	58
2	8245	8293	8342	8390	8439	8487	57
3	8536	8584	8632	8681	8729	8778	56
4	8826	8874	8923	8971	9020	9068	55
5	9117	9165	9213	9262	9310	9359	54
6	9407	9455	9504	9552	9601	9649	53
7	9698	9746	9794	9843	9891	9940	52
8	9988	0037	0085	0133	0182	0230	51
9	950279	0327	0375	0424	0472	0521	50
10	950569	0618	0666	0714	0763	0811	49
11	0860	0908	0957	1005	1053	1102	48
12	1150	1199	1247	1296	1344	1392	47
13	1441	1489	1538	1586	1634	1683	46
14	1731	1780	1828	1877	1925	1973	45
15	2022	2070	2119	2167	2216	2264	44
16	2312	2361	2409	2458	2506	2555	43
17	2603	2651	2700	2748	2797	2845	42
18	2894	2942	2990	3039	3087	3136	41
19	3184	3233	3281	3329	3378	3426	40
20	953475	3523	3572	3620	3668	3717	39
21	3765	3814	3862	3911	3959	4007	38
22	4056	4104	4153	4201	4250	4298	37
23	4346	4395	4443	4492	4540	4589	36
24	4637	4685	4734	4782	4831	4879	35
25	4928	4976	5024	5073	5121	5170	34
26	5218	5267	5315	5363	5412	5460	33
27	5509	5557	5606	5654	5703	5751	32
28	5799	5848	5896	5945	5993	6041	31
29	6090	6138	6187	6235	6284	6332	30
30	956381	6429	6477	6526	6574	6623	29
31	6671	6720	6768	6817	6865	6913	28
32	6962	7010	7059	7107	7156	7204	27
33	7252	7301	7349	7398	7446	7495	26
34	7543	7592	7640	7688	7737	7785	25
35	7834	7882	7931	7979	8027	8076	24
36	8124	8173	8221	8270	8318	8367	23
37	8415	8463	8512	8560	8609	8657	22
38	8706	8754	8802	8851	8899	8948	21
39	8996	9045	9093	9142	9190	9238	20
40	959287	9335	9384	9432	9481	9529	19
41	9578	9626	9674	9723	9771	9820	18
42	9868	9917	9965	0014	0062	0110	17
43	960159	0207	0256	0304	0353	0401	16
44	0450	0498	0546	0595	0643	0692	15
45	0740	0789	0837	0886	0934	0982	14
46	1031	1079	1128	1176	1225	1273	13
47	1322	1370	1418	1467	1515	1564	12
48	1612	1661	1709	1758	1806	1854	11
49	1903	1951	2000	2048	2097	2145	10
50	962194	2242	2290	2339	2387	2436	9
51	2484	2533	2581	2630	2678	2726	8
52	2775	2823	2872	2920	2969	3017	7
53	3066	3114	3162	3211	3259	3308	6
54	3356	3405	3453	3502	3550	3599	5
55	3647	3695	3744	3792	3841	3889	4
56	3938	3986	4035	4083	4131	4180	3
57	4228	4277	4325	4374	4422	4471	2
58	4519	4568	4616	4664	4713	4761	1
59	4810	4858	4907	4955	5004	5052	0
	60"	50"	40"	30"	20"	10"	
	Co-versed sine 2 degrees.						M

P. part { 1" 2" 3" 4" 5" 6" 7" 8" 9" / 5 10 15 19 24 29 34 39 44 }

M	Versed sine 88 degrees. 0″	10″	20″	30″	40″	50″	
0	965101	5149	5197	5246	5294	5343	59
1	5391	5440	5488	5537	5585	5633	58
2	5682	5730	5779	5827	5876	5924	57
3	5973	6021	6070	6118	6166	6215	56
4	6263	6312	6360	6409	6457	6506	55
5	6554	6603	6651	6699	6748	6796	54
6	6845	6893	6942	6990	7039	7087	53
7	7136	7184	7233	7281	7329	7378	52
8	7426	7475	7523	7572	7620	7669	51
9	7717	7765	7814	7862	7911	7959	50
10	968008	8056	8105	8153	8202	8250	49
11	8298	8347	8395	8444	8492	8541	48
12	8589	8638	8686	8735	8783	8832	47
13	8880	8928	8977	9025	9074	9122	46
14	9171	9219	9268	9316	9365	9413	45
15	9461	9510	9558	9607	9655	9704	44
16	9752	9801	9849	9898	9946	9995	43
17	970043	0091	0140	0188	0237	0285	42
18	0334	0382	0431	0479	0528	0576	41
19	0625	0673	0721	0770	0818	0867	40
20	970915	0964	1012	1061	1109	1158	39
21	1206	1255	1303	1352	1400	1448	38
22	1497	1545	1594	1642	1691	1739	37
23	1788	1836	1885	1933	1981	2030	36
24	2078	2127	2175	2224	2272	2321	35
25	2369	2418	2466	2515	2563	2611	34
26	2660	2708	2757	2805	2854	2902	33
27	2951	2999	3048	3096	3145	3193	32
28	3241	3290	3338	3387	3435	3484	31
29	3532	3581	3629	3678	3726	3775	30
30	973823	3872	3920	3968	4017	4065	29
31	4114	4162	4211	4259	4308	4356	28
32	4405	4453	4502	4550	4598	4647	27
33	4695	4744	4792	4841	4889	4938	26
34	4986	5035	5083	5132	5180	5229	25
35	5277	5325	5374	5422	5471	5519	24
36	5568	5616	5665	5713	5762	5810	23
37	5859	5907	5956	6004	6052	6101	22
38	6149	6198	6246	6295	6343	6392	21
39	6440	6489	6537	6586	6634	6683	20
40	976731	6780	6828	6876	6925	6973	19
41	7022	7070	7119	7167	7216	7264	18
42	7313	7361	7410	7458	7507	7555	17
43	7603	7652	7700	7749	7797	7846	16
44	7894	7943	7991	8040	8088	8137	15
45	8185	8234	8282	8331	8379	8427	14
46	8476	8524	8573	8621	8670	8718	13
47	8767	8815	8864	8912	8961	9009	12
48	9058	9106	9155	9203	9251	9300	11
49	9348	9397	9445	9494	9542	9591	10
50	979639	9688	9736	9785	9833	9882	9
51	9930	9979	0027	0075	0124	0172	8
52	980221	0269	0318	0366	0415	0463	7
53	0512	0560	0609	0657	0706	0754	6
54	0803	0851	0900	0948	0996	1045	5
55	1093	1142	1190	1239	1287	1336	4
56	1384	1433	1481	1530	1578	1627	3
57	1675	1724	1772	1821	1869	1917	2
58	1966	2014	2063	2111	2160	2208	1
59	2257	2305	2354	2402	2451	2499	0
	60″	50″	40″	30″	20″	10″	
	Co-versed sine 1 degree.						M

P. part { 1″ 2″ 3″ 4″ 5″ 6″ 7″ 8″ 9″ / 5 10 15 19 24 29 34 39 44

M	Versed sine 89 degrees. 0″	10″	20″	30″	40″	50″	
0	982548	2596	2645	2693	2741	2790	59
1	2838	2887	2935	2984	3032	3081	58
2	3129	3178	3226	3275	3323	3372	57
3	3420	3469	3517	3566	3614	3663	56
4	3711	3759	3808	3857	3905	3953	55
5	4002	4050	4099	4147	4196	4244	54
6	4293	4341	4390	4438	4487	4535	53
7	4584	4632	4680	4729	4777	4826	52
8	4874	4923	4971	5020	5068	5117	51
9	5165	5214	5262	5311	5359	5408	50
10	985456	5505	5553	5602	5650	5698	49
11	5747	5795	5844	5892	5941	5989	48
12	6038	6086	6135	6183	6232	6280	47
13	6329	6377	6426	6474	6523	6571	46
14	6620	6668	6716	6765	6813	6862	45
15	6911	6959	7007	7056	7104	7153	44
16	7201	7250	7298	7347	7395	7444	43
17	7492	7541	7589	7638	7686	7735	42
18	7783	7831	7880	7928	7977	8025	41
19	8074	8122	8171	8219	8268	8316	40
20	988365	8413	8462	8510	8559	8607	39
21	8656	8704	8753	8801	8850	8898	38
22	8946	8995	9043	9092	9140	9189	37
23	9237	9286	9334	9383	9431	9480	36
24	9528	9577	9625	9674	9722	9771	35
25	9819	9868	9916	9965	0013	0061	34
26	990110	0158	0207	0255	0304	0352	33
27	0401	0449	0498	0546	0595	0643	32
28	0692	0740	0789	0837	0886	0934	31
29	0983	1031	1080	1128	1177	1225	30
30	991273	1322	1370	1419	1467	1516	29
31	1564	1613	1661	1710	1758	1807	28
32	1855	1904	1952	2001	2049	2098	27
33	2146	2195	2243	2292	2340	2388	26
34	2437	2485	2534	2582	2631	2679	25
35	2728	2776	2825	2873	2922	2970	24
36	3019	3067	3116	3164	3213	3261	23
37	3310	3358	3407	3455	3504	3552	22
38	3601	3649	3697	3746	3794	3843	21
39	3891	3940	3988	4037	4085	4134	20
40	994182	4231	4279	4328	4376	4425	19
41	4473	4522	4570	4619	4667	4716	18
42	4764	4813	4861	4909	4958	5006	17
43	5055	5103	5152	5200	5249	5297	16
44	5346	5394	5443	5491	5540	5588	15
45	5637	5685	5734	5782	5831	5879	14
46	5928	5976	6025	6073	6122	6170	13
47	6218	6267	6315	6364	6412	6461	12
48	6509	6558	6606	6655	6703	6752	11
49	6800	6849	6897	6946	6994	7043	10
50	997091	7140	7188	7237	7285	7334	9
51	7382	7430	7479	7527	7576	7624	8
52	7673	7721	7770	7818	7867	7915	7
53	7964	8012	8061	8109	8158	8206	6
54	8255	8303	8352	8400	8449	8497	5
55	8546	8594	8643	8691	8739	8788	4
56	8836	8885	8933	8982	9030	9079	3
57	9127	9176	9224	9273	9321	9370	2
58	9418	9467	9515	9564	9612	9661	1
59	9709	9758	9806	9855	9903	9952	0
	60″	50″	40″	30″	10″	10″	
	Co-versed sine 0 degree.						M

P. part { 1″ 2″ 3″ 4″ 5″ 6″ 7″ 8″ 9″ / 5 10 15 19 24 29 34 39 44

TABLE XLII.

LOGARITHMIC DIFFERENCE.

Moon's hor. par.	App. Apparent Altitude of the Moon's Center. 0°	1°	2°	3°	4°	5°	6°
53′ 0″	0.000113	0.000042	9.999947	9.999839	9.999727	9.999612	9.999496
10	0113	0042	9946	9838	9725	9610	9494
20	0112	0041	9945	9836	9723	9608	9491
30	0112	0041	9944	9835	9722	9606	9489
40	0112	0040	9943	9833	9720	9604	9486
50	0112	0039	9942	9832	9718	9602	9484
54 0	0.000112	0.000039	9.999941	9.999831	9.999716	9.999599	9.999481
10	0112	0039	9940	9829	9714	9597	9479
20	0111	0038	9939	9829	9713	9595	9476
30	0111	0037	9938	9827	9711	9593	9474
40	0111	0037	9937	9826	9709	9591	9471
50	0111	0036	9936	9824	9707	9589	9469
55 0	0.000111	0.000036	9.999935	9.999823	9.999706	9.999587	9.999466
10	0111	0035	9934	9821	9704	9584	9464
20	0111	0035	9933	9820	9702	9582	9461
30	0111	0034	9932	9819	9700	9580	9459
40	0111	0033	9931	9817	9699	9578	9456
50	0110	0033	9930	9816	9697	9576	9454
56 0	0.000110	0.000032	9.999929	9.999815	9.999695	9.999574	9.999451
10	0110	0032	9928	9813	9693	9571	9449
20	0110	0031	9927	9812	9692	9569	9447
30	0110	0031	9927	9811	9690	9567	9444
40	0110	0030	9926	9809	9688	9565	9442
50	0110	0029	9925	9808	9686	9563	9439
57 0	0.000109	0.000029	9.999924	9.999807	9.999685	9.999561	9.999437
10	0109	0028	9923	9805	9683	9559	9434
20	0109	0028	9922	9804	9681	9557	9432
30	0109	0027	9921	9803	9680	9555	9429
40	0109	0027	9920	9802	9678	9553	9427
50	0109	0026	9919	9800	9676	9551	9424
58 0	0.000108	0.000025	9.999918	9.999799	9.999675	9.999549	9.999422
10	0108	0025	9917	9797	9673	9546	9419
20	0108	0024	9916	9796	9671	9544	9417
30	0108	0024	9915	9794	9669	9542	9414
40	0108	0023	9914	9793	9668	9540	9412
50	0108	0023	9913	9791	9666	9538	9409
59 0	0.000108	0.000022	9.999912	9.999790	9.999664	9.999536	9.999407
10	0107	0021	9911	9789	9662	9533	9404
20	0107	0021	9910	9787	9661	9531	9402
30	0107	0020	9909	9786	9659	9529	9399
40	0107	0020	9908	9785	9657	9527	9397
50	0107	0019	9907	9783	9655	9524	9394
60 0	0.000107	0.000018	9.999906	9.999782	9.999653	9.999523	9.999392
10	0106	0018	9905	9781	9652	9520	9389
20	0106	0017	9904	9779	9650	9518	9387
30	0106	0017	9903	9778	9648	9516	9384
40	0106	0016	9902	9777	9646	9514	9382
50	0106	0016	9901	9775	9645	9512	9379
61 0	0.000106	0.000015	9.999900	9.999774	9.999643	9.999510	9.999376
10	0105	0014	9899	9772	9641	9507	9374
20	0105	0014	9898	9771	9639	9505	9371
30	0105	0013	9897	9769	9638	9503	9369
40	0105	0013	9896	9768	9636	9501	9366
50	0104	0012	9895	9766	9634	9499	9364
62 0	0.000104	0.000011	9.999894	9.999765	9.999632	9.999497	9.999361

P. part. to sec. of par. alt. 1°	1″	3″	5″	7″	9″	alt. 3°	1″	3″	5″	7″	9″	alt. 5°	1″	3″	5″	7″	9″
	0	0	0	0	1		0	0	1	1	1		0	1	1	1	2

TABLE XLII.

LOGARITHMIC DIFFERENCE.

Moon's hor. parallax	Arg. Apparent Altitude of the Moon's Center. 7°	8°	9°	10°	11°	12°	Diff. to 1' of alt.
53′ 0″	9.999380	9.999264	9.999148	9.999032	9.998917	9.998803	1.92
10	9377	9261	9144	9028	8913	8798	1.93
20	9374	9257	9140	9024	8908	8794	1.93
30	9371	9254	9137	9021	8904	8789	1.94
40	9369	9251	9133	9017	8900	8784	1.95
50	9366	9247	9129	9013	8895	8779	1.96
54 0	9.999363	9.999244	9.999126	9.999009	9.998891	9.998774	1.96
10	9360	9241	9122	9005	8886	8770	1.97
20	9357	9238	9119	9001	8882	8765	1.97
30	9354	9234	9115	8997	8878	8760	1.98
40	9351	9231	9112	8993	8873	8756	1.98
50	9348	9228	9108	8989	8869	8751	1.99
55 0	9.999345	9.999225	9.999105	9.998985	9.998865	9.998746	2.00
10	9342	9222	9101	8981	8860	8742	2.00
20	9339	9219	9098	8977	8856	8737	2.01
30	9337	9216	9094	8973	8852	8732	2.02
40	9334	9212	9091	8969	8847	8727	2.02
50	9331	9209	9087	8965	8843	8722	2.03
56 0	9.999328	9.999206	9.999083	9.998961	9.998839	9.998718	2.03
10	9325	9203	9079	8957	8834	8713	2.04
20	9322	9199	9076	8953	8830	8708	2.05
30	9320	9196	9072	8949	8826	8704	2.05
40	9317	9193	9069	8945	8821	8699	2.06
50	9314	9190	9065	8941	8817	8694	2.07
57 0	9.999311	9.999186	9.999062	9.998937	9.998813	9.998690	2.07
10	9308	9183	9058	8934	8808	8685	2.08
20	9305	9180	9055	8930	8804	8680	2.08
30	9303	9176	9051	8926	8800	8676	2.09
40	9300	9173	9048	8922	8795	8671	2.10
50	9297	9170	9044	8918	8791	8666	2.10
58 0	9.999294	9.999167	9.999040	9.998914	9.998787	9.998662	2.11
10	9291	9163	9037	8910	8782	8657	2.11
20	9288	9160	9033	8906	8778	8652	2.12
30	9286	9157	9029	8902	8774	8648	2.13
40	9283	9154	9026	8898	8769	8643	2.13
50	9280	9151	9022	8894	8765	8638	2.14
59 0	9.999277	9.999147	9.999018	9.998890	9.998761	9.998634	2.14
10	9274	9144	9014	8886	8756	8629	2.15
20	9271	9141	9011	8882	8752	8624	2.16
30	9268	9137	9007	8878	8748	8619	2.16
40	9265	9134	9004	8874	8743	8615	2.17
50	9262	9131	9000	8870	8739	8610	2.17
60 0	9.999259	9.999128	9.998997	9.998866	9.998735	9.998605	2.18
10	9257	9124	8993	8862	8730	8600	2.19
20	9254	9121	8990	8858	8726	8596	2.19
30	9251	9118	8986	8854	8722	8591	2.20
40	9248	9115	8982	8850	8717	8586	2.21
50	9245	9111	8979	8846	8713	8581	2.21
61 0	9.999242	9.999108	9.998975	9.998842	9.998709	9.998577	2.22
10	9239	9105	8971	8838	8704	8572	2.22
20	9236	9102	8968	8834	8700	8567	2.23
30	9233	9098	8964	8830	8695	8562	2.24
40	9230	9095	8960	8826	8691	8558	2.24
50	9227	9092	8957	8822	8687	8553	2.25
62 0	9.999224	9.999089	9.998953	9.998818	9.998683	9.998548	2.25

P. part to sec. of par.	1″	2″	3″	4″	5″	6″	7″	8″	9″		1″	2″	3″	4″	5″	6″	7″	8″	9″
	0	1	1	1	2	2	2	3	3		0	1	1	2	2	3	3	3	4

TABLE XLII.

LOGARITHMIC DIFFERENCE.

Moon's hor. par.	Arg. Apparent Altitude of the Moon's Center.						Diff. to 1' of alt.
	13°	14°	15°	16°	17°	18°	
53' 0"	9.998688	9.998574	9.998460	9.998348	9.998236	9.998124	1.88
10	8683	8569	8454	8342	8230	8117	1.89
20	8678	8563	8449	8336	8223	8110	1.89
30	8673	8558	8443	8329	8217	8104	1.90
40	8668	8553	8437	8323	8210	8097	1.90
50	8663	8547	8432	8317	8203	8090	1.91
54 0	9.998658	9.998542	9.998426	9.998311	9.998197	9.998083	1.92
10	8653	8537	8420	8305	8190	8076	1.92
20	8648	8531	8414	8299	8184	8069	1.93
30	8643	8526	8409	8293	8177	8063	1.93
40	8638	8521	8403	8287	8171	8056	1.94
50	8633	8515	8397	8281	8164	8049	1.95
55 0	9.998628	9.998510	9.998391	9.998275	9.998158	9.998042	1.95
10	8623	8504	8386	8269	8151	8035	1.96
20	8618	8498	8380	8262	8145	8028	1.96
30	8613	8493	8374	8256	8138	8022	1.97
40	8607	8488	8368	8250	8132	8015	1.97
50	8602	8482	8363	8244	8125	8008	1.98
56 0	9.998597	9.998477	9.998357	9.998238	9.998119	9.998001	1.99
10	8592	8471	8351	8232	8112	7994	1.99
20	8587	8466	8345	8225	8106	7987	2.00
30	8582	8460	8339	8219	8099	7981	2.00
40	8577	8455	8334	8213	8093	7974	2.01
50	8572	8449	8328	8207	8086	7967	2.02
57 0	9.998567	9.998444	9.998322	9.998201	9.998080	9.997960	2.02
10	8562	8439	8316	8195	8073	7953	2.03
20	8557	8433	8311	8189	8067	7946	2.03
30	8552	8428	8305	8183	8060	7940	2.04
40	8546	8423	8299	8177	8054	7933	2.04
50	8541	8418	8294	8171	8047	7926	2.05
58 0	9.998536	9.998412	9.998288	9.998165	9.998041	9.997919	2.06
10	8531	8407	8282	8159	8034	7912	2.06
20	8526	8401	8276	8152	8028	7905	2.07
30	8521	8396	8271	8146	8021	7899	2.07
40	8516	8391	8265	8140	8015	7892	2.08
50	8511	8385	8259	8134	8008	7885	2.09
59 0	9.998506	9.998380	9.998253	9.998128	9.998002	9.997878	2.09
10	8501	8374	8247	8122	7995	7871	2.10
20	8496	8369	8241	8115	7989	7864	2.11
30	8491	8363	8235	8109	8982	7858	2.11
40	8486	8358	8230	8103	7976	7851	2.12
50	8481	8352	8224	8097	7969	7844	2.12
60 0	9.998476	9.998347	9.998218	9.998091	9.997963	9.997837	2.13
10	8471	8341	8212	8085	7956	7830	2.14
20	8466	8336	8206	8078	7950	7823	2.14
30	8461	8330	8201	8072	7943	7817	2.15
40	8455	8325	8195	8066	7937	7810	2.15
50	8450	8319	8189	8060	7930	7803	2.16
61 0	9.998445	9.998314	9.998183	9.998054	9.997924	9.997796	2.16
10	8440	8308	8177	8048	7917	7789	2.17
20	8435	8303	8171	8041	7911	7782	2.18
30	8430	8297	8165	8035	7904	7775	2.18
40	8425	8292	8160	8029	7898	7768	2.19
50	8420	8286	8154	8023	7891	7761	2.20
62 0	9.998415	9.998281	9.998148	9.998017	9.997885	9.997754	2.20

P. part. to sec. of par.	1"	2"	3"	4"	5"	6"	7"	8"	9"		1"	2"	3"	4"	5"	6"	7"	8"	9"
	1	1	2	2	3	3	4	4	5		1	1	2	3	3	4	5	5	6

TABLE XLII.

LOGARITHMIC DIFFERENCE.

Moon's hor. par.	Arg. Apparent Altitude of the Moon's Center. 19°	20°	21°	22°	23°	24°	Diff. to 1' of alt.
53′ 0″	9.998013	9.997902	9.997792	9.997684	9.997577	9.997468	1.82
10	8006	7895	7784	7675	7568	7459	1.82
20	7998	7887	7776	7667	7560	7450	1.83
30	7991	7880	7769	7659	7551	7442	1.83
40	7984	7872	7761	7651	7542	7433	1.84
50	7977	7865	7753	7643	7534	7424	1.84
54 0	9.997970	9.997857	9.997745	9.997635	9.997525	9.997416	1.85
10	7963	7850	7737	7626	7517	7406	1.86
20	7955	7842	7730	7618	7508	7397	1.86
30	7948	7835	7722	7610	7500	7388	1.87
40	7941	7827	7714	7602	7491	7380	1.87
50	7934	7820	7706	7594	7483	7371	1.88
55 0	9.997927	9.997812	9.997698	9.997586	9.997474	9.997362	1.88
10	7920	7805	7690	7577	7465	7353	1.89
20	7913	7797	7683	7569	7457	7344	1.90
30	7905	7790	7675	7561	7448	7335	1.90
40	7898	7782	7667	7553	7439	7326	1.91
50	7891	7775	7659	7545	7431	7318	1.91
56 0	9.997884	9.997767	9.997651	9.997537	9.997422	9.997309	1.92
10	7877	7760	7644	7528	7413	7300	1.92
20	7869	7752	7636	7520	7405	7291	1.93
30	7862	7745	7628	7512	7396	7282	1.93
40	7855	7737	7620	7504	7388	7273	1.94
50	7848	7730	7612	7495	7380	7264	1.95
57 0	9.997841	9.997722	9.997604	9.997487	9.997371	9.997255	1.95
10	7834	7715	7596	7479	7362	7246	1.96
20	7826	7707	7588	7470	7354	7237	1.96
30	7819	7700	7580	7462	7345	7228	1.97
40	7812	7692	7572	7454	7336	7220	1.97
50	7805	7685	7564	7446	7328	7211	1.98
58 0	9.997798	9.997677	9.997556	9.997438	9.997319	9.997202	1.99
10	7790	7670	7549	7429	7310	7193	1.99
20	7783	7662	7541	7421	7302	7184	2.00
30	7776	7655	7533	7413	7293	7175	2.00
40	7769	7647	7525	7405	7285	7166	2.01
50	7762	7640	7517	7396	7276	7157	2.02
59 0	9.997754	9.997632	9.997509	9.997388	9.997268	9.997148	2.02
10	7747	7625	7501	7380	7259	7139	2.03
20	7740	7617	7493	7372	7251	7130	2.03
30	7733	7610	7486	7363	7242	7122	2.04
40	7725	7602	7478	7355	7234	7113	2.04
50	7718	7594	7470	7347	7225	7104	2.05
60 0	9.997711	9.997586	9.997462	9.997339	9.997217	9.997095	2.05
10	7704	7579	7454	7330	7208	7086	2.06
20	7696	7571	7446	7322	7199	7077	2.06
30	7689	7564	7438	7314	7191	7068	2.07
40	7682	7556	7430	7306	7182	7060	2.07
50	7675	7549	7422	7297	7174	7051	2.08
61 0	9.997667	9.997541	9.997414	9.997290	9.997165	9.997042	2.08
10	7660	7534	7407	7281	7157	7033	2.09
20	7653	7526	7399	7273	7148	7024	2.10
30	7646	7519	7391	7265	7140	7015	2.10
40	7638	7511	7383	7257	7131	7006	2.11
50	7631	7504	7375	7248	7123	6997	2.11
62 0	9.997624	9.997496	9.997367	9.997240	9.997114	9.996988	2.12

P. part. to sec. of par.	1″	2″	3″	4″	5″	6″	7″	8″	9″	1″	2″	3″	4″	5″	6″	7″	8″	9″
	1	1	2	3	4	4	5	6	7	1	2	3	3	4	5	6	7	8

TABLE XLII.

LOGARITHMIC DIFFERENCE.

Moon's hor. par.	Arg. Apparent Altitude of the Moon's Center. 25°	26°	27°	28°	29°	30°	Diff. to 5' of alt.
53' 0"	9.997363	9.997257	9.997152	9.997048	9.996945	9.996844	1.73
10	7354	7247	7142	7038	6935	6833	1.74
20	7344	7238	7132	7028	6924	6822	1.74
30	7335	7228	7122	7017	6914	6812	1.75
40	7326	7218	7112	7007	6903	6801	1.75
50	7317	7209	7102	6997	6893	6790	1.76
54 0	9.997307	9.997199	9.997092	9.996987	9.996882	9.996779	1.76
10	7298	7190	7083	6977	6872	6768	1.77
20	7289	7180	7073	6966	6861	6757	1.77
30	7280	7171	7063	6956	6851	6746	1.78
40	7270	7161	7053	6946	6840	6735	1.78
50	7261	7152	7043	6936	6830	6724	1.79
55 0	9.997252	9.997142	9.997033	9.996926	9.996819	9.996713	1.79
10	7243	7132	7023	6915	6809	6703	1.80
20	7233	7123	7014	6905	6798	6692	1.80
30	7224	7113	7004	6895	6788	6681	1.81
40	7215	7103	6994	6885	6777	6670	1.81
50	7206	7094	6984	6875	6767	6659	1.82
56 0	9.997196	9.997084	9.996974	9.996864	9.996756	9.996648	1.83
10	7187	7075	6964	6854	6746	6638	1.83
20	7178	7065	6954	6844	6735	6627	1.84
30	7169	7056	6945	6834	6725	6616	1.84
40	7160	7046	6935	6824	6714	6605	1.85
50	7150	7037	6925	6813	6704	6594	1.85
57 0	9.997141	9.997027	9.996915	9.996803	9.996693	9.996583	1.86
10	7132	7018	6905	6793	6683	6573	1.86
20	7123	7008	6895	6783	6672	6562	1.87
30	7113	6999	6885	6772	6662	6551	1.87
40	7104	6989	6876	6762	6651	6540	1.88
50	7095	6980	6866	6752	6641	6529	1.88
58 0	9.997086	9.996970	9.996856	9.996742	9.996630	9.996518	1.89
10	7076	6961	6846	6731	6620	6508	1.89
20	7067	6951	6836	6721	6609	6497	1.90
30	7058	6942	6826	6711	6598	6486	1.91
40	7049	6932	6816	6701	6588	6475	1.91
50	7040	6923	6806	6690	6577	6464	1.92
59 0	9.997030	9.996913	9.996796	9.996680	9.996566	9.996453	1.92
10	7021	6904	6786	6670	6556	6442	1.93
20	7012	6894	6776	6660	6545	6431	1.94
30	7003	6884	6766	6649	6535	6420	1.94
40	6993	6875	6756	6639	6524	6409	1.95
50	6984	6865	6746	6629	6514	6398	1.95
60 0	9.996975	9.996855	9.996737	9.996619	9.996503	9.996387	1.96
10	6966	6846	6727	6608	6493	6377	1.96
20	6956	6836	6717	6598	6482	6366	1.97
30	6947	6827	6707	6588	6472	6355	1.97
40	6938	6817	6697	6578	6461	6344	1.98
50	6929	6808	6687	6567	6450	6333	1.98
61 0	9.996919	9.996798	9.996677	9.996557	9.996439	9.996322	1.99
10	6910	6788	6667	6547	6429	6312	1.99
20	6901	6779	6657	6537	6418	6301	2.00
30	6891	6769	6647	6526	6408	6290	2.00
40	6882	6759	6637	6516	6397	6279	2.01
50	6873	6759	6627	6506	6386	6268	2.02
62 0	9.996863	9.996740	9.996617	9.996496	9.996376	9.996257	2.02

P. part. to	1"	2"	3"	4"	5"	6"	7"	8"	9"	1"	2"	3"	4"	5"	6"	7"	8"	9"
sec. of par.	1	2	3	4	5	6	7	8	9	1	2	3	4	5	6	7	8	9

TABLE XLII.

LOGARITHMIC DIFFERENCE.

Moon's hor. par.	Arg. Apparent Altitude of the Moon's Center. 31°	32°	33°	34°	35°	36°	Diff. to 1′ of alt.
53′ 0″	9.996743	9.996643	9.996545	9.996447	9.996351	9.996256	1.62
10	6732	6632	6533	6435	6339	6243	1.63
20	6721	6620	6521	6423	6326	6231	1.63
30	6710	6609	6510	6411	6314	6218	1.64
40	6699	6598	6498	6399	6302	6205	1.64
50	6688	6586	6486	6386	6289	6193	1.65
54 0	9.996676	9.996575	9.996474	9.996374	9.996277	9.996180	1.65
10	6665	6563	6462	6362	6264	6167	1.66
20	6654	6552	6450	6350	6252	6154	1.67
30	6643	6540	6438	6338	6239	6142	1.67
40	6632	6529	6427	6326	6227	6129	1.68
50	6620	6517	6415	6314	6214	6116	1.68
55 0	9.996609	9.996506	9.996403	9.996302	9.996202	9.996103	1.69
10	6598	6495	6391	6289	6190	6091	1.69
20	6587	6483	6379	6277	6177	6078	1.70
30	6576	6472	6368	6265	6165	6065	1.70
40	6564	6460	6356	6253	6153	6052	1.71
50	6553	6449	6344	6241	6140	6040	1.71
56 0	9.996542	9.996437	9.996332	9.996229	9.996128	9.996027	1.72
10	6531	6426	6321	6217	6115	6014	1.72
20	6520	6414	6309	6205	6103	6002	1.73
30	6508	6403	6297	6193	6090	5989	1.73
40	6497	6391	6285	6181	6078	5976	1.74
50	6486	6380	6273	6169	6065	5963	1.74
57 0	9.996475	9.996368	9.996262	9.996157	9.996053	9.995951	1.75
10	6464	6357	6250	6144	6041	5938	1.75
20	6452	6345	6238	6132	6028	5926	1.76
30	6441	6334	6226	6120	6016	5913	1.76
40	6430	6322	6214	6108	6003	5900	1.77
50	6419	6311	6203	6096	5991	5887	1.77
58 0	9.996408	9.996299	9.996191	9.996084	9.995979	9.995875	1.78
10	6396	6288	6179	6072	5966	5862	1.78
20	6385	6276	6167	6060	5954	5849	1.79
30	6374	6265	6155	6047	5941	5836	1.79
40	6363	6253	6144	6035	5929	5824	1.80
50	6352	6242	6132	6023	5916	5811	1.80
59 0	9.996340	9.996230	9.996120	9.996011	9.995904	9.995798	1.81
10	6329	6219	6108	5999	5892	5785	1.81
20	6318	6207	6096	5987	5879	5773	1.82
30	6307	6196	6085	5975	5867	5760	1.82
40	6296	6184	6073	5963	5855	5747	1.83
50	6284	6173	6061	5951	5842	5735	1.83
60 0	9.996273	9.996161	9.996049	9.995939	9.995830	9.995722	1.84
10	6262	6150	6037	5926	5817	5709	1.84
20	6251	6138	6025	5914	5805	5696	1.85
30	6240	6126	6014	5902	5792	5684	1.85
40	6228	6115	6002	5890	5780	5671	1.86
50	6217	6103	5990	5878	5767	5658	1.86
61 0	9.996206	9.996091	9.995978	9.995866	9.995755	9.995645	1.87
10	6195	6080	5966	5853	5742	5633	1.87
20	6184	6068	5954	5841	5730	5620	1.88
30	6173	6057	5943	5829	5717	5607	1.88
40	6161	6045	5931	5817	5705	5594	1.89
50	6150	6034	5919	5805	5692	5582	1.89
62 0	9.996139	9.996022	9.995907	9.995793	9.995680	9.995569	1.90

P. part. to sec. of par.	1″	2″	3″	4″	5″	6″	7″	8″	9″	1″	2″	3″	4″	5″	6″	7″	8″	9″
	1	2	3	5	6	7	8	9	10	1	2	4	5	6	7	9	10	11

TABLE XLII.

LOGARITHMIC DIFFERENCE.

Moon's hor. par.	Arg. Apparent Altitude of the Moon's Center. 37°	38°	39°	40°	41°	42°	Diff. to 1′ of alt.
53′ 0″	9.996162	9.996069	9.995977	9.995887	9.995798	9.995711	1.50
10	6149	6056	5964	5873	5784	5697	1.51
20	6136	6042	5950	5859	5770	5682	1.51
30	6123	6029	5937	5846	5756	5668	1.52
40	6110	6016	5923	5832	5741	5653	1.52
50	6097	6003	5910	5818	5727	5639	1.53
54 0	9.996084	9.995989	9.995896	9.995804	9.995713	9.995624	1.53
10	6071	5976	5883	5790	5699	5610	1.54
20	6058	5963	5869	5776	5685	5596	1.54
30	6045	5950	5856	5762	5671	5581	1.55
40	6032	5936	5842	5748	5675	5567	1.55
50	6019	5923	5829	5735	5643	5552	1.56
55 0	9.996006	9.995910	9.995815	9.995721	9.995628	9.995538	1.56
10	5993	5896	5801	5707	5614	5523	1.57
20	5980	5883	5788	5693	5600	5509	1.57
30	5967	5870	5774	5679	5586	5494	1.58
40	5954	5856	5761	5665	5572	5480	1.58
50	5941	5843	5747	5652	5558	5465	1.59
56 0	9.995928	9.995830	9.995733	9.995638	9.995544	9.995451	1.59
10	5915	5816	5720	5624	5529	5437	1.59
20	5902	5803	5706	5610	5515	5422	1.60
30	5889	5790	5693	5596	5501	5408	1.60
40	5876	5776	5679	5582	5487	5394	1.61
50	5861	5763	5666	5568	5473	5379	1.61
57 0	9.995850	9.995750	9.995652	9.995554	9.995458	9.995365	1.62
10	5837	5736	5638	5541	5444	5350	1.62
20	5824	5723	5625	5527	5430	5336	1.63
30	5811	5710	5611	5513	5416	5321	1.63
40	5798	5696	5597	5499	5402	5307	1.64
50	5785	5683	5584	5485	5388	5292	1.64
58 0	9.995772	9.995670	9.995570	9.995471	9.995374	9.995278	1.65
10	5759	5656	5556	5457	5359	5263	1.65
20	5746	5643	5543	5443	5345	5249	1.66
30	5733	5630	5529	5429	5331	5234	1.66
40	5720	5616	5516	5416	5317	5220	1.67
50	5707	5603	5502	5402	5303	5205	1.67
59 0	9.995694	9.995590	9.995488	9.995388	9.995289	9.995191	1.68
10	5680	5576	5475	5374	5274	5177	1.68
20	5667	5563	5461	5360	5260	5163	1.68
30	5654	5550	5447	5346	5246	5148	1.69
40	5641	5536	5434	5332	5232	5134	1.69
50	5628	5523	5420	5318	5218	5119	1.70
60 0	9.995615	9.995510	9.995407	9.995304	9.995204	9.995105	1.70
10	5602	5496	5393	5291	5189	5090	1.71
20	5589	5483	5379	5277	5175	5076	1.71
30	5576	5470	5366	5263	5161	5061	1.72
40	5563	5456	5352	5249	5147	5047	1.72
50	5550	5443	5338	5235	5133	5032	1.73
61 0	9.995537	9.995430	9.995325	9.995221	9.995119	9.995018	1.73
10	5524	5416	5311	5207	5104	5003	1.74
20	5511	5403	5297	5193	5090	4989	1.74
30	5498	5390	5284	5180	5076	4974	1.75
40	5485	5376	5270	5166	5062	4960	1.75
50	5472	5363	5256	5152	5048	4945	1.76
62 0	9.995459	9.995350	9.995243	9.995138	9.995033	9.994931	1.76

P. part. to sec. of par.	1″	2″	3″	4″	5″	6″	7″	8″	9″	1″	2″	3″	4″	5″	6″	7″	8″	9″
	1	3	4	5	7	8	9	11	12	1	3	4	6	7	8	10	11	13

TABLE XLII.

LOGARITHMIC DIFFERENCE.

Moon's hor. par.	Arg. Apparent Altitude of the Moon's Center. 43°	44°	45°	46°	47°	48°	Diff. to 1' of alt.
53′ 0″	9.995625	9.995540	9.995456	9.995374	9.995294	9.995215	1.37
10	5610	5525	5441	5359	5279	5199	1.37
20	5595	5510	5426	5344	5263	5183	1.37
30	5581	5495	5410	5328	5247	5167	1.38
40	5566	5480	5395	5313	5231	5151	1.38
50	5551	5465	5380	5298	5216	5135	1.39
54 0	9.995536	9.995450	9.995365	9.995282	9.995200	9.995119	1.39
10	5522	5435	5350	5267	5184	5103	1.40
20	5507	5420	5334	5251	5168	5087	1.40
30	5492	5405	5319	5236	5153	5071	1.40
40	5478	5390	5304	5220	5137	5055	1.41
50	5463	5375	5289	5205	5121	5039	1.41
55 0	9.995448	9.995360	9.995274	9.995189	9.995105	9.995023	1.42
10	5434	5345	5258	5174	5089	5007	1.42
20	5419	5331	5243	5158	5074	4991	1.43
30	5404	5316	5228	5143	5058	4975	1.43
40	5390	5301	5213	5127	5042	4959	1.44
50	5375	5286	5197	5112	5026	4943	1.44
56 0	9.995360	9.995271	9.995182	9.995096	9.995011	9.994927	1.44
10	5346	5256	5167	5080	4995	4911	1.45
20	5331	5241	5152	5065	4979	4895	1.45
30	5316	5226	5136	5049	4964	4879	1.46
40	5302	5211	5121	5034	4948	4863	1.46
50	5287	5196	5106	5018	4932	4847	1.47
57 0	9.995272	9.995181	9.995091	9.995003	9.994916	9.994831	1.47
10	5258	5166	5076	4987	4901	4815	1.48
20	5243	5151	5060	4972	4885	4799	1.48
30.	5228	5136	5045	4956	4869	4783	1.48
40	5214	5121	5030	4941	4853	4767	1.49
50	5199	5106	5015	4925	4837	4751	1.49
58 0	9.995184	9.995091	9.995000	9.994910	9.994822	9.994735	1.50
10	5169	5076	4984	4894	4806	4719	1.50
20	5155	5061	4969	4879	4790	4703	1.51
30	5140	5046	4954	4863	4775	4687	1.51
40	5125	5031	4939	4848	4759	4671	1.51
50	5110	5016	4923	4832	4743	4655	1.52
59 0	9.995095	9.995001	9.994908	9.994817	9.994727	9.994639	1.52
10	5081	4986	4893	4791	4712	4623	1.53
20	5066	4971	4878	4786	4696	4607	1.53
30	5051	4956	4862	4770	4680	4591	1.53
40	5037	4941	4847	4755	4665	4575	1.54
50	5022	4926	4832	4739	4649	4559	1.54
60 0	9.995007	9.994911	9.994817	9.994724	9.994633	9.994543	1.55
10	4993	4896	4801	4708	4617	4527	1.55
20	4978	4881	4786	4693	4601	4511	1.56
30	4963	4866	4771	4677	4586	4495	1.56
40	4949	4851	4756	4663	4570	4479	1.57
50	4934	4836	4740	4646	4554	4463	1.57
61 0	9.994919	9.994821	9.994725	9.994631	9.994538	9.994447	1.57
10	4904	4806	4710	4615	4523	4431	1.58
20	4890	4791	4695	4600	4507	4415	1.58
30	4875	4776	4679	4584	4492	4399	1.59
40	4860	4761	4664	4569	4476	4383	1.59
50	4845	4746	4649	4553	4460	4367	1.59
62 0	9.994830	9.994731	9.994634	9.994538	9.994444	9.994351	1.60

P. part to Sec. of par.	1″	2″	3″	4″	5″	6″	7″	8″	9″		1″	2″	3″	4″	5″	6″	7″	8″	9″
	1	3	4	6	7	9	10	12	13		2	3	5	6	8	9	11	12	14

TABLE XLII.

LOGARITHMIC DIFFERENCE.

Moon's hor. par.	Arg. Apparent Altitude of the Moon's Center. 49°	50°	51°	52°	53°	54°	Diff. to 1' of alt.
53′ 0″	9.995137	9.995061	9.994986	9.994913	9.994842	9.994773	1.21
10	5121	5045	4969	4896	4825	4756	1.22
20	5105	5028	4952	4879	4808	4738	1.22
30	5089	5012	4936	4862	4791	4721	1.23
40	5072	4995	4919	4846	4774	4703	1.23
50	5056	4979	4902	4829	4757	4686	1.23
54 0	9.995040	9.994962	9.994886	9.994812	9.994739	9.994668	1.24
10	5024	4946	4869	4795	4722	4651	1.24
20	5008	4930	4852	4778	4705	4634	1.25
30	4991	4913	4836	4761	4688	4616	1.25
40	4975	4897	4819	4744	4671	4599	1.25
50	4959	4880	4802	4727	4654	4582	1.26
55 0	9.994943	9.994864	9.994786	9.994710	9.994637	9.994564	1.26
10	4926	4847	4769	4693	4619	4547	1.26
20	4910	4831	4752	4676	4602	4530	1.27
30	4894	4814	4736	4660	4585	4512	1.27
40	4878	4798	4719	4643	4568	4495	1.28
50	4861	4781	4702	4626	4551	4478	1.28
56 0	9.994845	9.994765	9.994686	9.994609	9.994534	9.994460	1.28
10	4829	4748	4669	4592	4516	4443	1.29
20	4813	4732	4652	4575	4499	4425	1.29
30	4796	4715	4636	4558	4482	4408	1.29
40	4780	4699	4619	4541	4465	4391	1.30
50	4764	4682	4602	4524	4448	4373	1.30
57 0	9.994748	9.994666	9.994586	9.994507	9.994431	9.994356	1.31
10	4731	4649	4569	4490	4414	4339	1.31
20	4715	4633	4552	4473	4396	4321	1.31
30	4699	4616	4536	4457	4379	4304	1.32
40	4683	4600	4519	4440	4362	4287	1.32
50	4666	4583	4502	4423	4341	4269	1.32
58 0	9.994650	9.994567	9.994486	9.994406	9.994328	9.994254	1.33
10	4634	4550	4469	4389	4311	4234	1.33
20	4618	4534	4452	4372	4294	4217	1.34
30	4601	4517	4436	4355	4276	4200	1.34
40	4585	4501	4419	4338	4259	4182	1.34
50	4569	4484	4402	4321	4242	4165	1.35
59 0	9.994553	9.994468	9.994385	9.994304	9.994225	9.994147	1.35
10	4536	4451	4369	4287	4208	4130	1.35
20	4520	4435	4352	4270	4191	4112	1.36
30	4504	4418	4335	4253	4174	4095	1.36
40	4488	4402	4319	4236	4156	4078	1.37
50	4472	4385	4302	4219	4139	4060	1.37
60 0	9.994455	9.994369	9.994285	9.994202	9.994122	9.994043	1.37
10	4439	4352	4269	4185	4105	4026	1.38
20	4423	4336	4252	4168	4088	4008	1.38
30	4407	4319	4235	4152	4071	3991	1.38
40	4390	4303	4219	4135	4054	3974	1.39
50	4374	4286	4202	4118	4036	3956	1.39
61 0	9.994358	9.994270	9.994185	9.994101	9.994019	9.993939	1.40
10	4342	4253	4168	4084	4002	3921	1.40
20	4325	4237	4152	4067	3985	3904	1.40
30	4309	4220	4135	4050	3968	3886	1.41
40	4293	4204	4118	4033	3951	3869	1.41
50	4277	4187	4101	4016	3934	3854	1.42
62 0	9.994260	9.994171	9.994084	9.993999	9.993916	9.993834	1.42

P. part. to sec. of par.	1″	2″	3″	4″	5″	6″	7″	8″	9″	1″	2″	3″	4″	5″	6″	7″	8″	9″
	2	3	5	7	8	10	12	13	15	2	3	5	7	9	10	12	14	15

TABLE XLII.

LOGARITHMIC DIFFERENCE.

Moon's hor. par.	Arg. Apparent Altitude of the Moon's Center. 55°	56°	57°	58°	59°	60°	Diff. to 1' of alt.
53′ 0″	9.994705	9.994639	9.994574	9.994512	9.994451	9.994392	1.04
10	4687	4621	4556	4494	4432	4374	1.04
20	4670	4603	4538	4476	4414	4355	1.05
30	4652	4586	4520	4458	4396	4337	1.05
40	4635	4568	4502	4439	4377	4318	1.05
50	4617	4550	4484	4421	4359	4300	1.06
54 0	9.994599	9.994532	9.994466	9.994403	9.994341	9.994281	1.06
10	4582	4514	4448	4385	4322	4263	1.06
20	4564	4496	4430	4366	4304	4244	1.07
30	4547	4479	4412	4348	4285	4225	1.07
40	4529	4461	4394	4330	4267	4207	1.07
50	4512	4443	4376	4312	4249	4188	1.08
55 0	9.994494	9.994425	9.994358	9.994294	9.994230	9.994169	1.08
10	4476	4408	4340	4275	4212	4151	1.08
20	4459	4390	4322	4257	4194	4132	1.09
30	4441	4372	4304	4239	4175	4114	1.09
40	4423	4354	4286	4221	4157	4095	1.09
50	4406	4337	4268	4203	4139	4077	1.10
56 0	9.994388	9.994319	9.994250	9.994184	9.994120	9.994058	1.10
10	4371	4301	4232	4166	4102	4039	1.10
20	4353	4283	4214	4148	4083	4021	1.11
30	4336	4265	4196	4130	4065	4002	1.11
40	4318	4247	4178	4112	4046	3984	1.11
50	4301	4230	4160	4093	4028	3965	1.12
57 0	9.994283	9.994212	9.994142	9.994075	9.994009	9.993946	1.12
10	4265	4194	4124	4057	3991	3928	1.12
20	4248	4176	4106	4039	3973	3909	1.13
30	4230	4158	4088	4021	3954	3891	1.13
40	4213	4140	4070	4002	3936	3872	1.14
50	4195	4123	4052	3984	3918	3854	1.14
58 0	9.994177	9.994105	9.994034	9.993966	9.993899	9.993835	1.14
10	4160	4087	4016	3948	3881	3816	1.15
20	4142	4069	3998	3930	3863	3798	1.15
30	4125	4051	3980	3911	3844	3779	1.15
40	4107	4033	3960	3893	3826	3760	1.16
50	4090	4016	3944	3875	3808	3742	1.16
59 0	9.994072	9.993998	9.993926	9.993857	9.993789	9.993723	1.16
10	4054	3980	3908	3839	3771	3705	1.17
20	4037	3962	3890	3820	3753	3686	1.17
30	4019	3944	3872	3802	3734	3668	1.17
40	4002	3927	3854	3784	3716	3649	1.18
50	3984	3909	3836	3766	3697	3630	1.18
60 0	9.993966	9.993891	9.993818	9.993748	9.993679	9.993612	1.18
10	3949	3873	3800	3729	3660	3593	1.19
20	3931	3855	3782	3711	3642	3574	1.19
30	3913	3837	3764	3693	3623	3556	1.19
40	3896	3820	3746	3675	3605	3537	1.20
50	3878	3802	3728	3657	3587	3518	1.20
61 0	9.993860	9.993784	9.993710	9.993638	9.993568	9.993500	1.20
10	3843	3766	3692	3620	3550	3481	1.21
20	3825	3748	3674	3602	3532	3462	1.21
30	3808	3731	3656	3584	3513	3444	1.21
40	3790	3713	3638	3566	3495	3425	1.22
50	3773	3695	3620	3548	3477	3406	1.22
62 0	9.993755	9.993677	9.993602	9.993529	9.993458	9.993388	1.22

P. part. to sec. of par.	1″	2″	3″	4″	5″	6″	7″	8″	9″	1″	2″	3″	4″	5″	6″	7″	8″	9″
	2	4	5	7	9	11	12	14	16	2	4	6	7	9	11	13	15	17

TABLE XLII.

LOGARITHMIC DIFFERENCE.

Moon's hor. par.	Arg. Apparent Altitude of the Moon's Center. 61°	62°	63°	64°	65°	66°	Diff. to 1' of alt.
53′ 0″	9.994335	9.994279	9.994224	9.994172	9.994122	9.994073	0.87
10	4316	4260	4205	4153	4102	4053	0.88
20	4297	4241	4186	4134	4083	4034	0.88
30	4279	4222	4167	4114	4063	4014	0.88
40	4260	4203	4148	4095	4044	3995	0.88
50	4241	4184	4129	4076	4025	3975	0.89
54 0	9.994222	9.994165	9.994110	9.994057	9.994005	9.993956	0.89
10	4204	4146	4091	4038	3986	3936	0.89
20	4185	4127	4072	4018	3966	3917	0.89
30	4166	4109	4053	3999	3947	3897	0.90
40	4147	4090	4034	3980	3927	3877	0.90
50	4129	4071	4015	3961	3908	3858	0.90
55 0	9.994110	9.994052	9.993996	9.993941	9.993888	9.993838	0.91
10	4091	4033	3977	3922	3869	3819	0.91
20	4072	4014	3957	3903	3850	3799	0.91
30	4054	3995	3938	3884	3830	3780	0.91
40	4035	3976	3919	3864	3811	3760	0.92
50	4016	3957	3900	3845	3791	3740	0.92
56 0	9.993997	9.993938	9.993881	9.993826	9.993772	9.993721	0.92
10	3978	3919	3862	3807	3753	3701	0.92
20	3959	3900	3843	3787	3733	3682	0.93
30	3940	3882	3824	3768	3714	3662	0.93
40	3922	3863	3805	3749	3695	3642	0.93
50	3903	3844	3786	3730	3675	3623	0.93
57 0	9.993885	9.993825	9.993766	9.993710	9.993656	9.993603	0.94
10	3866	3806	3747	3691	3636	3584	0.94
20	3847	3787	3728	3672	3617	3564	0.94
30	3828	3768	3709	3652	3598	3545	0.94
40	3810	3749	3690	3633	3578	3525	0.95
50	3791	3730	3671	3614	3559	3505	0.95
58 0	9.993772	9.993711	9.993652	9.993594	9.993539	9.993486	0.95
10	3753	3692	3632	3575	3520	3466	0.96
20	3735	3673	3613	3556	3500	3447	0.96
30	3716	3654	3594	3537	3481	3427	0.96
40	3697	3635	3575	3517	3461	3407	0.97
50	3678	3616	3556	3498	3442	3388	0.97
59 0	9.993660	9.993597	9.993537	9.993479	9.993422	9.993368	0.97
10	3641	3579	3518	3459	3403	3349	0.97
20	3622	3560	3498	3440	3384	3329	0.98
30	3603	3541	3479	3421	3364	3310	0.98
40	3584	3522	3460	3402	3345	3290	0.98
50	3566	3503	3441	3382	3325	3270	0.98
60 0	9.993547	9.993484	9.993422	9.993363	9.993306	9.993251	0.99
10	3528	3465	3403	3344	3286	3231	0.99
20	3509	3446	3384	3324	3267	3212	0.99
30	3490	3427	3365	3305	3248	3192	0.99
40	3472	3408	3346	3286	3228	3172	1.00
50	3453	3389	3327	3267	3209	3153	1.00
61 0	9.993434	9.993370	9.993308	9.993247	9.993189	9.993133	1.00
10	3415	3351	3289	3228	3170	3114	1.00
20	3396	3332	3270	3209	3150	3094	1.01
30	3378	3313	3250	3189	3131	3075	1.01
40	3359	3294	3231	3170	3111	3055	1.01
50	3340	3275	3212	3151	3092	3035	1.02
62 0	9.993321	9.993256	9.993193	9.993131	9.993072	9.993016	1.02

P. part. to inc. of par.	1″	2″	3″	4″	5″	6″	7″	8″	9″	1″	2″	3″	4″	5″	6″	7″	8″	9″
	2	4	6	8	9	11	13	15	17	2	4	6	8	10	12	14	16	17

TABLE XLII.

LOGARITHMIC DIFFERENCE.

Moon's hor. par.	Arg. Apparent Altitude of the Moon's Center. 67°	68°	69°	70°	71°	72°	Diff. to 1' of alt.
53' 0"	9.994027	9.993982	9.993939	9.993898	9.993858	9.993822	0.68
10	4007	3962	3919	3878	3838	3802	0.68
20	3987	3942	3899	3858	3817	3781	0.69
30	3968	3922	3879	3838	3797	3761	0.69
40	3948	3902	3859	3818	3777	3740	0.69
50	3928	3882	3839	3798	3757	3720	0.69
54 0	9.993908	9.993862	9.993818	9.993777	9.993737	9.993699	0.70
10	3889	3843	3798	3757	3716	3679	0.70
20	3869	3823	3778	3737	3696	3658	0.70
30	3849	3803	3758	3717	3676	3638	0.70
40	3830	3783	3738	3697	3656	3617	0.71
50	3810	3763	3718	3677	3636	3597	0.71
55 0	9.993790	9.993743	9.993698	9.993656	9.993615	9.993576	0.71
10	3770	3724	3678	3636	3595	3556	0.71
20	3751	3704	3658	3616	3575	3536	0.72
30	3731	3684	3638	3596	3555	3515	0.72
40	3711	3664	3618	3576	3535	3495	0.72
50	3692	3644	3598	3556	3514	3475	0.72
56 0	9.993672	9.993624	9.993578	9.993535	9.993494	9.993455	0.72
10	3652	3605	3558	3515	3474	3434	0.73
20	3632	3585	3538	3495	3454	3414	0.73
30	3613	3565	3518	3475	3434	3394	0.73
40	3593	3545	3498	3455	3413	3373	0.73
50	3573	3525	3478	3435	3393	3353	0.73
57 0	9.993553	9.993505	9.993458	9.993415	9.993373	9.993332	0.74
10	3534	3486	3438	3394	3353	3312	0.74
20	3514	3466	3418	3374	3332	3292	0.74
30	3494	3446	3398	3354	3312	3272	0.74
40	3474	3426	3378	3334	3292	3251	0.74
50	3455	3406	3358	3314	3272	3231	0.75
58 0	9.993435	9.993386	9.993338	9.993294	9.993251	9.993211	0.75
10	3415	3366	3318	3273	3231	3190	0.75
20	3396	3346	3298	3253	3211	3170	0.75
30	3376	3327	3278	3233	3191	3149	0.76
40	3356	3307	3258	3213	3170	3129	0.76
50	3336	3287	3238	3193	3150	3108	0.76
59 0	9.993317	9.993267	9.993218	9.993173	9.993130	9.993088	0.76
10	3297	3247	3198	3152	3110	3068	0.76
20	3277	3227	3178	3132	3089	3047	0.77
30	3258	3207	3158	3112	3069	3027	0.77
40	3238	3187	3138	3092	3049	3007	0.77
50	3218	3167	3118	3072	3029	2986	0.77
60 0	9.993198	9.993147	9.993098	9.993052	9.993008	9.992966	0.77
10	3179	3128	3078	3031	2988	2946	0.78
20	3159	3108	3058	3011	2968	2925	0.78
30	3139	3088	3038	2991	2948	2905	0.78
40	3119	3068	3018	2971	2927	2885	0.78
50	3100	3048	2998	2951	2907	2864	0.79
61 0	9.993080	9.993028	9.992978	9.992931	9.992887	9.992844	0.79
10	3060	3008	2958	2910	2867	2824	0.79
20	3040	2988	2938	2890	2846	2803	0.79
30	3021	2968	2918	2870	2826	2783	0.79
40	3001	2948	2898	2850	2806	2762	0.80
50	2981	2928	2878	2830	2786	2742	0.80
62 0	9.992961	9.992908	9.992858	9.992810	9.992765	9.992721	0.80

P. part. to sec. of par.	1"	2"	3"	4"	5"	6"	7"	8"	9"
	2	4	6	8	10	12	14	16	18
	1"	2"	3"	4"	5"	6"	7"	8"	9"
	2	4	6	8	10	12	14	16	18

TABLE XLII.

LOGARITHMIC DIFFERENCE.

Moon's hor. par.	Arg. Apparent Altitude of the Moon's Center.						
	73°	74°	75°	76°	77°	78°	79°
53′ 0″	9.993788	9.993753	9.993721	9.993691	9.993664	9.993640	9.993617
10	3767	3732	3700	3670	3643	3619	3596
20	3746	3712	3679	3649	3623	3598	2575
30	3726	3691	3659	3629	3602	3577	3554
40	3705	3671	3638	3608	3581	3556	3533
50	3684	3650	3617	3587	3560	3535	3512
54 0	9.993664	9.993630	9.993597	9.993567	9.993540	9.993514	9.993491
10	3643	3609	3576	3546	3519	3493	3470
20	3623	3588	3555	3525	3498	3472	3449
30	3602	3568	3535	3505	3477	3451	3428
40	3582	3547	3514	3484	3456	3430	3407
50	3561	3526	3493	3463	3436	3409	3386
55 0	9.993541	9.993506	9.993473	9.993443	9.993415	9.993388	9.993365
10	3520	3485	3452	3422	3394	3368	3344
20	3500	3464	3431	3401	3373	3347	3323
30	3479	3444	3411	3380	3352	3326	3302
40	3459	3423	3390	3360	3331	3305	3281
50	3438	3402	3369	3339	3310	3284	3260
56 0	9.993418	.993382	9.993349	9.993318	9.993289	9.993263	9.993239
10	3397	3361	3328	3297	3269	3242	3218
20	3377	3340	3307	3276	3248	3221	3197
30	3356	3320	3287	3255	3227	3200	3176
40	3336	3299	3266	3234	3206	3179	3155
50	3315	3278	3245	3214	3185	3158	3134
57 0	9.993295	9.993258	9.993225	9.993193	9.993164	9.993137	9.993113
10	3274	3237	3204	3172	3144	3117	3092
20	3254	3217	3183	3151	3123	3096	3071
30	3233	3196	3163	3131	3102	3075	3050
40	3213	3176	3142	3110	3081	3054	3029
50	3192	3155	3121	3089	3060	3033	3008
58 0	9.993172	9.993135	9.993101	9.993068	9.993039	9.993012	9.992987
10	3151	3114	3080	3047	3018	2991	2966
20	3131	3094	3059	3027	2997	2970	2945
30	3110	3073	3039	3006	2977	2949	2924
40	3090	3053	3018	2985	2956	2928	2902
50	3069	3032	2997	2964	2935	2907	2881
59 0	9.993049	9.993012	9.992977	9.992944	9.992914	9.992886	9.992860
10	3028	2991	2956	2923	2893	2865	2839
20	3008	2970	2935	2902	2872	2844	2818
30	2987	2950	2915	2881	2851	2823	2797
40	2967	2929	2894	2860	2830	2802	2776
50	2946	2908	2873	2840	2809	2781	2755
60 0	9.992926	9.992888	9.992853	9.992819	9.992788	9.992760	9.992734
10	2905	2867	2832	2798	2768	2739	2713
20	2885	2847	2811	2777	2747	2718	2692
30	2864	2826	2791	2756	2726	2697	2671
40	2844	2806	2770	2736	2705	2676	2650
50	2823	2785	2749	2715	2684	2655	2629
61 0	9.992803	9.992765	9.992729	9.992694	9.992663	9.992634	9.992608
10	2782	2744	2708	2673	2642	2614	2587
20	2762	2723	2687	2653	2621	2593	2566
30	2741	2703	2666	2632	2600	2572	2545
40	2721	2682	2645	2611	2580	2551	2524
50	2700	2661	2625	2590	2559	2530	2503
62 0	9.992680	9.992641	9.992604	9.992570	9.992538	9.992509	9.992482

P. part to sec. of par.	1″	2″	3″	4″	5″	6″	7″	8″	9″
	2	4	6	8	10	12	15	17	19

TABLE XLII.

LOGARITHMIC DIFFERENCE.

Moon's hor. par.	Arg. Apparent Altitude of the Moon's Center.						
	80°	81°	82°	83°	84°	85°	86°
53′ 0″	9.993595	9.993577	9.993560	9.993545	9.993532	9.993521	9.993512
10	3574	3556	3539	3523	3511	3500	3491
20	3553	3535	3518	3502	3489	3478	3469
30	3532	3513	3496	3481	3468	3457	3448
40	3511	3492	3475	3460	3447	3436	3427
50	3490	3471	3454	3439	3426	3415	3405
54 0	9.993469	9.993450	9.993433	9.993418	9.993405	9.993393	9.993384
10	3448	3429	3412	3397	3383	3372	3362
20	3427	3408	3390	3375	3362	3351	3341
30	3406	3387	3369	3354	3341	3329	3319
40	3385	3365	3348	3333	3320	3308	3298
50	3364	3344	3327	3312	3298	3287	3277
55 0	9.993343	9.973323	9.993306	9.993290	9.993277	9.993265	9.993255
10	3322	3302	3284	3269	3256	3244	3234
20	3301	3281	3263	3247	3234	3223	3213
30	3280	3260	3242	3226	3213	3201	3191
40	3259	3238	3221	3205	3192	3180	3170
50	3238	3217	3200	3184	3171	3159	3149
56 0	9.993217	9.993196	9.993178	9.993162	9.993149	9.993137	9.993127
10	3196	3175	3157	3141	3128	3116	3106
20	3175	3154	3136	3120	3106	3095	3085
30	3154	3133	3115	3098	3085	3073	3063
40	3133	3111	3094	3077	3064	3052	3042
50	3111	3090	3072	3056	3042	3031	3021
57 0	9.993090	9.993069	9.993051	9.993035	9.993021	9.993009	9.992999
10	3069	3048	3030	3013	2999	2988	2978
20	3048	3027	3009	2992	2978	2967	2957
30	3027	3006	2988	2971	2957	2946	2935
40	3006	2985	2966	2950	2936	2924	2914
50	2985	2964	2945	2928	2914	2903	2893
58 0	9.992964	9.992943	9.992924	9.992907	9.992893	9.992881	9.992871
10	2943	2921	2903	2886	2871	2860	2850
20	2922	2900	2881	2865	2850	2839	2829
30	2901	2879	2860	2843	2829	2817	2807
40	2880	2858	2839	2822	2807	2796	2786
50	2858	2837	2818	2801	2786	2775	2765
59 0	9.992837	9.992816	9.992797	9.992780	9.992765	9.992753	9.992743
10	2816	2795	2775	2758	2744	2732	2422
20	2795	2773	2754	2737	2722	2711	2701
30	2774	2752	2733	2716	2701	2690	2679
40	2752	2731	2712	2695	2680	2668	2658
50	2731	2710	2691	2673	2659	2647	2637
60 0	9.992510	9.992689	9.992669	9.992652	9.992638	9.992625	9.992615
10	2689	2668	2648	2631	2616	2604	2594
20	2668	2647	2627	2610	2595	2583	2573
30	2647	2626	2606	2588	2574	2561	2551
40	2626	2604	2585	2567	2552	2540	2530
50	2605	2583	2563	2546	2531	2519	2508
61 0	9.992584	9.992562	9.992542	9.992525	9.992510	9.992497	9.992487
10	2563	2541	2521	2503	2488	2476	2465
20	2542	2520	2500	2482	2467	2455	2444
30	2520	2499	2479	2461	2446	2433	2423
40	2499	2478	2457	2440	2424	2412	2401
50	2478	2456	2436	2418	2403	2391	2380
62 0	9.992457	9.992435	9.992415	9.992397	9.992382	9.992369	9.992358

P. part to sec. of par.									
	1″	2″	3″	4″	5″	6″	7″	8″	9″
	2	4	6	8	11	13	15	17	19

TABLE XLII.

LOGARITHMIC DIFFERENCE.

Moon's hor. parallax.	Arg. App. Alt. of Moon's Center. 87°	88°	89°	90°
53′ 0″	9.993505	9.993499	9.993497	9.993495
10	3483	3477	3475	3474
20	3462	3456	3454	3452
30	3441	3435	3432	3431
40	3419	3413	3411	3409
50	3398	3392	3389	3388
54 0	9.993377	9.993371	9.993368	9.993366
10	3351	3349	3346	3345
20	3334	3328	3325	3323
30	3313	3306	3303	3301
40	3291	3285	3282	3280
50	3270	3264	3261	3258
55 0	9.993248	9.993242	9.993239	9.993237
10	3227	3221	3218	3215
20	3205	3199	3197	3194
30	3184	3178	3175	3172
40	3163	3156	3154	3151
50	3141	3135	3132	3129
56 0	9.993120	9.993114	9.993111	9.993108
10	3098	3092	3090	3087
20	3077	3071	3068	3065
30	3056	3049	3047	3044
40	3034	3028	3025	3023
50	3013	3007	3004	3001
57 0	9.992991	9.992985	9.992982	9.992980
10	2970	2964	2961	2959
20	2949	2943	2940	2938
30	2927	2921	2918	2917
40	2906	2900	2897	2895
50	2885	2879	2876	2874
58 0	9.992863	9.992857	9.992854	9.992853
10	2842	2836	2833	2831
20	2821	2814	2812	2810
30	2799	2791	2792	2788
40	2778	2771	2769	2767
50	2757	2750	2748	2746
59 0	9.992735	9.992729	9.992726	9.992725
10	2714	2707	2705	2703
20	2693	2686	2683	2682
30	2672	2664	2662	2660
40	2650	2643	2640	2639
50	2629	2622	2619	2627
60 0	9.992607	9.992600	9.992597	9.992596
10	2586	2579	2576	2574
20	2565	2558	2555	2553
30	2543	2537	2533	2531
40	2522	2515	2512	2510
50	2501	2494	2490	2488
61 0	9.992479	9.992472	9.992469	9.992467
10	2458	2452	2448	2445
20	2436	2430	2426	2424
30	2415	2408	2405	2402
40	2393	2387	2385	2381
50	2372	2366	2362	2359
62 0	9.992350	9.992344	9.992340	9.992338

P. part. to sec. of par.	1″	2″	3″	4″	5″	6″	7″	8″	9″
	2	4	6	9	11	13	15	17	19

TABLE XLIII.

Correction of the Logarithmic Difference, when the Moon's Distance from the Sun is observed.

App. alt. of Sun's center.	Cor. Subtract.	Ap. alt. S's cen.	Cor. Subtract.	Ap. alt. S's cen.	Cor. Subtract.
0° 0′	140	6°	11	40°	12
0 10	126	7	9	41	13
0 20	114	8	8	42	13
0 30	104	9	7	44	13
0 40	94	10	7	46	14
0 50	85	12	7	48	14
1 0	77	14	7	49	14
1 10	[illegible]	16	7	50	15
1 20	64	18	7	52	15
1 30	58	20	7	54	15
1 40	53	21	8	55	16
1 50	49	22	8	56	16
2 0	45	24	8	58	16
2 10	41	25	8	60	16
2 20	38	26	9	62	16
2 30	35	28	9	63	17
2 40	33	29	10	65	17
2 50	30	30	10	68	17
3 0	28	32	10	69	18
3 20	25	33	11	70	18
3 40	22	34	11	75	18
4 0	19	36	11	80	18
4 30	16	37	12	85	18
5 0	14	38	12	90	18

TABLE XLIV.

Correction of the Log. Difference, when the Moon's Distance from a Star is observed.

App. alt. of the Star.	Cor. Subtract.	App. alt. of the Star.	Cor. Subtract.	Ap. alt. of Star	Cor. Subtract.
0° 0′	140	2° 30′	35	12°	3
0 10	126	2 40	32	13	2
0 20	114	2 50	29	14	2
0 30	103	3 0	27	15	1
0 40	94	3 20	24	16	1
0 50	85	3 40	20	17	1
1 0	77	4 0	18	18	1
1 10	70	4 30	15	19	1
1 20	64	5 0	13	20	1
1 30	58	6 0	9	21	1
1 40	53	7 0	7	22	1
1 50	48	8 0	6	23	1
2 0	44	9 0	5	25	1
2 10	41	10 0	4	26	0
2 20	37	11 0	3	90	0

TABLE XLV.

PROPORTIONAL LOGARITHMS.

	0° 0′	0° 1′	0° 2′	0° 3′	0° 4′	0° 5′	0° 6′	0° 7′	0° 8′
0″		2.2553	1.9542	1.7782	1.6532	1.5563	1.4771	1.4102	1.3522
1	4.0334	2.2481	1.9506	1.7757	1.6514	1.5548	1.4759	1.4091	1.3513
2	3.7324	2.2410	1.9470	1.7733	1.6496	1.5534	1.4747	1.4081	1.3504
3	3.5563	2.2341	1.9435	1.7710	1.6478	1.5520	1.4735	1.4071	1.3495
4	3.4313	2.2272	1.9400	1.7686	1.6460	1.5505	1.4723	1.4060	1.3486
5	3.3344	2.2205	1.9365	1.7662	1.6442	1.5491	1.4711	1.4050	1.3477
6	3.2553	2.2139	1.9331	1.7639	1.6425	1.5477	1.4699	1.4040	1.3468
7	3.1883	2.2073	1.9296	1.7616	1.6407	1.5463	1.4687	1.4030	1.3459
8	3.1303	2.2009	1.9262	1.7592	1.6390	1.5449	1.4676	1.4020	1.3450
9	3.0792	2.1946	1.9228	1.7570	1.6372	1.5435	1.4664	1.4010	1.3441
10	3.0334	2.1883	1.9195	1.7546	1.6355	1.5420	1.4652	1.3999	1.3432
11	2.9920	2.1821	1.9161	1.7524	1.6337	1.5406	1.4640	1.3989	1.3423
12	2.9542	2.1761	1.9128	1.7501	1.6320	1.5393	1.4629	1.3979	1.3415
13	2.9195	2.1701	1.9096	1.7478	1.6303	1.5379	1.4617	1.3969	1.3406
14	2.8873	2.1642	1.9063	1.7456	1.6286	1.5365	1.4605	1.3959	1.3397
15	2.8573	2.1584	1.9031	1.7434	1.6269	1.5351	1.4594	1.3949	1.3388
16	2.8293	2.1526	1.8999	1.7411	1.6252	1.5337	1.4582	1.3939	1.3379
17	2.8030	2.1469	1.8967	1.7389	1.6235	1.5323	1.4571	1.3929	1.3370
18	2.7782	2.1413	1.8935	1.7368	1.6218	1.5310	1.4559	1.3919	1.3362
19	2.7546	2.1358	1.8904	1.7345	1.6201	1.5296	1.4548	1.3909	1.3353
20	2.7324	2.1303	1.8873	1.7324	1.6184	1.5283	1.4536	1.3899	1.3344
21	2.7112	2.1249	1.8842	1.7302	1.6168	1.5269	1.4525	1.3890	1.3336
22	2.6910	2.1196	1.8811	1.7281	1.6151	1.5255	1.4513	1.3880	1.3327
23	2.6717	2.1143	1.8781	1.7259	1.6134	1.5242	1.4502	1.3870	1.3318
24	2.6532	2.1091	1.8751	1.7238	1.6118	1.5229	1.4491	1.3860	1.3310
25	2.6355	2.1040	1.8720	1.7216	1.6102	1.5215	1.4479	1.3850	1.3301
26	2.6184	2.0989	1.8690	1.7195	1.6085	1.5202	1.4468	1.3841	1.3293
27	2.6021	2.0939	1.8661	1.7175	1.6069	1.5189	1.4457	1.3831	1.3284
28	2.5862	2.0889	1.8631	1.7153	1.6053	1.5175	1.4446	1.3821	1.3275
29	2.5710	2.0840	1.8602	1.7133	1.6037	1.5162	1.4435	1.3812	1.3267
30	2.5563	2.0792	1.8573	1.7112	1.6021	1.5149	1.4424	1.3802	1.3259
31	2.5420	2.0744	1.8544	1.7091	1.6004	1.5136	1.4412	1.3792	1.3250
32	2.5283	2.0696	1.8516	1.7071	1.5988	1.5123	1.4401	1.3783	1.3241
33	2.5149	2.0649	1.8487	1.7050	1.5973	1.5110	1.4390	1.3773	1.3233
34	2.5019	2.0603	1.8459	1.7030	1.5957	1.5097	1.4379	1.3763	1.3224
35	2.4893	2.0557	1.8431	1.7010	1.5941	1.5084	1.4368	1.3754	1.3216
36	2.4771	2.0512	1.8403	1.6990	1.5925	1.5071	1.4357	1.3745	1.3208
37	2.4652	2.0466	1.8375	1.6969	1.5909	1.5058	1.4346	1.3735	1.3199
38	2.4536	2.0422	1.8347	1.6949	1.5894	1.5045	1.4335	1.3725	1.3191
39	2.4424	2.0378	1.8320	1.6930	1.5878	1.5032	1.4325	1.3716	1.3183
40	2.4313	2.0334	1.8293	1.6910	1.5862	1.5019	1.4313	1.3706	1.3174
41	2.4206	2.0291	1.8266	1.6890	1.5847	1.5006	1.4303	1.3697	1.3166
42	2.4102	2.0248	1.8239	1.6871	1.5832	1.4994	1.4292	1.3688	1.3158
43	2.3999	2.0206	1.8212	1.6851	1.5816	1.4981	1.4281	1.3678	1.3149
44	2.3899	2.0164	1.8186	1.6832	1.5801	1.4968	1.4270	1.3669	1.3141
45	2.3802	2.0122	1.8159	1.6812	1.5786	1.4956	1.4260	1.3660	1.3133
46	2.3706	2.0081	1.8133	1.6793	1.5770	1.4943	1.4249	1.3650	1.3124
47	2.3613	2.0040	1.8107	1.6774	1.5755	1.4931	1.4238	1.3641	1.3116
48	2.3522	2.0000	1.8081	1.6755	1.5740	1.4918	1.4228	1.3632	1.3108
49	2.3432	1.9960	1.8055	1.6736	1.5725	1.4906	1.4217	1.3622	1.3099
50	2.3344	1.9920	1.8030	1.6717	1.5710	1.4893	1.4206	1.3613	1.3091
51	2.3259	1.9881	1.8004	1.6698	1.5695	1.4881	1.4196	1.3604	1.3083
52	2.3174	1.9842	1.7979	1.6679	1.5680	1.4869	1.4185	1.3595	1.3075
53	1.3091	1.9803	1.7954	1.6660	1.5665	1.4856	1.4175	1.3585	1.3067
54	2.3010	1.9765	1.7929	1.6642	1.5651	1.4844	1.4165	1.3576	1.3059
55	2.2930	1.9727	1.7904	1.6623	1.5636	1.4832	1.4154	1.3567	1.3050
56	2.2852	1.9689	1.7879	1.6605	1.5621	1.4820	1.4143	1.3558	1.3042
57	2.2775	1.9652	1.7855	1.6587	1.5607	1.4808	1.4133	1.3549	1.3034
58	2.2700	1.9615	1.7830	1.6568	1.5592	1.4795	1.4122	1.3540	1.3026
59	2.2626	1.9579	1.7805	1.6550	1.5577	1.4783	1.4112	1.3531	1.3018
60	2.2553	1.9542	1.7782	1.6532	1.5563	1.4771	1.4102	1.3522	1.3010

TABLE XLV.

PROPORTIONAL LOGARITHMS.

	0° 9′	0° 10′	0° 11′	0° 12′	0° 13′	0° 14′	0° 15′	0° 16′	0° 17′
0″	1.3010	1.2553	1.2139	1.1761	1.1413	1.1091	1.0792	1.0512	1.0248
1	1.3002	1.2545	1.2132	1.1755	1.1408	1.1086	1.0787	1.0507	1.0244
2	1.2994	1.2538	1.2125	1.1749	1.1402	1.1081	1.0782	1.0502	1.0240
3	1.2986	1.2531	1.2119	1.1743	1.1397	1.1076	1.0777	1.0498	1.0235
4	1.2978	1.2524	1.2112	1.1737	1.1391	1.1071	1.0772	1.0493	1.0231
5	1.2970	1.2517	1.2106	1.1731	1.1385	1.1066	1.0768	1.0489	1.0227
6	1.2962	1.2510	1.2099	1.1725	1.1380	1.1061	1.0763	1.0484	1.0223
7	1.2954	1.2502	1.2093	1.1719	1.1374	1.1055	1.0758	1.0480	1.0218
8	1.2946	1.2495	1.2086	1.1713	1.1369	1.1050	1.0753	1.0475	1.0214
9	1.2939	1.2488	1.2080	1.1707	1.1363	1.1045	1.0749	1.0471	1.0210
10	1.2931	1.2481	1.2073	1.1701	1.1358	1.1040	1.0744	1.0466	1.0206
11	1.2923	1.2474	1.2067	1.1695	1.1352	1.1035	1.0739	1.0462	1.0201
12	1.2915	1.2467	1.2061	1.1689	1.1347	1.1030	1.0734	1.0458	1.0197
13	1.2907	1.2459	1.2054	1.1683	1.1341	1.1025	1.0729	1.0453	1.0193
14	1.2899	1.2452	1.2047	1.1677	1.1336	1.1020	1.0725	1.0448	1.0189
15	1.2891	1.2445	1.2041	1.1671	1.1331	1.1015	1.0720	1.0444	1.0185
16	1.2883	1.2438	1.2035	1.1665	1.1325	1.1009	1.0715	1.0440	1.0180
17	1.2875	1.2431	1.2028	1.1659	1.1319	1.1004	1.0710	1.0435	1.0176
18	1.2868	1.2424	1.2022	1.1654	1.1314	1.0999	1.0706	1.0431	1.0172
19	1.2860	1.2417	1.2015	1.1648	1.1309	1.0994	1.0701	1.0426	1.0168
20	1.2852	1.2410	1.2009	1.1642	1.1303	1.0989	1.0696	1.0422	1.0164
21	1.2845	1.2403	1.2003	1.1636	1.1298	1.0984	1.0692	1.0418	1.0160
22	1.2837	1.2396	1.1996	1.1630	1.1292	1.0979	1.0687	1.0413	1.0155
23	1.2829	1.2389	1.1990	1.1624	1.1287	1.0974	1.0682	1.0408	1.0151
24	1.2821	1.2382	1.1984	1.1619	1.1282	1.0969	1.0678	1.0404	1.0147
25	1.2814	1.2375	1.1977	1.1613	1.1276	1.0964	1.0673	1.0400	1.0143
26	1.2806	1.2368	1.1971	1.1607	1.1271	1.0959	1.0668	1.0395	1.0139
27	1.2798	1.2362	1.1965	1.1601	1.1266	1.0954	1.0663	1.0391	1.0135
28	1.2791	1.2355	1.1958	1.1595	1.1260	1.0947	1.0659	1.0386	1.0130
29	1.2783	1.2348	1.1952	1.1589	1.1255	1.0944	1.0654	1.0382	1.0126
30	1.2775	1.2341	1.1946	1.1584	1.1249	1.0939	1.0649	1.0378	1.0122
31	1.2768	1.2334	1.1939	1.1578	1.1244	1.0934	1.0645	1.0373	1.0118
32	1.2760	1.2327	1.1933	1.1572	1.1238	1.0929	1.0640	1.0369	1.0114
33	1.2753	1.2320	1.1927	1.1566	1.1233	1.0924	1.0635	1.0365	1.0110
34	1.2745	1.2313	1.1920	1.1560	1.1228	1.0919	1.0631	1.0360	1.0106
35	1.2737	1.2306	1.1914	1.1555	1.1222	1.0914	1.0626	1.0356	1.0102
36	1.2730	1.2300	1.1908	1.1549	1.1217	1.0909	1.0621	1.0352	1.0098
37	1.2722	1.2293	1.1902	1.1543	1.1212	1.0904	1.0617	1.0347	1.0093
38	1.2715	1.2286	1.1895	1.1537	1.1206	1.0899	1.0612	1.0343	1.0089
39	1.2707	1.2279	1.1889	1.1532	1.1201	1.0894	1.0608	1.0339	1.0085
40	1.2700	1.2272	1.1883	1.1526	1.1196	1.0889	1.0603	1.0334	1.0081
41	1.2692	1.2265	1.1877	1.1520	1.1191	1.0884	1.0598	1.0330	1.0077
42	1.2685	1.2259	1.1871	1.1515	1.1186	1.0880	1.0594	1.0326	1.0073
43	1.2677	1.2252	1.1864	1.1509	1.1180	1.0875	1.0589	1.0321	1.0069
44	1.2670	1.2245	1.1858	1.1503	1.1175	1.0870	1.0584	1.0317	1.0065
45	1.2663	1.2239	1.1852	1.1498	1.1170	1.0865	1.0580	1.0313	1.0061
46	1.2655	1.2232	1.1846	1.1492	1.1164	1.0860	1.0575	1.0308	1.0057
47	1.2648	1.2225	1.1840	1.1486	1.1159	1.0855	1.0571	1.0304	1.0053
48	1.2640	1.2218	1.1834	1.1481	1.1154	1.0850	1.0566	1.0300	1.0049
49	1.2633	1.2212	1.1828	1.1475	1.1148	1.0845	1.0561	1.0295	1.0044
50	1.2626	1.2205	1.1822	1.1469	1.1143	1.0840	1.0557	1.0291	1.0040
51	1.2618	1.2198	1.1816	1.1464	1.1138	1.0835	1.0552	1.0287	1.0036
52	1.2611	1.2192	1.1809	1.1458	1.1133	1.0830	1.0548	1.0282	1.0032
53	1.2603	1.2185	1.1803	1.1452	1.1128	1.0826	1.0543	1.0278	1.0028
54	1.2596	1.2178	1.1797	1.1447	1.1123	1.0821	1.0539	1.0274	1.0024
55	1.2589	1.2172	1.1791	1.1441	1.1117	1.0816	1.0534	1.0269	1.0020
56	1.2582	1.2165	1.1785	1.1435	1.1112	1.0811	1.0529	1.0265	1.0016
57	1.2574	1.2159	1.1779	1.1430	1.1107	1.0806	1.0525	1.0261	1.0012
58	1.2567	1.2152	1.1773	1.1424	1.1102	1.0801	1.0520	1.0257	1.0008
59	1.2560	1.2145	1.1767	1.1419	1.1096	1.0796	1.0516	1.0252	1.0004
60	1.2553	1.2139	1.1761	1.1413	1.1091	1.0792	1.0512	1.0248	1.0000

0°18'	0°19'	0°20'	0°21'	0°22'	0°23'	0°24'	0°25'	0°26'	0°27'	0°28'
0000	9765	9542	9331	9128	8935	8751	8573	8403	8239	8081
9996	9761	9539	9327	9125	8932	8748	8570	8400	8236	8078
9992	9757	9535	9323	9122	8929	8745	8567	8397	8234	8076
9988	9754	9532	9320	9119	8926	8742	8565	8395	8231	8073
9984	9750	9528	9317	9115	8923	8739	8562	8392	8228	8071
9980	9746	9524	9313	9112	8920	8736	8559	8389	8225	8068
9976	9742	9521	9310	9109	8917	8733	8556	8386	8223	8066
9972	9738	9517	9306	9105	8913	8730	8553	8383	8220	8063
9968	9735	9513	9303	9102	8910	8727	8550	8381	8217	8060
9964	9731	9510	9300	9099	8907	8724	8547	8378	8215	8058
9960	9727	9506	9296	9096	8904	8721	8544	8375	8212	8055
9956	9723	9503	9293	9092	8901	8718	8541	8372	8209	8053
9952	9720	9499	9289	9089	8898	8715	8539	8370	8207	8050
9948	9716	9495	9286	9086	8895	8712	8536	8367	8204	8047
9944	9712	9492	9282	9082	8891	8709	8533	8364	8202	8045
9940	9708	9488	9279	9079	8888	8706	8530	8361	8199	8042
9936	9704	9485	9276	9076	8885	8703	8527	8358	8196	8040
9932	9701	9481	9272	9073	8882	8700	8524	8356	8194	8037
9928	9697	9478	9269	9070	8879	8697	8522	8353	8191	8035
9924	9693	9474	9265	9066	8876	8694	8519	8350	8188	8032
9920	9689	9470	9262	9063	8873	8691	8516	8347	8186	8030
9916	9686	9467	9259	9060	8870	8688	8513	8345	8183	8027
9912	9682	9463	9255	9056	8867	8685	8510	8342	8180	8024
9908	9678	9460	9252	9053	8864	8682	8507	8339	8178	8022
9905	9675	9456	9249	9050	8861	8679	8504	8337	8175	8020
9901	9671	9453	9245	9047	8857	8676	8501	8334	8172	8017
9897	9667	9449	9242	9044	8854	8673	8498	8331	8170	8014
9893	9664	9446	9238	9041	8851	8670	8496	8328	8167	8012
9889	9660	9442	9235	9037	8848	8667	8493	8326	8164	8009
9885	9656	9439	9231	9034	8845	8664	8490	8323	8162	8007
9881	9652	9435	9228	9031	8842	8661	8487	8320	8159	8004
9877	9648	9431	9225	9027	8839	8658	8484	8317	8157	8002
9873	9645	9428	9221	9024	8836	8655	8481	8315	8154	7999
9869	9641	9425	9218	9021	8833	8652	8479	8312	8152	7997
9865	9637	9421	9215	9018	8830	8649	8476	8309	8149	7994
9861	9634	9417	9211	9015	8827	8646	8473	8306	8146	7991
9858	9630	9414	9208	9012	8824	8643	8470	8304	8144	7989
9854	9626	9410	9205	9008	8820	8640	8467	8301	8141	7986
9850	9623	9407	9201	9005	8817	8637	8464	8298	8138	7984
9846	9619	9404	9198	9002	8814	8635	8462	8296	8136	7981
9842	9615	9400	9195	8999	8811	8632	8459	8293	8133	7979
9838	9612	9396	9191	8995	8808	8629	8456	8290	8130	7976
9834	9608	9393	9188	8992	8805	8626	8453	8288	8128	7974
9830	9604	9389	9185	8989	8802	8623	8450	8285	8125	7971
9826	9601	9386	9181	8986	8799	8620	8448	8282	8122	7969
9823	9597	9383	9178	8983	8796	8617	8445	8279	8120	7966
9819	9593	9379	9175	8980	8793	8614	8442	8277	8117	7964
9815	9590	9375	9171	8976	8790	8611	8439	8274	8115	7961
9811	9586	9372	9168	8973	8787	8608	8437	8271	8112	7959
9807	9582	9368	9165	8970	8784	8605	8434	8268	8109	7956
9803	9579	9365	9161	8967	8781	8602	8431	8266	8107	7954
9800	9575	9362	9158	8964	8778	8599	8428	8263	8104	7951
9796	9571	9358	9155	8960	8775	8596	8425	8260	8102	7949
9792	9568	9355	9151	8957	8772	8593	8422	8258	8099	7946
9788	9564	9351	9148	8954	8769	8591	8420	8255	8097	7944
9784	9560	9348	9145	8951	8766	8588	8417	8252	8094	7941
9780	9557	9344	9141	8948	8763	8585	8414	8250	8091	7939
9777	9553	9341	9138	8945	8760	8582	8411	8247	8089	7936
9773	9549	9337	9135	8942	8757	8579	8408	8244	8086	7934
9769	9546	9334	9132	8938	8754	8576	8406	8241	8084	7931

TABLE XLV.

PROPORTIONAL LOGARITHMS.

	0°30'	0°31'	0°32'	0°33'	0°34'	0°35'	0°36'	0°37'	0°38'	0°39'	0°40'	0°41'
0"	7782	7639	7501	7368	7238	7112	6990	6871	6755	6642	6532	6425
1	7779	7637	7499	7365	7236	7110	6988	6869	6753	6640	6530	6423
2	7776	7634	7496	7363	7234	7108	6986	6867	6751	6638	6528	6421
3	7774	7632	7494	7361	7231	7106	6984	6865	6749	6637	6527	6420
4	7772	7630	7492	7359	7229	7104	6982	6863	6747	6635	6525	6418
5	7769	7627	7490	7356	7227	7102	6980	6861	6745	6633	6523	6416
6	7767	7625	7488	7354	7225	7100	6978	6859	6743	6631	6521	6414
7	7764	7623	7485	7352	7223	7097	6976	6857	6741	6629	6519	6412
8	7762	7620	7483	7350	7221	7095	6974	6855	6739	6627	6517	6411
9	7760	7618	7481	7348	7219	7093	6972	6853	6738	6625	6516	6409
10	7757	7616	7478	7345	7216	7091	6970	6851	6736	6623	6514	6407
11	7755	7613	7476	7343	7214	7089	6968	6849	6734	6621	6512	6405
12	7753	7611	7474	7341	7212	7087	6966	6847	6732	6620	6510	6404
13	7750	7609	7472	7339	7210	7085	6964	6845	6730	6618	6508	6402
14	7748	7606	7469	7337	7208	7083	6962	6843	6728	6616	6507	6400
15	7745	7604	7467	7335	7206	7081	6960	6841	6726	6614	6505	6398
16	7743	7602	7465	7332	7204	7079	6958	6839	6724	6612	6503	6397
17	7740	7599	7463	7330	7202	7077	6956	6837	6722	6610	6501	6395
18	7738	7597	7461	7328	7200	7075	6954	6836	6721	6609	6500	6393
19	7736	7595	7458	7326	7197	7073	6952	6834	6719	6607	6498	6391
20	7733	7592	7456	7324	7195	7071	6950	6832	6717	6605	6496	6390
21	7731	7590	7454	7322	7193	7069	6948	6830	6715	6603	6494	6388
22	7729	7588	7452	7319	7191	7067	6946	6828	6713	6601	6492	6386
23	7726	7586	7449	7317	7189	7065	6944	6826	6711	6599	6490	6384
24	7724	7583	7447	7315	7187	7063	6942	6824	6709	6598	6489	6383
25	7722	7581	7445	7313	7185	7061	6940	6823	6707	6596	6487	6381
26	7719	7579	7443	7311	7183	7059	6938	6820	6705	6594	6485	6379
27	7717	7577	7441	7309	7181	7057	6936	6818	6704	6592	6484	6377
28	7714	7574	7438	7306	7179	7054	6934	6816	6702	6590	6482	6376
29	7712	7572	7436	7304	7177	7052	6932	6814	6700	6588	6480	6374
30	7710	7570	7434	7302	7175	7050	6930	6812	6698	6587	6478	6372
31	7707	7567	7432	7300	7172	7048	6928	6810	6696	6585	6476	6370
32	7705	7565	7429	7298	7170	7046	6926	6808	6694	6583	6474	6369
33	7703	7563	7427	7296	7168	7044	6924	6807	6692	6581	6473	6367
34	7700	7560	7425	7293	7166	7042	6922	6805	6690	6579	6471	6365
35	7698	7558	7423	7291	7164	7040	6920	6803	6689	6577	6469	6363
36	7696	7556	7421	7289	7162	7038	6918	6801	6687	6576	6467	6362
37	7693	7553	7418	7287	7160	7036	6916	6799	6685	6574	6465	6360
38	7691	7551	7416	7285	7158	7034	6914	6797	6683	6572	6464	6358
39	7688	7549	7414	7283	7156	7032	6912	6795	6681	6570	6462	6357
40	7686	7546	7411	7281	7153	7030	6910	6793	6679	6568	6460	6355
41	7683	7544	7409	7278	7151	7028	6908	6791	6677	6566	6458	6353
42	7681	7542	7407	7276	7149	7026	6906	6789	6676	6565	6457	6351
43	7679	7540	7405	7274	7147	7024	6904	6787	6674	6563	6455	6349
44	7676	7537	7403	7272	7145	7022	6902	6785	6672	6561	6453	6348
45	7674	7535	7401	7270	7143	7020	6900	6784	6670	6559	6451	6346
46	7672	7533	7398	7268	7141	7018	6898	6782	6668	6557	6449	6344
47	7669	7531	7396	7266	7139	7016	6896	6780	6666	6556	6448	6342
48	7667	7528	7394	7264	7137	7014	6894	6778	6664	6554	6446	6341
49	7665	7526	7392	7261	7135	7012	6892	6776	6662	6552	6444	6339
50	7662	7524	7389	7259	7133	7010	6890	6774	6660	6550	6442	6337
51	7660	7522	7387	7257	7131	7008	6888	6772	6659	6548	6441	6336
52	7658	7519	7385	7255	7128	7006	6886	6770	6657	6546	6439	6334
53	7655	7517	7383	7253	7126	7004	6884	6768	6655	6545	6437	6332
54	7653	7515	7381	7251	7124	7002	6882	6766	6653	6543	6435	6331
55	7651	7512	7378	7248	7122	7000	6880	6764	6651	6541	6434	6329
56	7648	7510	7376	7246	7120	6998	6878	6762	6649	6539	6432	6327
57	7646	7508	7374	7244	7118	6996	6877	6761	6648	6538	6430	6325
58	7644	7506	7372	7242	7116	6994	6875	6759	6646	6536	6428	6323
59	7641	7503	7370	7240	7114	6992	6873	6757	6644	6534	6426	6322
60	7639	7501	7368	7238	7112	6990	6871	6755	6642	6532	6425	6320

TABLE XLV.

PROPORTIONAL LOGARITHMS.

	0°42′	0°43′	0°44′	0°45′	0°46′	0°47′	0°48′	0°49′	0°50′	0°51′	0°52′	0°53′
0″	6320	6218	6118	6021	5925	5832	5740	5651	5563	5477	5393	5310
1	6318	6216	6116	6019	5923	5830	5739	5649	5561	5475	5391	5308
2	6317	6214	6115	6017	5922	5828	5737	5648	5560	5474	5390	5307
3	6315	6213	6113	6016	5920	5827	5736	5646	5559	5473	5389	5306
4	6313	6211	6111	6014	5919	5825	5734	5645	5557	5471	5387	5304
5	6311	6209	6110	6012	5917	5824	5733	5643	5556	5470	5386	5303
6	6310	6208	6108	6011	5916	5823	5731	5642	5554	5469	5384	5302
7	6308	6206	6106	6009	5914	5821	5730	5640	5553	5467	5383	5300
8	6306	6204	6105	6008	5912	5819	5728	5639	5551	5465	5381	5299
9	6305	6203	6103	6006	5911	5818	5727	5637	5550	5464	5380	5298
10	6303	6201	6102	6004	5909	5816	5725	5636	5548	5463	5379	5296
11	6301	6199	6100	6003	5908	5815	5724	5634	5547	5461	5377	5295
12	6300	6198	6099	6001	5906	5813	5722	5633	5546	5460	5376	5294
13	6298	6196	6097	6000	5905	5812	5721	5631	5544	5458	5374	5292
14	6296	6194	6095	5998	5903	5810	5719	5630	5543	5457	5373	5291
15	6294	6193	6094	5997	5902	5809	5718	5629	5541	5456	5372	5290
16	6293	6191	6092	5995	5900	5807	5716	5627	5540	5454	5370	5288
17	6291	6189	6090	5993	5898	5805	5715	5626	5538	5453	5369	5287
18	6289	6188	6089	5992	5897	5804	5713	5624	5537	5452	5368	5285
19	6287	6186	6087	5990	5895	5802	5712	5623	5535	5450	5366	5284
20	6286	6184	6085	5988	5894	5801	5710	5621	553[illegible]	[illegible]9	5365	5283
21	6284	6183	6084	5987	5892	5800	5709	5620	5533	5447	5364	5281
22	6282	6181	6082	5985	5890	5798	5707	5618	5531	5446	5362	5280
23	6281	6179	6080	5984	5889	5796	5706	5617	5530	5444	5361	5278
24	6279	6178	6079	5982	5888	5795	5704	5615	5528	5443	5359	5277
25	6277	6176	6077	5980	5886	5793	5703	5614	5527	5441	5358	5276
26	6275	6174	6075	5979	5884	5792	5701	5612	5525	5440	5356	5274
27	6274	6173	6074	5977	5883	5790	5700	5611	5524	5439	5355	5273
28	6272	6171	6072	5976	5881	5789	5698	5609	5522	5437	5354	5272
29	6270	6169	6071	5974	5880	5787	5697	5608	5521	5436	5352	5270
30	6269	6168	6069	5973	5878	5786	5695	5607	5520	5435	5351	5269
31	6267	6166	6067	5971	5876	5784	5694	5605	5518	5433	5350	5268
32	6265	6164	6066	5969	5875	5783	5693	5604	5517	5432	5348	5266
33	6264	6163	6064	5968	5874	5781	5691	5602	5516	5430	5347	5265
34	6262	6161	6062	5966	5872	5779	5689	5601	5514	5429	5345	5264
35	6260	6159	6061	5964	5870	5778	5688	5599	5512	5427	5344	5262
36	6259	6158	6059	5963	5869	5777	5686	5598	5511	5426	5343	5261
37	6257	6156	6058	5961	5867	5775	5685	5596	5510	5425	5341	5260
38	62[illegible]5	6154	6056	5960	5866	5773	5683	5595	5508	5423	5340	5258
39	6254	6153	6055	5958	5864	5772	5682	5594	5507	5422	5339	5257
40	6252	6151	6053	5957	5862	5770	5680	5592	5505	5420	5337	5255
41	6250	6149	6051	5955	5861	5769	5679	5590	5504	5419	5336	5254
42	6248	6148	6050	5954	5860	5768	5677	5589	5503	5418	5335	5253
43	6247	6146	6048	5952	5858	5766	5676	5587	5501	5416	5333	5251
44	6245	6144	6046	5950	5856	5764	5674	5586	5500	5415	5332	5250
45	6243	6143	6045	5949	5855	5763	5673	5585	5498	5414	5331	5249
46	6242	6141	6043	5947	5853	5761	5671	5583	5497	5412	5329	5247
47	6240	6139	6041	5945	5852	5760	5670	5582	5495	5411	5328	5246
48	6238	6138	6040	5944	5850	5758	5669	5580	5494	5409	5326	5245
49	6236	6136	6038	5942	5849	5757	5667	5579	5492	5408	5325	5243
50	6235	6134	6037	5941	5847	5755	5665	5577	5491	5406	5323	5242
51	6233	6133	6035	5939	5846	5754	5664	5576	5490	5405	5322	5241
52	6231	6131	6033	5938	5844	5752	5662	5574	5488	5404	5321	5239
53	6230	6130	6032	5936	5842	5751	5661	5573	5487	5402	5319	5238
54	6228	6128	6030	5935	5841	5749	5660	5572	5486	5401	5318	5237
55	6226	6126	6028	5933	5839	5748	5658	5570	5484	5399	5317	5235
56	6225	6125	6027	5931	5838	5746	5656	5569	5483	5398	5315	5234
57	6223	6123	6025	5930	5836	5745	5655	5567	5481	5397	5314	5233
58	6221	6121	6024	5928	5835	5743	5654	5566	5480	5395	5312	5231
59	6220	6120	6022	5927	5833	5742	5652	5564	5478	5394	5311	5230
60	6218	6118	6021	5925	5832	5740	5651	5563	5477	5393	5310	5229

TABLE XLV.

PROPORTIONAL LOGARITHMS.

	0°54′	0°55′	0°56′	0°57′	0°58′	0°59′	1°0′	1°1′	1°2′	1°3′	1°4′	1°5′
0″	5229	5149	5071	4994	4918	4844	4771	4699	4629	4559	4491	4424
1	5227	5148	5069	4992	4917	4843	4770	4698	4627	4558	4490	4422
2	5226	5146	5068	4991	4916	4842	4769	4697	4626	4557	4489	4421
3	5225	5145	5067	4990	4915	4841	4768	4696	4625	4556	4488	4420
4	5223	5144	5065	4989	4913	4839	4766	4694	4624	4555	4486	4419
5	5222	5142	5064	4987	4912	4838	4765	4693	4623	4553	4485	4418
6	5221	5141	5063	4986	4911	4837	4764	4692	4622	4552	4484	4417
7	5219	5140	5062	4985	4910	4835	4763	4691	4620	4551	4483	4416
8	5218	5138	5060	4984	4908	4834	4761	4690	4619	4550	4482	4415
9	5217	5137	5059	4983	4907	4833	4760	4689	4618	4549	4481	4414
10	5215	5136	5058	4981	4906	4832	4759	4687	4617	4548	4479	4412
11	5214	5134	5056	4980	4905	4831	4758	4686	4616	4547	4478	4411
12	5213	5133	5055	4979	4903	4830	4757	4685	4615	4546	4477	4410
13	5211	5132	5054	4977	4902	4828	4755	4684	4613	4544	4476	4409
14	5210	5130	5053	4976	4901	4827	4754	4683	4612	4543	4475	4408
15	5209	5129	5051	4975	4900	4826	4753	4682	4611	4542	4474	4407
16	5207	5128	5050	4973	4898	4824	4752	4680	4610	4541	4473	4406
17	5206	5127	5049	4972	4897	4823	4751	4679	4609	4540	4472	4405
18	5205	5125	5048	4971	4896	4822	4750	4678	4608	4539	4471	4404
19	5203	5124	5046	4970	4895	4821	4748	4677	4606	4537	4469	4402
20	5202	[illegible]	5045	4968	4893	4820	4747	4676	4605	4536	4468	4401
21	5201	5122	5044	4967	4892	4819	4746	4675	4604	4535	4467	4400
22	5199	5120	5042	4966	4891	4817	4745	4673	4603	4534	4466	4399
23	5198	5119	5041	4965	4890	4816	4743	4672	4602	4533	4465	4398
24	5197	5118	5040	4964	4889	4815	4742	4671	4601	4532	4464	4397
25	5195	5116	5038	4962	4887	4813	4741	4670	4600	4530	4463	4396
26	5194	5115	5037	4961	4886	4812	4740	4669	4598	4529	4461	4395
27	5193	5114	5036	4960	4885	4811	4739	4668	4597	4528	4460	4394
28	5191	5112	5035	4958	4883	4810	4737	4666	4596	4527	4459	4392
29	5190	5111	5033	4957	4882	4809	4736	4665	4595	4526	4458	4391
30	5189	5110	5032	4956	4881	4808	4735	4664	4594	4525	4457	4390
31	5187	5108	5031	4955	4880	4806	4734	4663	4593	4524	4456	4389
32	5186	5107	5029	4953	4878	4805	4733	4661	4591	4523	4455	4388
33	5185	5106	5028	4952	4877	4804	4732	4660	4590	4522	4454	4387
34	5183	5104	5027	4951	4876	4802	4730	4659	4589	4520	4452	4386
35	5182	5103	5026	4950	4875	4801	4729	4658	4588	4519	4451	4385
36	5181	5102	5025	4949	4874	4800	4728	4657	4587	4518	4450	4384
37	5179	5100	5023	4947	4872	4799	4727	4656	4586	4517	4449	4382
38	5178	5099	5022	4946	4871	4798	4725	4654	4585	4516	4448	4381
39	5177	5098	5021	4945	4870	4797	4724	4653	4584	4515	4447	4380
40	5175	5097	5019	4943	4869	4795	4723	4652	4582	4513	4446	4379
41	5174	5095	5018	4942	4867	4794	4722	4651	4581	4512	4445	4378
42	5173	5094	5017	4941	4866	4793	4721	4650	4580	4511	4444	4377
43	5172	5093	5015	4940	4865	4792	4719	4648	4579	4510	4443	4376
44	5170	5091	5014	4938	4864	4790	4718	4647	4578	4509	4441	4375
45	5169	5090	5013	4937	4863	4789	4717	4646	4577	4508	4440	4374
46	5167	5089	5012	4936	4861	4788	4716	4645	4575	4507	4439	4372
47	5166	5087	5010	4934	4860	4787	4715	4644	4574	4506	4438	4371
48	5165	5086	5009	4933	4859	4786	4714	4643	4573	4505	4437	4370
49	5163	5085	5008	4932	4858	4784	4712	4641	4572	4503	4436	4369
50	5162	5084	5006	4931	4856	4783	4711	4640	4571	4502	4435	4368
51	5161	5082	5005	4930	4855	4782	4710	4639	4570	4501	4434	4367
52	5159	5081	5004	4928	4854	4781	4709	4638	4568	4500	4432	4366
53	5158	5080	5003	4927	4853	4779	4708	4637	4567	4499	4431	4365
54	5157	5079	5002	4926	4852	4778	4707	4636	4566	4498	4430	4364
55	5155	5077	5000	4924	4850	4777	4705	4634	4565	4496	4429	4363
56	5154	5076	4999	4923	4849	4776	4704	4633	4564	4495	4428	4362
57	5153	5075	4998	4922	4848	4775	4703	4632	4563	4494	4427	4361
58	5152	5073	4995	4921	4846	4773	4702	4631	4561	4493	4426	4359
59	5150	5072	4995	4919	4845	4772	4700	4630	4560	4492	4425	4358
60	5149	5071	4994	4918	4844	4771	4699	4629	4559	4491	4424	4357

TABLE XLV.

PROPORTIONAL LOGARITHMS.

	1° 6'	1° 7'	1° 8'	1° 9'	1° 10'	1° 11'	1° 12'	1° 13'	1° 14'	1° 15'	1° 16'	1° 17'
0"	4357	4292	4228	4164	4102	4040	3979	3919	3860	3802	3745	3688
1	4356	4291	4226	4163	4101	4039	3978	3918	3859	3801	3744	3687
2	4355	4290	4225	4162	4100	4038	3977	3917	3858	3800	3743	3686
3	4354	4289	4224	4161	4099	4037	3976	3917	3857	3799	3742	3685
4	4353	4287	4223	4160	4098	4036	3975	3916	3856	3798	3741	3684
5	4352	4286	4222	4159	4097	4035	3974	3915	3855	3797	3740	3683
6	4351	4285	4221	4158	4096	4034	3973	3914	3855	3796	3739	3682
7	4349	4284	4220	4157	4094	4033	3972	3913	3854	3795	3738	3681
8	4348	4283	4219	4156	4093	4032	3971	3912	3853	3794	3737	3680
9	4347	4282	4218	4155	4092	4031	3970	3911	3852	3793	3736	3679
10	4346	4281	4217	4154	4091	4030	3969	3910	3851	3792	3735	3678
11	4345	4280	4216	4153	4090	4029	3968	3709	3850	3791	3734	3677
12	4344	4279	4215	4152	4089	4028	3967	3908	3849	3791	3733	3677
13	4343	4278	4214	4151	4088	4027	3966	3907	3848	3790	3732	3676
14	4342	4277	4213	4150	4087	4026	3965	3906	3847	3789	3731	3675
15	4341	4276	4212	4149	4086	4025	3964	3905	3846	3788	3730	3674
16	4340	4275	4211	4147	4085	4024	3963	3904	3845	3787	3729	3673
17	4339	4274	4210	4146	4084	4023	3962	3903	3844	3786	3728	3672
18	4338	4273	4209	4145	4083	4022	3961	3902	3843	3785	3727	3671
19	4336	4271	4207	4144	4082	4021	3960	3901	3848	3784	3726	3670
20	4335	4270	4206	4143	4081	4020	3959	3900	3841	3783	3725	3669
21	4334	4269	4205	4142	4080	4019	3958	3899	3840	3782	3725	3668
22	4333	4268	4204	4141	4079	4018	3957	3898	3839	3781	3724	3667
23	4332	4267	4203	4140	4078	4017	3956	3897	3838	3780	3723	3666
24	4331	4266	4202	4139	4077	4016	3955	3896	3837	3779	3722	3665
25	4330	4265	4201	4138	4076	4015	3954	3895	3836	3778	3721	3664
26	4329	4264	4200	4137	4075	4014	3953	3894	3835	3777	3720	3663
27	4328	4263	4199	4136	4074	4013	3952	3893	3834	3776	3719	3663
28	4327	4262	4198	4135	4073	4012	3951	3892	3833	3775	3718	3662
29	4326	4261	4197	4134	4072	4011	3950	3891	3832	3774	3717	3661
30	4325	4260	4196	4133	4071	4010	3949	3890	3831	3773	[illegible]	3660
31	4323	4258	4195	4132	4070	4009	3948	3889	3830	3772	3715	3659
32	4322	4257	4194	4131	4069	4008	3947	3888	3829	3771	3714	3658
33	4321	4256	4193	4130	4068	4007	3946	3887	3828	3770	3713	3657
34	4320	4255	4191	4129	4067	4006	3945	3886	3827	3769	3712	3656
35	4319	4254	4190	4128	4066	4005	3944	3885	3826	3768	3711	3655
36	4318	4253	4189	4127	4065	4004	3943	3884	3825	3768	3710	3654
37	4317	4252	4188	4126	4064	4003	3942	3883	3824	3767	3709	3653
38	4316	4251	4187	4125	4063	4002	3941	3882	3823	3766	3708	3652
39	4315	4250	4186	4124	4062	4001	3940	3881	3822	3765	3708	3651
40	4313	4249	4185	4122	4061	4000	3939	3880	3821	3764	3707	3650
41	4312	4248	4184	4121	4060	3999	3938	3879	3820	3763	3706	3649
42	4311	4247	4183	4120	4059	3998	3937	3878	3820	3762	3705	3649
43	4310	4246	4182	4119	4057	3997	3936	3877	3819	3761	3704	3648
44	4309	4245	4181	4118	4056	3996	3935	3876	3818	3760	3703	3647
45	4308	4244	4180	4117	4055	3995	3934	3875	3817	3759	3702	3646
46	4307	4242	4179	4116	4054	3993	3933	3874	3816	3758	3701	3645
47	4306	4241	4178	4115	4053	3992	3932	3873	3815	3757	3700	3644
48	4305	4240	4177	4114	4052	3991	3931	3872	3814	3756	3699	3643
49	4304	4239	4176	4113	4051	3990	3930	3871	3813	3755	3698	3642
50	4303	4238	4175	4112	4050	3989	3929	3870	3812	3754	3697	3641
51	4302	4237	4174	4111	4049	3988	3928	3869	3811	3753	3696	3640
52	4300	4236	4173	4110	4048	3987	3927	3868	3810	3752	3695	3639
53	4299	4235	4172	4109	4047	3986	3926	3867	3809	3751	3694	3638
54	4298	4234	4171	4108	4046	3985	3925	3866	3808	3750	3693	3637
55	4297	4233	4169	4107	4045	3984	3924	3865	3807	3749	3692	3636
56	4296	4232	4168	4106	4044	3983	3923	3864	3806	3748	3691	3635
57	4295	4231	4167	4105	4043	3982	3922	3863	3805	3747	3691	3635
58	4294	4230	4166	4104	4042	3981	3921	3862	3804	3746	3690	3634
59	4293	4229	4165	4103	4041	3980	3920	3861	3803	3745	3689	3633
60	4292	4228	4164	4102	4040	3979	3919	3860	3802	3745	3688	3632

TABLE XLV.

PROPORTIONAL LOGARITHMS.

	1°18′	1°19′	1°20′	1°21′	1°22′	1°23′	1°24′	1°25′	1°26′	1°27′	1°28′	1°29′
0″	3632	3576	3522	3468	3415	3362	3310	3259	3208	3158	3108	3059
1	3631	3575	3521	3467	3414	3361	3309	3258	3207	3157	3107	3058
2	3630	3574	3520	3466	3413	3360	3308	3257	3206	3156	3106	3057
3	3629	3574	3519	3465	3412	3359	3307	3256	3205	3155	3105	3056
4	3628	3573	3518	3464	3411	3358	3306	3255	3204	3154	3105	3056
5	3627	3572	3517	3463	3410	3358	3306	3254	3203	3153	3104	3055
6	3626	3571	3516	3463	3409	3357	3305	3253	3203	3153	3103	3054
7	3625	3570	3515	3462	3408	3356	3304	3253	3202	3152	3102	3053
8	3624	3569	3514	3461	3407	3355	3303	3252	3201	3151	3102	3058
9	3623	3568	3514	3460	3407	3354	3302	3251	3200	3150	3101	3052
10	3622	3567	3513	3459	3406	3353	3301	3250	3199	3149	3100	3051
11	3621	3566	3512	3458	3405	3352	3300	3249	3198	3148	3099	3050
12	3621	3565	3511	3457	3404	3351	3300	3248	3198	3148	3098	3049
13	3620	3564	3510	3456	3403	3351	3299	3247	3197	3147	3097	3048
14	3619	3563	3509	3455	3402	3350	3298	3247	3196	3146	3096	3047
15	3618	3563	3508	3454	3401	3349	3297	3246	3195	3145	3096	3047
16	3617	3562	3507	3454	3400	3348	3296	3245	3194	3144	3095	3046
17	3616	3561	3506	3453	3400	3347	3295	3244	3193	3143	3094	3045
18	3615	3560	3506	3452	3399	3346	3294	3243	3193	3143	3093	3044
19	3614	3559	3505	3451	3398	3345	3294	3242	3192	3142	3092	3043
20	3613	3558	3504	3450	3397	3344	3293	3241	3191	3141	3091	3043
21	3612	3557	3503	3449	3396	3344	3292	3241	3190	3140	3091	3042
22	3611	3556	3502	3448	3395	3343	3291	3240	3189	3139	3090	3041
23	3610	3555	3501	3447	3394	3342	3290	3239	3188	3138	3089	3040
24	3610	3555	3500	3446	3393	3341	3289	3238	3188	3138	3088	3039
25	3609	3554	3499	3445	3393	3340	3288	3237	3187	3137	3087	3038
26	3608	3553	3498	3445	3392	3339	3287	3236	3186	3136	3086	3038
27	3607	3552	3497	3444	3391	3338	3287	3236	3185	3135	3086	3037
28	3606	3551	3496	3443	3390	3338	3286	3235	3184	3134	3085	3036
29	3605	3550	3496	3442	3389	3337	3285	3234	3183	3135	3084	3035
30	3604	3549	3495	3441	3388	3336	3284	3233	3183	3133	3083	3034
31	3603	3548	3494	3440	3387	3335	3283	3232	3182	3132	3082	3034
32	3602	3547	3493	3439	3386	3334	3282	3231	3181	3131	3082	3033
33	3601	3546	3492	3438	3386	3333	3281	3231	3180	3130	3081	3032
34	3600	3545	3491	3438	3385	3332	3281	3230	3179	3129	3080	3031
35	3599	3544	3490	3437	3384	3331	3280	3229	3178	3128	3079	3030
36	3598	3544	3489	3436	3383	3331	3279	3228	3178	3128	3078	3030
37	3597	3543	3488	3435	3382	3330	3278	3227	3177	3127	3078	3029
38	3596	3542	3487	3434	3381	3329	3277	3226	3176	3126	3077	3028
39	3596	3541	3487	3433	3380	3328	3276	3225	3175	3125	3076	3027
40	3595	3540	3486	3432	3379	3327	3276	3225	3174	3124	3975	3026
41	3594	3529	3485	3431	3178	3326	3275	3224	3173	3123	3074	3026
42	3593	3538	3484	3431	3178	3325	3274	3223	3173	3123	3073	3025
43	3592	3537	3483	3430	3377	3325	3273	3222	3172	3122	3073	3024
44	3591	5536	3482	3429	3376	3324	3272	3221	3171	3121	3072	3023
45	3590	3535	3481	3428	3375	3323	3271	3220	3170	3120	3071	3022
46	3589	3534	3480	3427	3374	3322	3270	3219	3169	3119	3070	3022
47	3588	3533	3479	3426	3373	3321	3270	3219	3168	3119	3069	3021
48	3587	3533	3479	3425	3372	3320	3269	3218	3168	3118	3069	3020
49	3586	3532	3478	3424	3171	3319	3268	3217	3167	3117	3068	3019
50	3585	3531	3477	3423	3371	3318	3267	3216	3166	3116	3067	3018
51	3585	3530	3476	3423	3370	3318	3266	3215	3165	3115	3066	3018
52	3584	3529	3475	3422	3169	3317	3265	3214	3164	3114	3065	3017
53	3583	3528	3474	3421	3368	3316	3264	3214	3163	3114	3064	3016
54	3582	3527	3473	3420	3367	3315	3264	3213	3163	3113	3064	3015
55	3581	3526	3472	3419	3166	3314	3263	3212	3162	3112	3063	3014
56	3580	3525	3471	3418	3365	3313	3262	3211	3161	3111	3062	3013
57	3579	3525	3471	3417	3365	3313	3261	3210	3160	3110	3061	3013
58	3578	3524	3470	3416	[illegible]	3312	3260	3209	3159	3109	3060	3012
59	3577	3523	3469	3415	3363	3311	3259	3209	3158	3109	3060	3011
60	3576	3522	3468	3115	3352	3310	3259	3208	3158	3108	3059	3010

TABLE XLV.

PROPORTIONAL LOGARITHMS.

	1°30′	1°31′	1°32′	1°33′	1°34′	1°35′	1°36′	1°37′	1°38′	1°39′	1°40′	1°41′
0″	3010	2962	2915	2868	2821	2775	2730	2685	2640	2596	2553	2510
1	3009	2961	2914	2867	2821	2775	2729	2684	2640	2596	2552	2509
2	3009	2961	2913	2866	2820	2774	2728	2683	2639	2595	2551	2508
3	3008	2960	2912	2866	2819	2773	2728	2683	2638	2594	2551	2507
4	3007	2959	2912	2865	2818	2772	2727	2682	2637	2593	2550	2507
5	3006	2958	2911	2864	2818	2772	2726	2681	2637	2593	2549	2506
6	3005	2958	2910	2863	2817	2771	2725	2681	2636	2592	2548	2505
7	3005	2957	2909	2862	2816	2770	2725	2680	2635	2591	2548	2504
8	3004	2956	2908	2862	2815	2769	2724	2679	2634	2590	2547	2504
9	3003	2955	2908	2861	2815	2769	2723	2678	2634	2590	2546	2503
10	3002	2954	2907	2860	2814	2768	2722	2678	2633	2589	2545	2502
11	3001	2954	2906	2859	2813	2767	2722	2677	2632	2588	2545	2502
12	3001	2953	2905	2859	2812	2766	2721	2676	2632	2588	2544	2501
13	3000	2952	2905	2858	2811	2766	2720	2675	2631	2587	2543	2500
14	1999	2951	2904	2857	2811	2765	2719	2675	2630	2586	2543	2499
15	2998	2950	2903	2856	2810	2764	2719	2674	2619	2585	2542	2499
16	2997	2950	2902	2855	2809	2763	2718	2673	2629	2585	2541	2498
17	2997	2949	2901	2855	2808	2762	2717	2672	2628	2584	2540	2497
18	2996	2948	2901	2854	2808	2762	2716	2672	2627	2583	2540	2497
19	2995	2947	2900	2853	2807	2761	2716	2671	2626	2582	2519	2496
20	2994	2946	2899	2852	2806	2760	2715	2670	2626	2582	2538	2495
21	2993	2946	2898	2852	2805	2760	2714	2669	2625	2581	2538	2494
22	2993	2945	2898	2851	2804	2759	2713	2669	2624	2580	2537	2494
23	2992	2944	2897	2850	2804	2758	2713	2668	2623	2580	2536	2493
24	2991	2943	2896	2849	2803	2757	2712	2667	2623	2579	2535	2472
25	2990	2942	2895	2848	2802	2756	2711	2666	2622	2578	2535	2492
26	2989	2942	2894	2848	2801	2756	2710	2666	2621	2577	2534	2491
27	2989	2941	2894	2847	2801	2755	2710	2665	2621	2577	2533	2490
28	2988	2940	2893	2846	2800	2754	2709	2664	2620	2576	2532	2489
29	2987	2939	2892	2845	2799	2753	2708	2663	2619	2575	2532	2489
30	2986	2939	2891	2845	2798	2753	2707	2663	2618	2574	2531	2488
31	2985	2938	2890	2844	2798	2752	2707	2662	2618	2574	2530	2487
32	2985	2937	2870	2843	2797	2751	2706	2661	2617	2573	2530	2487
33	2984	2936	2889	2842	2796	2750	2705	2660	2616	2572	2529	2486
34	2983	2935	2888	2842	2795	2750	2704	2660	2615	2572	2528	2485
35	2982	2934	2887	2841	2795	2749	2704	2659	2615	2571	2527	2484
36	1981	2934	2887	2840	2794	2748	2703	2658	2614	2570	2527	2484
37	2981	2933	2886	2839	2793	2747	2702	2657	2613	2569	2526	6483
38	2980	2932	2885	2838	2792	2747	2701	2657	2612	2569	2525	2482
39	2979	2931	2884	2838	2792	2746	2701	2656	2612	2568	2525	2482
40	2978	2931	2883	2837	2791	2745	2700	2655	2611	2567	2524	2481
41	2977	2930	2883	2836	2790	2744	2699	2654	2610	2566	2523	2480
42	2977	2929	2882	2835	2789	2744	2698	2654	2610	2566	2522	2480
43	2976	2928	2881	2834	2788	2743	2698	2653	2609	2565	2522	2479
44	2975	2927	2880	2834	2788	2742	2697	2652	2608	2564	2521	2478
45	2974	2927	2880	2833	2787	2741	2696	2652	2607	2564	2520	2477
46	2973	2926	2879	2832	2786	2741	2695	2651	2607	2563	2520	2477
47	2973	2925	2878	2831	2785	2740	2695	2650	2606	2562	2519	2476
48	2972	2924	2877	2831	2785	2739	2694	2649	2605	2561	2518	2475
49	2971	2923	2876	2830	2784	2738	2693	2649	2604	2561	2517	2474
50	2970	2923	2876	2829	2783	2737	2692	2648	2604	2560	2517	2474
51	2969	2922	2875	2828	2782	2737	2692	2647	2603	2559	2516	2473
52	2969	2921	2874	2828	2782	2736	2691	2646	2602	2558	2515	2472
53	2968	2920	2873	2827	2781	2735	2690	2646	2601	2558	2514	2472
54	2967	2920	2873	2826	2780	2735	2689	2645	2601	2557	2514	2471
55	2966	2919	2872	2825	2779	2734	2689	2644	2600	2556	2513	2470
56	1965	2918	2871	2824	2778	2733	2688	2643	2599	2556	2512	2470
57	2965	2917	2870	2824	2778	2732	2687	2643	2599	2555	2512	2469
58	2964	2916	2869	2823	2777	2731	2686	2642	2598	2554	2511	2468
59	2963	2916	2869	2822	2776	2731	2686	2641	2597	2553	2510	2467
60	2962	2915	2863	2821	2775	2730	2685	2640	2596	2553	2510	2467

TABLE XLV.

PROPORTIONAL LOGARITHMS.

	1°42′	1°43′	1°44′	1°45′	1°46′	1°47′	1°48′	1°49′	1°50′	1°51′	1°52′	1°53′
0″	2467	2424	2382	2341	2300	2259	2218	2178	2139	2099	2061	2022
1	2466	2424	2382	2340	2299	2258	2218	2178	2138	2099	2060	2021
2	2465	2423	2381	2339	2298	2257	2217	2177	2137	2098	2059	2021
3	2465	2422	2380	2339	2298	2257	2216	2176	2137	2098	2059	2020
4	2464	2421	2380	2338	2297	2256	2216	2176	2136	2097	2058	2019
5	2463	2421	2379	2337	2296	2255	2215	2175	2135	2096	2057	2019
6	2462	2420	2378	2337	2296	2255	2214	2174	2135	2096	2057	2018
7	2462	2419	2378	2336	2295	2254	2214	2174	2134	2095	2056	2017
8	2461	2419	2377	2335	2294	2253	2213	2173	2133	2094	2055	2017
9	2460	2418	2376	2335	2294	2253	2212	2172	2133	2094	2055	2016
10	2460	2417	2375	2334	2293	2252	2212	2172	2132	2093	2054	2016
11	2459	2417	2375	2333	2292	2251	2211	2171	2132	2092	2053	2015
12	2458	2416	2374	2333	2291	2251	2210	2170	2131	2092	2053	2014
13	2457	2415	2373	2332	2291	2250	2210	2170	2130	2091	2052	2014
14	2457	2414	2373	2331	2290	2249	2209	2169	2130	2090	2051	2013
15	2456	2414	2372	2331	2289	2249	2208	2169	2129	2090	2051	2012
16	2455	2413	2371	2330	2289	2248	2208	2168	2128	2089	2050	2012
17	2455	2412	2371	2329	2288	2247	2207	2167	2128	2088	2050	2011
18	2454	2412	2370	2328	2287	2247	2206	2167	2127	2088	2049	2010
19	2453	2411	2369	2328	2287	2246	2206	2166	2126	2087	2048	2010
20	2452	2410	2368	2327	2286	2245	2205	2165	2126	2086	2048	2009
21	2452	2410	2368	2326	2285	2245	2204	2165	2125	2086	2047	2009
22	2451	2409	2367	2326	2285	2244	2204	2164	2124	2085	2046	2008
23	2450	2408	2366	2325	2284	2243	2203	2163	2124	2084	2046	2007
24	2450	2408	2366	2324	2283	2243	2202	2163	2123	2084	2045	2007
25	2449	2407	2365	2324	2283	2242	2202	2162	2122	2083	2044	2006
26	2448	2406	2364	2323	2282	2241	2201	2161	2122	2083	2044	2005
27	2448	2405	2364	2322	2281	2241	2200	2161	2121	2082	2043	2005
28	2447	2405	2363	2321	2281	2240	2200	2160	2120	2081	2042	2004
29	2446	2404	2362	2321	2280	2239	2199	2159	2120	2081	2042	2004
30	2445	2403	2362	2320	2279	2239	2198	2159	2119	2080	2041	2003
31	2445	2403	2361	2319	2279	2238	2198	2158	2118	2079	2041	2002
32	2444	2402	2360	2319	2278	2237	2197	2157	2118	2079	2040	2001
33	2443	2401	2359	2318	2377	2237	2196	2157	2117	2078	2039	2001
34	2443	2400	2359	2317	2276	2236	2196	2156	2116	2077	2039	2000
35	2442	2400	2358	2317	2276	2235	2195	2155	2116	2077	2038	2000
36	2441	2399	2357	2316	2275	2235	2194	2155	2115	2076	2037	1999
37	2440	2398	2357	2315	2274	2234	2194	2154	2114	2075	2037	1998
38	2440	1398	2356	2315	2274	2233	2193	2153	2114	2075	2036	1998
39	2437	2397	2355	2314	2273	2233	2192	2153	2113	2074	2035	1997
40	2438	2396	2355	2313	2272	2232	2192	2152	2113	2073	2035	1996
41	2438	2396	2354	2313	2272	2231	2191	2152	2112	2073	2034	1996
42	2437	2395	2353	2312	2271	2231	2190	2151	2111	2072	2033	1995
43	2436	2394	2353	2311	2270	2230	2190	2150	2111	2071	2033	1994
44	2436	2394	2352	2311	2270	2229	2189	2149	2110	2071	2032	1994
45	2435	2393	2351	2310	2269	2229	2188	2149	2109	2070	2032	1993
46	2434	2392	2350	2309	2268	2228	2188	2148	2109	2070	2031	1993
47	2433	2391	2350	2308	2268	2227	2187	2147	2108	2069	2030	1992
48	2433	2391	2349	2308	2267	2227	2186	2147	2107	2068	2030	1991
49	2432	2390	2348	2307	2266	2226	2186	2146	2107	2068	2029	1991
50	2431	2389	2348	2306	2266	2225	2185	2145	2106	2067	2028	1990
51	2431	2389	2347	2306	2265	2225	2184	2145	2105	2066	2028	1989
52	2430	2388	2346	2305	2264	2224	2184	2144	2105	2066	2027	1989
53	2429	2387	2346	2304	2264	2223	2183	2143	2104	2065	2026	1988
54	2429	2387	2345	2304	2263	2223	2182	2143	2103	2064	2026	1987
55	2428	2386	2344	2303	2262	2222	2182	2142	2103	2064	2025	1987
56	2427	2385	2344	2302	2262	2221	2181	2141	2102	2063	2024	1986
57	2426	2384	2343	2302	2261	2220	2180	2141	2101	2062	2024	1986
58	2426	2384	2342	2301	2260	2220	2180	2140	2101	2062	2023	1985
59	2425	2383	2341	2300	2260	2219	2179	2139	2100	2061	2023	1984
60	2424	2382	2341	2300	2259	2218	2178	2139	2099	2061	2022	1984

TABLE XLV.

PROPORTIONAL LOGARITHMS.

	1° 54′	1° 55′	1° 56′	1° 57′	1° 58′	1° 59′	2° 0′	2° 1′	2° 2′	2° 3′	2° 4′
0″	1984	1946	1908	1871	1834	1797	1761	1725	1689	1654	1619
1	1983	1945	1907	1870	1833	1797	1760	1724	1688	1653	1618
2	1982	1944	1907	1870	1833	1796	1760	1724	1688	1652	1617
3	1982	1944	1906	1869	1832	1795	1759	1723	1687	1652	1617
4	1981	1943	1906	1868	1831	1795	1758	1722	1687	1651	1616
5	1980	1943	1905	1868	1831	1794	1758	1722	1686	1651	1616
6	1980	1942	1904	1867	1830	1794	1757	1721	1686	1650	1615
7	1979	1941	1904	1867	1830	1793	1757	1721	1685	1650	1614
8	1979	1941	1903	1866	1829	1792	1756	1720	1684	1649	1614
9	1978	1940	1903	1865	1828	1792	1755	1719	1684	1648	1613
10	1977	1939	1902	1865	1828	1791	1755	1719	1683	1648	1613
11	1977	1939	1901	1864	1827	1791	1754	1718	1683	1647	1612
12	1976	1938	1901	1863	1827	1790	1754	1718	1682	1647	1612
13	1975	1938	1900	1863	1826	1789	1753	1717	1681	1646	1611
14	1975	1937	1899	1862	1825	1789	1752	1716	1681	1645	1610
15	1974	1936	1899	1862	1825	1788	1752	1716	1680	1645	1610
16	1973	1936	1898	1861	1824	1787	1751	1715	1680	1644	1609
17	1973	1935	1898	1860	1823	1787	1751	1715	1679	1644	1609
18	1972	1934	1897	1860	1823	1786	1750	1714	1678	1643	1608
19	1972	1934	1896	1859	1822	1786	1749	1713	1678	1642	1607
20	1971	1933	1896	1858	1822	1785	1749	1713	1677	1642	1607
21	1970	1933	1895	1858	1821	1785	1748	1712	1677	1641	1606
22	1970	1932	1894	1857	1820	1784	1748	1712	1676	1641	1606
23	1969	1931	1894	1857	1820	1783	1747	1711	1675	1640	1605
24	1968	1931	1893	1856	1819	1783	1746	1711	1675	1640	1605
25	1968	1930	1893	1855	1819	1782	1746	1710	1674	1639	1604
26	1967	1929	1892	1855	1818	1781	1745	1709	1674	1638	1603
27	1967	1929	1891	1854	1817	1781	1745	1709	1673	1638	1603
28	1966	1928	1891	1854	1817	1780	1744	1708	1673	1637	1602
29	1965	1927	1890	1853	1816	1780	1743	1708	1672	1637	1602
30	1965	1927	1889	1852	1816	1779	1743	1707	1671	1636	1601
31	1964	1926	1889	1851	1815	1778	1742	1706	1671	1635	1600
32	1963	1926	1888	1851	1814	1778	1742	1706	1670	1635	1600
33	1963	1925	1888	1850	1814	1777	1741	1705	1670	1634	1599
34	1962	1924	1887	1850	1813	1777	1740	1705	1669	1634	1599
35	1961	1924	1886	1849	1812	1776	1740	1704	1668	1633	1598
36	1961	1923	1886	1849	1812	1775	1739	1703	1668	1633	1598
37	1960	1922	1885	1848	1811	1775	1739	1703	1667	1632	1597
38	1960	1922	1884	1847	1811	1774	1738	1702	1667	1631	1596
39	1959	1921	1884	1847	1810	1774	1737	1702	1666	1631	1596
40	1958	1921	1883	1846	1809	1773	1737	1701	1665	1630	1595
41	1958	1920	1883	1846	1809	1772	1736	1700	1665	1630	1595
42	1957	1919	1882	1845	1808	1772	1736	1700	1664	1629	1594
43	1956	1919	1881	1844	1808	1771	1735	1699	1664	1628	1593
44	1956	1918	1881	1844	1807	1771	1734	1699	1663	1628	1593
45	1955	1918	1880	1843	1806	1770	1734	1698	1663	1627	1592
46	1955	1917	1879	1842	1806	1769	1733	1697	1662	1627	1592
47	1954	1916	1879	1842	1805	1769	1733	1697	1661	1626	1591
48	1953	1916	1878	1841	1805	1768	1732	1696	1661	1626	1591
49	1952	1915	1878	1841	1804	1768	1731	1696	1660	1625	1590
50	1952	1914	1877	1840	1803	1767	1731	1695	1660	1624	1589
51	1951	1914	1876	1839	1803	1766	1730	1694	1659	1624	1589
52	1951	1913	1876	1839	1802	1766	1730	1694	1658	1623	1588
53	1950	1912	1875	1838	1801	1765	1729	1693	1658	1623	1588
54	1950	1912	1875	1838	1801	1765	1728	1693	1657	1622	1587
55	1949	1911	1874	1837	1800	1764	1728	1692	1657	1621	1586
56	1948	1911	1873	1836	1800	1763	1727	1691	1656	1621	1586
57	1948	1910	1873	1836	1799	1763	1727	1691	1655	1620	1585
58	1947	1909	1872	1835	1798	1762	1726	1690	1655	1620	1585
59	1946	1909	1871	1834	1798	1761	1725	1690	1654	1619	1584
60	1946	1908	1871	1834	1797	1761	1725	1689	1654	1619	1584

TABLE XLV.

PROPORTIONAL LOGARITHMS.

	2° 5′	2° 6′	2° 7′	2° 8′	2° 9′	2° 10′	2° 11′	2° 12′	2° 13′	2° 14′	2° 15′
0″	1584	1549	1515	1481	1447	1413	1380	1347	1314	1282	1249
1	1583	1548	1514	1480	1446	1413	1379	1346	1314	1281	1249
2	1582	1548	1514	1479	1446	1412	1379	1346	1313	1281	1248
3	1582	1547	1513	1479	1445	1412	1378	1345	1313	1280	1248
4	1581	1547	1512	1478	1445	1411	1378	1345	1312	1279	1247
5	1581	1546	1512	1478	1444	1410	1377	1344	1311	1279	1247
6	1580	1546	1511	1477	1443	1410	1377	1344	1311	1278	1246
7	1580	1545	1511	1477	1443	1409	1376	1343	1310	1278	1246
8	1579	1544	1510	1476	1442	1409	1376	1343	1310	1277	1245
9	1578	1544	1510	1476	1442	1408	1375	1342	1309	1277	1245
10	1578	1543	1509	1475	1441	1408	1374	1341	1309	1276	1244
11	1577	1543	1508	1474	1441	1407	1374	1341	1308	1276	1243
12	1577	1542	1508	1474	1440	1407	1373	1340	1308	1275	1243
13	1576	1542	1507	1473	1440	1406	1373	1340	1307	1275	1242
14	1575	1541	1507	1473	1439	1405	1372	1339	1307	1274	1242
15	1575	1540	1506	1472	1438	1405	1372	1339	1306	1274	1241
16	1574	1540	1506	1472	1438	1404	1371	1338	1305	1273	1241
17	1574	1539	1505	1471	1437	1404	1371	1338	1305	1272	1240
18	1573	1539	1504	1470	1437	1403	1370	1337	1304	1272	1240
19	1573	1538	1504	1470	1436	1403	1369	1337	1304	1271	1239
20	1572	1538	1503	1469	1436	1402	1369	1336	1303	1271	1239
21	1571	1537	1503	1469	1435	1402	1368	1335	1303	1270	1238
22	1571	1536	1502	1468	1434	1401	1368	1335	1302	1270	1238
23	1570	1536	1502	1468	1434	1400	1367	1334	1302	1269	1237
24	1570	1535	1501	1467	1433	1400	1367	1334	1301	1269	1237
25	1569	1535	1500	1466	1433	1399	1366	1333	1301	1268	1236
26	1569	1534	1500	1466	1432	1399	1366	1333	1300	1268	1235
27	1568	1534	1499	1465	1432	1398	1365	1332	1300	1267	1235
28	1567	1533	1499	1465	1431	1398	1365	1332	1299	1267	1234
29	1567	1532	1498	1464	1431	1397	1364	1331	1298	1266	1234
30	1566	1532	1498	1464	1430	1397	1363	1331	1298	1266	1233
31	1566	1531	1497	1463	1429	1396	1363	1330	1297	1265	1233
32	1565	1531	1496	1463	1429	1395	1362	1329	1297	1264	1232
33	1565	1530	1496	1462	1428	1395	1362	1329	1296	1264	1232
34	1564	1529	1495	1461	1428	1394	1361	1328	1296	1263	1231
35	1563	1529	1495	1461	1427	1394	1361	1328	1295	1263	1231
36	1563	1528	1494	1460	1427	1393	1360	1327	1295	1262	1230
37	1562	1528	1494	1460	1426	1393	1360	1327	1294	1262	1230
38	1562	1527	1493	1459	1426	1392	1359	1326	1294	1261	1229
39	1561	1527	1493	1459	1425	1392	1359	1326	1293	1261	1229
40	1560	1526	1492	1458	1424	1391	1358	1325	1292	1260	1228
41	1560	1525	1491	1457	1424	1390	1357	1325	1292	1260	1227
42	1559	1525	1491	1457	1423	1390	1357	1324	1291	1259	1227
43	1559	1524	1490	1456	1423	1389	1356	1323	1291	1258	1226
44	1558	1524	1490	1456	1422	1389	1356	1323	1290	1258	1226
45	1558	1523	1489	1455	1422	1388	1355	1322	1290	1257	1225
46	1557	1523	1489	1455	1421	1388	1355	1322	1289	1257	1225
47	1556	1522	1488	1454	1420	1387	1354	1321	1289	1256	1224
48	1556	1522	1487	1454	1420	1387	1354	1321	1288	1256	1224
49	1555	1521	1487	1453	1419	1386	1353	1320	1288	1255	1223
50	1555	1520	1486	1452	1419	1386	1352	1320	1287	1255	1223
51	1554	1520	1486	1452	1418	1385	1352	1319	1287	1254	1222
52	1554	1519	1485	1451	1418	1384	1351	1319	1286	1254	1222
53	1553	1518	1485	1451	1417	1384	1351	1318	1285	1253	1221
54	1552	1518	1484	1450	1417	1383	1350	1317	1285	1253	1221
55	1552	1518	1483	1450	1416	1383	1350	1317	1284	1252	1220
56	1551	1517	1483	1449	1415	1382	1349	1316	1284	1251	1219
57	1551	1516	1482	1449	1415	1382	1349	1316	1283	1251	1219
58	1550	1516	1482	1448	1414	1381	1348	1315	1283	1250	1218
59	1550	1515	1481	1447	1414	1381	1347	1315	1282	1250	1218
60	1549	1515	1481	1447	1413	1380	1347	1314	1282	1249	1217

TABLE XLV.

PORTIONAL LOGARITHMS.

2° 18′	2° 19′	2° 20′	2° 21′	2° 22′	2° 23′	2° 2
1154	1123	1091	1061	1030	0999	096
1153	1122	1091	1060	1029	0999	096
1153	1121	1090	1059	1029	0998	096
1152	1121	1090	1059	1028	0998	096
1152	1120	1089	1058	1028	0997	096
1151	1120	1089	1058	1027	0997	096
1151	1119	1088	1057	1027	0996	096
1150	1119	1088	1057	1026	0996	096
1150	1118	1087	1056	1026	0995	096
1149	1118	1087	1056	1025	0995	096
1149	1117	1086	1055	1025	0994	096
1148	1117	1086	1055	1024	0994	096
1148	1116	1085	1054	1024	0993	096
1147	1116	1085	1054	1023	0993	096
1147	1115	1084	1053	1023	0992	096
1146	1115	1084	1053	1022	0992	096
1146	1114	1083	1052	1022	0991	096
1145	1114	1083	1052	1021	0991	096
1145	1113	1082	1051	1021	0990	096
1144	1113	1082	1051	1020	0990	096
1143	1112	1081	1050	1020	0989	095
1143	1112	1081	1050	1019	0989	095
1142	1111	1080	1049	1019	0988	095
1142	1111	1080	1049	1018	0988	095
1141	1110	1079	1048	1018	0987	095
1141	1110	1079	1048	1017	0987	095
1140	1109	1078	1047	1017	0986	095
1140	1109	1078	1047	1016	0986	095
1139	1108	1077	1046	1016	0985	095
1139	1107	107[illegible]	1046	1015	0985	095
1138	1107	1076	1045	1015	0984	095
1138	1106	1075	1045	1014	0984	095
1137	1106	1075	1044	1014	0983	095
1137	1105	1074	1044	1013	0983	095
1136	1105	1074	1043	1013	0982	095
1136	1104	1073	1043	1012	0982	095
1135	1104	1073	1042	1012	0981	095
1135	1103	1072	1042	1011	0981	095
1134	1103	1072	1041	1010	0980	095
1134	1102	1071	1041	1010	0980	095
1133	1102	1071	1040	1009	0979	094
1132	1101	1070	1039	1009	0979	094
1132	1101	1070	1039	1008	0978	094
1131	1100	1069	1038	1008	0978	094
1131	1100	1069	1038	1007	0977	094
1130	1099	1068	1037	1007	0977	094
1130	1099	1068	1037	1006	0976	094
1129	1098	1067	1036	1006	0976	094
1129	1098	1067	1036	1005	0975	094
1128	1097	1066	1035	1005	0975	094
1128	1097	1086	1035	1004	0974	094
112[illegible]	1096	1065	1034	1004	0974	094
1127	1096	1065	1034	1003	0973	094
1126	1095	1064	1033	1003	0973	094
1126	1095	1064	1033	1002	0972	094
1125	1094	1063	1032	1002	0972	094
1125	1093	1063	1032	1001	0971	094
1124	1093	1062	1031	1001	0971	094
1124	1092	1062	1031	1000	0970	094
1123	1092	1061	1030	1000	0970	094

TABLE XLV.

PROPORTIONAL LOGARITHMS.

	2°27′	2°28′	2°29′	2°30′	2°31′	2°32′	2°33′	2°34′	2°35′	2°36′	2°37′
0″	0880	0850	0821	0792	0763	0734	0706	0678	0649	0621	0594
1	0879	0850	0820	0791	0763	0734	0705	0677	0649	0621	0593
2	0879	0849	0820	0791	0762	0733	0705	0677	0648	0621	0592
3	0878	0849	0819	0790	0762	0733	0704	0676	0648	0620	0592
4	0878	0848	0819	0790	0761	0732	0704	0676	0648	0620	0592
5	0877	0848	0818	0789	0761	0732	0703	0675	0647	0619	0591
6	0877	0847	0818	0789	0760	0731	0703	0675	0647	0619	0591
7	0876	0847	0817	0788	0760	0731	0702	0674	0646	0618	0590
8	0876	0846	0817	0788	0759	0730	0702	0674	0646	0618	0590
9	0875	0846	0816	0787	0759	0730	0702	0673	0645	0617	0590
10	0875	0845	0816	0787	0758	0729	0701	0673	0645	0617	0589
11	0874	0845	0815	0787	0758	0729	0701	0672	0644	0616	0589
12	0874	0844	0815	0786	0757	0729	0700	0672	0644	0616	0588
13	0873	0844	0814	0786	0757	0728	0700	0671	0643	0615	0588
14	0873	0843	0814	0785	0756	0728	0699	0671	0643	0615	0587
15	0872	0843	0814	0785	0756	0727	0699	0670	0642	0615	0587
16	0872	0842	0813	0784	0755	0727	0698	0670	0642	0614	0586
17	0871	0842	0813	0784	0755	0726	0698	0669	0641	0614	0586
18	0871	0841	0812	0783	0754	0726	0697	0669	0641	0613	0585
19	0870	0841	0812	0783	0754	0725	0697	0669	0641	0613	0585
20	0870	0840	0811	0782	0753	0725	0696	0668	0640	0612	0584
21	0869	0840	0811	0782	0753	0724	0696	0668	0640	0612	0584
22	0869	0839	0810	0781	0752	0724	0695	0667	0639	0611	0584
23	0868	0839	0810	0781	0752	0723	0695	0667	0639	0611	0583
24	0868	0838	0809	0780	0751	0723	0694	0666	0638	0610	0583
25	0867	0838	0809	0780	0751	0722	0694	0666	0638	0610	0582
26	0867	0837	0808	0779	0750	0722	0693	0665	0637	0609	0582
27	0866	0837	0808	0779	0750	0721	0693	0665	0637	0609	0581
28	0866	0836	0807	0778	0750	0721	0693	0664	0636	0608	0581
29	0865	0836	0807	0778	0749	0720	0692	0664	0636	0608	0580
30	0865	0835	0806	0777	0749	0720	0692	0663	0635	0608	0580
31	0864	0835	0806	0777	0748	0720	0691	0663	0635	0607	0579
32	0864	0834	0805	0776	0748	0719	0691	0662	0634	0607	0579
33	0863	0834	0805	0776	0747	0719	0690	0662	0634	0606	0579
34	0863	0833	0804	0775	0747	0718	0690	0662	0634	0606	0578
35	0862	0833	0804	0775	0746	0718	0689	0661	0633	0605	0578
36	0862	0833	0803	0774	0746	0717	0689	0661	0633	0605	0577
37	0861	0832	0803	0774	0745	0717	0688	0660	0632	0604	0577
38	0861	0832	0802	0773	0745	0716	0688	0660	0632	0604	0576
39	0860	0831	0802	0773	0744	0716	0687	0659	0631	0603	0576
40	0860	0831	0801	0773	0744	0715	0687	0659	0631	0603	0575
41	0859	0830	0801	0772	0743	0715	0686	0658	0630	0602	0575
42	0859	0830	0801	0772	0743	0714	0686	0658	0630	0602	0574
43	0858	0829	0800	0771	0742	0714	0685	0657	0629	0601	0574
44	0858	0829	0800	0771	0742	0713	0685	0657	0629	0601	0573
45	0857	0828	0799	0770	0741	0713	0685	0656	0628	0601	0573
46	0857	0828	0799	0770	0741	0712	0684	0656	0628	0600	0573
47	0856	0827	0798	0769	0740	0712	0684	0655	0627	0600	0572
48	0856	0827	0798	0769	0740	0711	0683	0655	0627	0599	0572
49	0855	0826	0797	0768	0739	0711	0683	0655	0627	0599	0571
50	0855	0826	0797	0768	0739	0711	0682	0654	0626	0598	0571
51	0855	0825	0796	0767	0739	0710	0682	0654	0626	0598	0570
52	0854	0825	0796	0767	0738	0710	0681	0653	0625	0597	0570
53	0854	0824	0795	0766	0738	0709	0681	0653	0625	0597	0569
54	0853	0824	0795	0766	0737	0709	0680	0652	0624	0596	0569
55	0853	0823	0794	0765	0737	0708	0680	0652	0624	0596	0568
56	0852	0823	0794	0765	0736	0708	0679	0651	0623	0596	0568
57	0852	0822	0793	0764	0736	0707	0679	0651	0623	0595	0568
58	0851	0822	0793	0764	0735	0707	0678	0650	0622	0595	0567
59	0851	0821	0792	0763	0735	0706	0678	0650	0622	0594	0567
60	0850	0821	0792	0763	0734	0706	0678	0649	0621	0594	0566

TABLE XLV.

PROPORTIONAL LOGARITHMS.

	2°38′	2°39′	2°40′	2°41′	2°42′	2°43′	2°44′	2°45′	2°46′	2°47′	2°48′
0″	0566	0539	0512	0484	0458	0431	0404	0378	0352	0320	0300
1	0566	0538	0511	0484	0457	0430	0404	0377	0351	0325	0299
2	0565	0538	0511	0484	0457	0430	0403	0377	0351	0325	0299
3	0565	0537	0510	0483	0456	0430	0403	0377	0350	0324	0298
4	0564	0537	0510	0483	0456	0429	0402	0376	0350	0324	0298
5	0564	0536	0509	0482	0455	0429	0402	0376	0349	0323	0297
6	0563	0536	0509	0482	0455	0428	0402	0375	0349	0323	0297
7	0563	0536	0508	0481	0454	0428	0401	0375	0349	0322	0297
8	0562	0535	0508	0481	0454	0427	0401	0374	0348	0322	0296
9	0562	053[illegible]	0507	0480	0454	0427	0400	0374	0348	0322	0296
10	0562	0534	0507	0480	0453	0426	0400	0373	0347	0321	0295
11	0561	0534	0507	0479	0453	0426	0399	0373	0347	0321	0295
12	0561	0533	0506	0479	0452	0426	0399	0373	0346	0320	0294
13	0560	0533	0506	0479	0452	0425	0399	0372	0346	0320	0294
14	0560	0532	0505	0478	0451	0425	0398	0372	0346	0319	0294
15	0559	0532	0505	0478	0451	0424	0398	0371	0345	0319	0293
16	0559	0531	0504	0477	0450	0424	0397	0371	0345	0319	0293
17	0558	0531	0504	0477	0450	0423	0397	0370	0344	0318	0292
18	0558	0531	0503	0476	0450	0423	0396	0370	0344	0318	0292
19	0557	0530	0503	0476	0449	0422	0396	0370	0343	0317	0291
20	0557	0530	0502	0475	0449	0422	0395	0369	0343	0317	0291
21	0557	0529	0502	0475	0448	0422	0395	0369	0342	0316	0291
22	0556	0529	0502	0475	0448	0421	0395	0368	0342	0316	0290
23	0556	0528	0501	0474	0447	0421	0394	0368	0342	0316	0290
24	0555	0528	0501	0474	0447	0420	0394	0367	0341	0315	0289
25	0555	0527	0500	0473	0446	0420	0393	0367	0341	0315	0289
26	0554	0527	0500	0473	0446	0419	0393	0366	0340	0314	0288
27	0554	0526	0499	0472	0446	0419	0392	0366	0340	0314	0288
28	0553	0526	0499	0472	0445	0418	0392	0366	0339	0313	0288
29	0553	0526	0498	0471	0445	0418	0391	0365	0339	0313	0287
30	0552	0525	0498	0471	0444	0418	0391	0365	0339	0313	0287
31	0552	0525	0497	0471	0444	0417	0391	0364	0338	0312	0286
32	0551	0524	0497	0470	0443	0417	0390	0364	0338	0312	0286
33	0551	0524	0497	0470	0443	0416	0390	0363	0337	0311	0285
34	0551	0523	0496	0469	0442	0416	0389	0363	0337	0311	0285
35	0550	0523	0496	0469	0442	0415	0389	0363	0336	0310	0285
36	0550	0522	0495	0468	0442	0415	0388	0362	0336	0310	0284
37	0549	0522	0495	0468	0441	0414	0388	0362	0336	0310	0284
38	0549	0521	0494	0467	0441	0414	0388	0361	0335	0309	0283
39	0548	0521	0494	0467	0440	0414	0387	0361	0335	0309	0283
40	0548	0521	0493	0466	0440	0413	0387	0360	0334	0308	0282
41	0547	0520	0493	0466	0439	0413	0386	0360	0334	0308	0282
42	0547	0520	0493	0466	0439	0412	0386	0359	0333	0307	0282
43	0546	0519	0492	0465	0438	0412	0385	0359	0333	0307	0281
44	0546	0519	0492	0465	0438	0411	0385	0359	0332	0306	0281
45	0546	0518	0491	0464	0438	0411	0384	0358	0332	0306	0280
46	0545	0518	0491	0464	0437	0410	0384	0358	0332	0306	0280
47	0545	0517	0490	0463	0437	0410	0384	0357	0331	0305	0279
48	0544	0517	0490	0463	0436	0410	0383	0357	0331	0305	0279
49	0544	0516	0489	0462	0436	0409	0383	0356	0330	0304	0279
50	0543	0516	0489	0462	0435	0409	0382	0356	0330	0304	0278
51	0543	0516	0489	0462	0435	0408	0382	0356	0329	0304	0278
52	0542	0515	0488	0461	0434	0408	0381	0355	0329	0303	0277
53	0542	0515	0488	0461	0434	0407	0381	0355	0329	0303	0277
54	0541	0514	0487	0460	0434	0407	0381	0354	0328	0302	0276
55	0541	0514	0487	0460	0433	0406	0380	0354	0328	0302	0276
56	0541	0513	0486	0459	0433	0406	0380	0353	0327	0301	0276
57	0540	0513	0486	0459	0432	0406	0379	0353	0327	0301	0275
58	0540	0512	0485	0458	0432	0405	0379	0352	0326	0300	0275
59	0539	0512	0485	0458	0431	0405	0378	0352	0326	0300	0274
60	0539	0512	0484	0458	0431	0404	0378	0352	0326	0300	0274

TABLE XLV.

PROPORTIONAL LOGARITHMS.

	2°49′	2°50′	2°51′	2°52′	2°53′	2°54′	2°55′	2°56′	2°57′	2°58′	2°59′
0″	0274	0248	0223	0197	0172	0147	0122	0098	0073	0049	0024
1	0273	0248	0222	0197	0172	0147	0122	0097	0073	0048	0024
2	0273	0247	0222	0197	0171	0146	0121	0097	0072	0048	0023
3	0273	0247	0221	0196	0171	0146	0121	0096	0072	0047	0023
4	0272	0246	0221	0196	0171	0146	0121	0096	0071	0047	0023
5	0272	0246	0221	0195	0170	0145	0120	0096	0071	0046	0022
6	0271	0246	0220	0195	0170	0145	0120	0095	0071	0046	0022
7	0271	0245	0220	0194	0169	0144	0119	0095	0070	0046	0021
8	0270	0245	0219	0194	0169	0144	0119	0094	0070	0045	0021
9	0270	0244	0219	0194	0169	0143	0119	0094	0069	0045	0021
10	0270	0244	0218	0193	0168	0143	0118	0093	0069	0044	0020
11	0269	0244	0218	0193	0168	0143	0118	0093	0068	0044	0020
12	0269	0243	0218	0192	0167	0142	0117	0093	0068	0044	0019
13	0268	0243	0217	0192	0167	0142	0117	0092	0068	0043	0019
14	0268	0242	0217	0192	0166	0141	0117	0092	0067	0043	0018
15	0267	0242	0216	0191	0166	0141	0116	0091	0067	0042	0018
16	0267	0241	0216	0191	0166	0141	0116	0091	0066	0042	0018
17	0267	0241	0216	0190	0165	0140	0115	0091	0066	0042	0017
18	0266	0241	0215	0190	0165	0140	0115	0090	0066	0041	0017
19	0266	0240	0215	0189	0164	0139	0114	0090	0065	0041	0016
20	0265	0240	0214	0189	0164	0139	0114	0089	0065	0040	0016
21	0265	0239	0214	0189	0163	0139	0114	0089	0064	0040	0016
22	0264	0239	0213	0188	0163	0138	0113	0089	0064	0040	0015
23	0264	0238	0213	0188	0163	0138	0113	0088	0064	0039	0015
24	0264	0238	0213	0187	0162	0137	0112	0088	0063	0039	0015
25	0263	0238	0212	0187	0162	0137	0112	0087	0063	0038	0014
26	0263	0237	0212	0186	0161	0136	0112	0087	0062	0038	0014
27	0262	0237	0211	0186	0161	0136	0111	0087	0062	0038	0013
28	0262	0236	0211	0186	0161	0136	0111	0086	0062	0037	0013
29	0261	0236	0210	0185	0160	0135	0110	0086	0061	0037	0012
30	0261	0235	0210	0185	0160	0135	0110	0085	0061	0036	0012
31	0261	0235	0210	0184	0159	0134	0110	0085	0060	0036	0012
32	0260	0235	0209	0184	0159	0134	0109	0084	0060	0035	0011
33	0260	0234	0209	0184	0158	0134	0109	0084	0060	0035	0011
34	0259	0234	0208	0183	0158	0133	0108	0084	0059	0035	0010
35	0259	0233	0208	0183	0158	0133	0108	0083	0059	0034	0010
36	0258	0233	0208	0182	0157	0132	0107	0083	0058	0034	0010
37	0258	0232	0207	0182	0157	0132	0107	0082	0058	0033	0009
38	0258	0232	0207	0181	0156	0131	0107	0082	0057	0033	0009
39	0257	0232	0206	0181	0156	0131	0106	0082	0057	0033	0008
40	0257	0231	0206	0181	0156	0131	0106	0081	0057	0032	0008
41	0256	0231	0205	0180	0155	0130	0105	0081	0056	0032	0008
42	0256	0230	0205	0180	0155	0130	0105	0080	0056	0031	0007
43	0255	0230	0205	0179	0154	0129	0105	0080	0055	0031	0007
44	0255	0230	0204	0179	0154	0129	0104	0080	0055	0031	0006
45	0255	0229	0204	0179	0153	0129	0104	0079	0055	0030	0006
46	0254	0229	0203	0178	0153	0128	0103	0079	0054	0030	0006
47	0254	0228	0203	0178	0153	0128	0103	0078	0054	0029	0005
48	0253	0228	0202	0177	0152	0127	0103	0078	0053	0029	0005
49	0253	0227	0202	0177	0152	0127	0102	0077	0053	0029	0004
50	0252	0227	0202	0176	0151	0126	0102	0077	0053	0028	0004
51	0252	0227	0201	0176	0151	0126	0101	0077	0052	0028	0004
52	0251	0226	0201	0176	0151	0126	0101	0076	0052	0027	0003
53	0251	0226	0200	0175	0150	0125	0100	0076	0051	0027	0003
54	0251	0225	0200	0175	0150	0125	0100	0075	0051	0027	0002
55	0250	0225	0200	0174	0149	0124	0100	0075	0051	0026	0002
56	0250	0224	0199	0174	0149	0124	0099	0075	0050	0026	0002
57	0250	0224	0199	0174	0148	0124	0099	0074	0050	0025	0001
58	0249	0224	0198	0173	0148	0123	0098	0074	0049	0025	0001
59	0249	0223	0198	0173	0148	0123	0098	0073	0049	0025	0000
60	0248	0223	0197	0172	0147	0122	0098	0073	0049	0024	0000

TABLE XLVI.

The Right Ascensions and Declinations of the principal Fixed Stars, adapted to the beginning of the year 1793.

Names of Stars.	Mag.	Mean Right Ascension in Time. h.	′	″	an. var ″	Degrees. °	′	″	an. var ″	Declination. °	′	″		an. var. ″
γ Pegasi	2	0	2	35	3.06	0	38	50	45.9	14	1	15	N.	+20.0
β Hydræ	3	0	14	8	2.68	3	31	59	40.2	78	25	27	S.	−20.1
α Phœnicis	2.3	0	16	1	2.99	4	0	11	44.8	43	25	32	S.	−20.0
δ Andromedæ	3	0	28	17	3.17	7	4	12	47.5	29	41	39	N.	+19.9
Schedar	3	0	28	50	3.31	7	12	37	49.7	55	23	59	N.	+19.9
β Ceti	2	0	33	11	3.01	8	17	51	45.1	19	7	32	S.	−19.8
γ Cassiopeiæ	3	0	44	20	3.52	11	4	53	52.8	59	35	35	N.	+19.7
Alruccabah p. star	2.3	0	50	44	12.42	12	41	3	186.3	88	12	8	N.	+19.6
Mirach	2	0	58	10	3.31	14	32	29	49.6	34	31	53	N.	+19.4
δ Cassiopeiæ	3	1	12	22	3.77	18	5	35	56.5	59	9	12	N.	+19.1
Achernar	1	1	30	0	2.25	22	29	55	33.7	58	17	33	S.	−18.5
ε Cassiopeiæ	3	1	39	40	4.16	24	54	54	62.4	62	38	30	N.	+18.2
β Arietis	3.4	1	43	13	3.28	25	48	18	49.2	19	47	12	N.	+18.1
Almaach	2	1	51	15	3.62	27	48	44	54.3	41	19	47	N.	+17.7
α Piscium	3	1	51	22	3.09	27	50	20	46.4	1	45	33	N.	+17.7
α Arietis	3	1	55	32	3.14	28	53	0	50.1	22	28	32	N.	+17.5
ο Ceti	Var	2	8	54	3.03	32	13	33	45.4	3	55	16	S.	−17.0
δ Ceti	3	2	28	54	3.07	37	13	30	46.0	0	34	16	S.	−16.0
ε Ceti	3	2	30	34	2.89	37	38	28	43.4	12	45	22	S.	−15.9
γ Ceti	3	2	32	36	3.11	38	9	0	46.6	2	21	29	N.	+15.8
γ Persei	3	2	49	54	4.35	42	28	33	65.8	52	40	59	N.	+14.8
Menkar	2	2	51	28	3.11	42	52	3	46.6	3	16	17	N.	+14.7
Algol	Var	2	54	45	3.85	43	41	21	57.8	40	8	47	N.	+14.5
Algenib	2	3	9	38	4.21	47	24	35	63.1	49	6	40	N.	+13.6
ε Eridani	3	3	23	14	2.89	50	48	36	43.3	10	10	2	S.	−12.7
δ Persei	3	3	28	15	4.21	52	3	45	63.2	47	6	39	N.	+12.3
δ Eridani	3	3	33	22	2.87	53	20	26	43.1	10	28	12	S.	−12.0
η Pleiadum	3	3	35	12	3.54	53	48	3	53.1	23	27	13	N.	+11.9
ζ Persei	3	3	41	9	3.74	55	17	10	56.1	31	15	20	N.	+11.4
ε Persei	3	3	44	1	3.98	56	0	8	59.7	39	23	40	N.	+11.2
γ Eridani	3	3	48	23	2.79	57	5	55	41.9	14	6	27	S.	−10.9
γ Tauri	3	4	8	1	3.39	62	0	21	50.9	15	6	57	N.	+ 9.4
ε Tauri	3	4	16	32	3.48	64	8	4	52.2	18	42	34	N.	+ 8.8
Aldebaran	1	4	24	3	3.43	66	0	50	51.3	16	4	48	N.	+ 8.2
β Eridani	3	4	57	42	2.95	74	25	30	44.3	5	21	53	S.	− 5.4
Capella	1	5	1	25	4.41	75	21	19	66.2	45	46	15	N.	+ 5.1
Rigel	1	5	4	35	2.87	76	8	52	43.0	8	27	9	S.	− 4.8
β Tauri	2	5	13	13	3.78	78	18	14	56.7	28	25	1	N.	+ 4.3
Bellatrix	2	5	14	3	3.21	78	30	44	48.2	6	8	53	N.	+ 4.0
η Orionis	3	5	14	5	3.01	78	31	15	45.2	2	16	1	S.	− 4.0
δ Orionis	2	5	21	27	3.07	80	21	47	46.0	0	27	50	S.	− 3.4
α Leporis	3	5	23	37	2.65	80	54	18	39.7	17	58	57	S.	− 3.2
ζ Tauri	3	5	25	17	3.57	81	19	10	53.7	21	0	8	N.	+ 3.0
ε Orionis	2	5	25	44	3.04	81	25	55	41.6	1	20	48	S.	− 3.0
ζ Orionis	2	5	30	20	3.01	82	35	6	45.4	2	3	12	S.	− 2.6
α Columbæ	2	5	32	10	2.17	83	2	34	32.6	34	15	14	S.	− 2.4
κ Orionis	2.3	5	37	58	2.85	84	29	31	42.7	9	45	53	S.	− 1.9
β Columbæ	3	5	43	45	2.11	85	55	53	31.7	35	55	36	S.	− 1.4
Betelgeuse	1	5	41	18	3.24	85	19	28	48.6	7	22	57	N.	+ 1.6
ζ Canis Majoris	2.3	6	12	21	2.31	93	5	46	34.6	29	58	53	S.	+ 1.1
β Canis Majoris	2.3	6	13	37	2.65	93	23	52	39.7	17	51	56	S.	+ 1.2
Canopus	1	6	19	22	1.33	94	50	26	20.0	52	35	16	S.	+ 1.7
Sirius	1	6	36	1	2.65	99	0	21	39.7	16	26	25	S.	+ 4.3
ε Canis Majoris	3	6	50	30	2.36	102	37	30	35.4	28	42	2	S.	+ 4.6
ζ Geminorum	3	6	51	49	3.57	102	57	10	53.6	20	51	38	N.	− 4.5
δ Canis Majoris	2	6	59	59	2.44	104	59	43	36.6	26	4	12	S.	+ 5.2
δ Geminorum	3	7	7	44	3.60	106	56	6	54.0	22	25	1	N.	− 5.8
π Navis	3	7	9	50	2.11	107	27	34	31.7	36	41	0	S.	+ 6.0
η Canis Majoris	2	7	15	54	2.37	108	58	37	35.6	28	54	33	S.	+ 6.5
Castor	1.2	7	21	22	3.85	110	20	37	57.8	32	19	30	N.	− 6.9

TABLE XLVI.

The Right Ascensions and Declinations of the principal Fixed Stars, adapted to the beginning of the year 1793.

Names of Stars.	Mag.	Mean Right Ascension in Time. h.	′	″	an.va. ″	Mean Right Ascension in Degrees. °	′	″	an.va. ″	Declination. °	′	″		an. var. ″
Procyon - -	1.2	7	28	27	3.14	112	6	47	47.0	5	45	3	N.	− 7.5
Pollux - -	2.1	7	32	37	3.67	113	9	21	55.1	28	30	42	N.	− 7.9
ζ Navis - - -	2	7	56	19	2.11	119	4	48	31.7	39	21	36	S.	+ 9.2
γ Navis - - - -	2	8	3	10	1.85	120	47	34	27.8	46	41	54	S.	+10.3
ε Navis - - -	2.3	8	18	16	1.25	124	34	2	18.8	58	39	18	S.	+11.4
δ Navis - - - -	2.1	8	39	0	1.65	129	44	58	24.9	53	57	21	S.	+13.8
ι Ursæ Majoris -	3	8	44	58	4.21	131	14	36	61.4	48	50	38	N.	−13.2
Acubens - -	3	8	47	8	3.12	131	47	8	49.5	12	39	6	N.	−13.4
λ Navis - - -	2.1	9	0	24	2.21	135	6	0	33.1	42	36	53	S.	+14.2
β Navis - -	1	9	10	54	0.75	137	41	29	11.2	68	52	2	S.	+14.8
χ Navis - - -	2.3	9	11	44	1.86	137	55	53	27.9	54	2	49	S.	+15.1
Alphard - -	2	9	17	34	2.93	139	21	6	44.0	7	46	7	S.	+15.2
θ Ursæ Majoris -	3	9	19	0	4.21	139	44	57	63.1	52	32	1	N.	−15.3
ε Leonis - - -	3	9	34	4	3.44	143	31	5	51.6	24	43	9	N	−16.1
μ Leonis - -	3	9	40	58	3.46	145	14	35	51.4	26	58	30	N.	−16.5
η Leonis -	3	9	56	1	3.29	149	0	18	49.4	17	46	4	N.	−17.2
Regulus - -	1	9	57	20	3.20	149	19	54	48.1	12	58	21	N.	−17.2
ζ Leonis - -	3	10	5	8	3.37	151	17	7	50.5	24	26	34	N.	−17.6
γ Leonis - - -	3	10	8	31	3.31	152	7	49	49.7	20	53	10	N.	−17.7
θ Navis - - -	2.1	10	35	37	2.11	158	54	30	31.7	63	18	47	S.	+18.7
μ Navis - - -	3	10	37	4	2.30	159	16	7	34.5	58	36	3	S.	+18.7
β Ursæ Majoris -	2	10	49	14	3.71	162	18	29	55.7	57	29	17	N.	−19.1
Dubhe - - -	2	10	50	49	3.85	162	41	19	57.8	62	53	9	N.	−19.2
δ Leonis - - -	2.3	11	3	4	3.21	165	45	53	48.1	21	39	30	N.	−19.4
β Leonis - - -	2	11	38	29	3.06	174	37	16	45.9	15	43	42	N.	−19.9
β Virginis - -	3	11	39	51	3.12	174	58	38	46.8	2	53	18	N.	−20.0
γ Ursæ Majoris	2	11	42	53	3.22	175	42	46	48.3	54	50	42	N.	−20.0
δ Centauri - -	2.3	11	57	43	3.06	179	25	39	45.9	49	34	1	S.	+20.0
δ Crucis - - -	3	12	4	16	3.11	181	3	59	46.7	57	33	50	S.	+20.0
δ Ursæ Majoris -	3	12	5	6	3.01	181	16	23	45.4	58	11	3	N.	−20.0
Algorab - -	3	12	5	10	3.08	181	17	31	46.2	26	21	32	S.	+20.0
α Crucis - - -	1	12	15	14	3.34	183	48	38	48.6	62	47	6	S.	+20.0
γ Crucis - - -	2	12	19	47	3.24	184	56	46	48.6	55	57	2	S.	+20.0
γ Centauri - -	2.3	12	30	12	3.27	187	33	58	49.0	47	49	8	S.	+19.9
γ Virginis - - - -	3	12	31	13	3.07	187	48	10	46.1	0	18	35	S.	+19.9
β Crucis - - -	2	12	35	46	3.41	188	56	26	51.2	58	31	59	S.	+19.8
Alioth - - -	2	12	44	53	2.67	191	13	52	40.1	57	5	14	N.	−19.7
δ Virginis - -	3	12	45	53	3.05	191	28	9	45.8	4	37	43	N.	−19.7
Cor. Caroli - -	3	12	46	25	2.85	191	35	0	42.8	39	26	21	N.	−19.6
Vindemiatrix -	3	12	51	53	3.01	192	58	13	45.1	12	4	36	N.	−19.5
Spica Virginis -	1	13	14	18	3.15	198	34	33	47.3	10	4	35	S.	+19.0
ζ Ursæ Majoris -	2	13	15	32	2.41	198	53	0	36.5	56	0	41	N.	−19.0
ε Centauri - -	2.3	13	26	14	3.71	201	43	21	55.7	52	24	17	S.	+18.6
Benetnach - -	2	13	39	23	2.40	204	50	48	36.0	50	22	7	N.	−18.2
ζ Centauri - -	3	13	43	43	3.68	205	40	48	55.2	46	15	37	S.	+18.1
η Bootis - - -	3	13	44	42	2.87	206	12	21	43.0	19	26	50	N.	−18.0
β Centauri - -	1.2	13	49	23	4.11	207	20	44	61.6	59	21	49	S.	+17.8
α Draconis - -	3	13	58	49	1.63	209	41	53	24.4	65	22	20	N.	−17.4
Arcturus - -	1	14	6	15	2.72	211	33	57	40.8	20	55	58	N.	−17.1
η Centauri - - -	2.3	14	22	20	3.75	215	16	50	56.3	41	14	53	S.	+16.3
γ Bootis - - -	3	14	23	41	2.43	215	56	5	36.1	39	53	10	N.	−16.2
α Centauri - -	1	14	26	1	4.45	216	30	8	66.7	59	58	44	S.	+16.1
α Leporis - -	3	14	28	16	3.92	217	3	53	58.8	46	29	21	S.	+16.0
α Libræ - - -	2.3	14	39	21	3.32	219	51	46	49.3	15	10	12	S.	+15.4
β Ursæ Minoris -	3	14	51	29	0.31	222	52	13	4.9	75	0	19	N.	−14.7
Zubenelg - -	2.3	15	5	54	3.22	226	28	28	48.3	8	16	28	S.	+13.8
β Serpentis - -	3	15	24	56	2.87	231	13	57	43.0	15	54	39	N.	−12.6
Alphacca - -	2	15	25	55	2.53	231	28	52	37.9	27	21	16	N.	−12.5
α Serpentis -	2.3	15	34	5	2.94	233	31	20	44.0	7	5	14	N.	−11.9
β Serpentis - -	3	15	36	38	2.76	234	9	33	41.4	16	4	52	N.	−11.8

TABLE XLVI.

The Right Ascensions and Declinations of the principal Fixed Stars, adapted to the beginning of the year 1793.

Names of Stars.	Mag.	Mean Right Ascension in Time. h.	′	″	an. va. ″	Degrees. °	′	″	an. va. ″	Declination. °	′	″		an. var. ″
γ Serpentis - -	3	15	46	54	2.75	236	43	34	41.3	16	21	47	N.	−11.0
β Scorpii - - -	3	15	48	7	3.51	237	1	52	52.9	22	1	8	S.	+10.9
δ Scorpii - - -	2	15	53	26	3.47	238	21	28	52.1	10	13	29	S.	+10.5
δ Ophiuchi - -	3	16	3	33	3.14	240	52	44	47.1	3	8	49	S.	+ 9.8
ε Ophiuchi - -	3	16	7	31	3.16	241	50	42	47.4	4	10	37	S.	+ 9.5
γ Herculis - -	3	16	12	48	2.65	243	11	55	39.7	19	39	0	N.	− 9.1
Antares - -	1	16	16	44	3.64	244	11	3	54.7	25	57	22	S.	+ 8.7
β Herculis - -	3	16	21	21	3.54	245	20	12	38.1	21	57	7	N.	− 8.4
ζ Ophiuchi - -	2.3	16	25	47	3.29	246	26	41	49.4	10	8	3	S.	+ 8.0
α Trianguli Austra.	2.3	16	26	55	6.21	246	43	44	93.1	68	37	11	S.	+ 8.0
ε Scorpii - -	3	16	36	49	3.91	249	12	22	58.7	33	53	50	S.	+ 7.1
μ Scorpii - -	3	16	37	53	4.04	249	28	21	60.6	37	40	25	S.	+ 7.1
ζ Scorpii - - -	3	16	40	4	4.21	250	0	59	63.1	41	59	1	S.	+ 6.9
η Ophiuchi - -	2.3	16	58	31	3.43	254	37	48	51.5	15	27	19	S.	+ 5.3
Ras. Algethi -	2	17	5	11	2.73	256	18	10	40.9	14	38	18	N.	− 4.8
δ Herculis - -	3	17	6	28	2.47	256	38	0	37.0	25	5	46	N.	− 4.6
α Aræ - - -	3	17	15	53	4.62	258	58	3	69.3	49	41	22	S.	+ 3.9
λ Scorpii - - -	2.3	17	19	35	4.06	259	53	19	60.9	36	56	4	S.	+ 3.5
θ Scorpii - - -	2.3	17	22	20	4.30	260	37	2	64.5	42	59	49	S.	+ 3.3
Ras. Alhague -	2	17	25	10	2.77	261	19	53	41.5	12	43	28	N.	− 3.0
Raitaben - -	3	17	51	48	1.39	267	56	58	20.9	51	31	10	N.	− 0.7
γ Sagittarii - -	3	18	7	44	3.84	271	55	55	57.6	30	53	55	S.	− 0.7
ε Sagittarii - -	3	18	10	37	3.99	272	36	41	59.9	34	27	42	S.	− 0.9
Vega - - -	1	18	29	56	2.03	277	28	55	30.4	38	35	55	N.	+ 2.6
σ Sagittarii - -	2.3	18	42	26	3.73	280	36	37	56.0	26	32	14	S.	− 3.7
β Lyræ - - -	2.3	18	42	26	2.21	280	36	34	33.2	33	8	3	N.	+ 3.2
δ Lyræ - - -	3	18	47	17	2.10	281	49	8	31.5	36	18	45	N.	+ 4.1
ζ Sagittarii - -	3	18	49	26	3.83	282	21	28	57.1	30	9	14	S.	− 4.3
ε Aquilæ - -	3.4	18	50	14	2.73	282	33	36	40.9	14	43	2	N.	+ 4.3
γ Lyræ - - -	3	18	51	12	2.25	282	48	1	33.7	32	25	0	N.	+ 4.4
α Sagittarii - -	3.4	19	9	31	4.19	287	22	44	62.8	49	59	9	S.	− 6.0
δ Draconis - -	3	19	12	37	0.04	288	6	47	0.6	67	17	50	N.	+ 6.2
δ Aquilæ - - -	3	19	15	3	3.01	288	45	49	45.2	2	42	53	N.	+ 6.4
β Cygni - - -	3	19	21	22	2.42	290	35	34	36.3	27	32	7	N.	+ 7.0
γ Aquilæ - -	3	19	16	25	2.84	294	6	21	42.8	10	6	12	N.	+ 8.2
δ Cygni - - -	3	19	38	30	1.87	294	17	35	28.1	44	37	58	N.	+ 8.3
Altair - - -	1.2	19	40	41	2.92	295	10	9	43.8	8	19	48	N.	+ 8.5
β Aquilæ - -	3	19	45	8	2.94	296	17	6	44.1	5	54	57	N.	+ 8.9
α 1 Capricorni - -	3	20	6	9	3.33	301	32	22	49.9	13	8	14	S.	−10.5
α 2 Capricorni - -	3	20	6	33	3.33	301	38	17	50.0	13	10	30	S.	−10.5
α Pavonis - -	3.4	20	9	10	4.84	302	17	31	72.6	57	22	50	S.	−10.7
β Capricorni - -	3	20	9	22	3.39	302	20	29	50.8	15	25	22	S.	−10.7
γ Cygni - - -	3	20	14	48	2.15	303	41	57	32.3	39	36	12	N.	+11.1
Deneb - - -	2	20	34	22	2.03	308	31	37	30.5	44	32	31	N.	+12.5
ε Cygni - - -	3	20	37	48	2.40	309	27	5	36.0	33	31	59	N.	+12.7
α Canis Minoris -	3.4	21	5	27	3.01	316	21	40	45.1	4	24	10	N.	+14.5
Alderaimin -	3	21	13	36	1.43	318	23	58	21.4	61	42	45	N.	+15.0
β Aquarii - - -	3	21	20	40	3.17	320	9	57	47.6	6	28	31	S.	−15.4
γ Capricorni -	3	21	28	36	3.33	322	8	53	50.0	17	35	22	S.	−15.8
ε Pegasi - - -	3	21	34	0	2.95	323	30	0	44.2	8	56	2	N.	+16.2
δ Capricorni - -	3	21	35	35	3.31	323	53	49	49.7	17	3	20	S.	−16.2
α Gruis - - -	2	21	55	6	3.85	328	46	34	57.7	47	57	8	S.	−17.1
α Aquarii - - -	3	21	55	8	3.07	328	47	4	46.0	1	19	7	S.	−17.1
γ Aquarii - -	3	22	10	58	3.10	332	44	27	46.5	2	25	26	S.	−17.8
ζ Pegasi - - -	3	22	31	7	2.97	337	46	44	44.8	9	45	26	N.	+18.6
Fomalhaut -	1	22	46	10	3.33	341	33	14	49.9	30	43	51	S.	−19.0
Scheat - - -	2	22	53	45	2.87	343	26	8	43.1	26	57	44	N.	+19.2
Markab - -	2	22	54	27	2.96	343	36	49	44.5	14	5	43	N.	+19.2
α Andromedæ -	2	23	57	43	3.07	359	25	43	46.0	27	56	55	N.	+20.0
β Cassiopeiæ - -	2.3	23	58	10	3.05	359	32	26	45.8	58	0	35	N.	+20.0

TABLE XLVII. Part First of Preceſſion in Right Aſcenſion for complete Years.

Y.	Precef. ′ ″	Y.	Precef. ′ ″	Y.	Precef. ′ ″	Y.	Precef. ′ ″	Y.	Precef. ′ ″	Y.	Precef. ′ ″	Y.	Preceſſion. ° ′ ″
1	0 46.0	9	6 54.2	17	13 2.3	25	19 10.5	33	25 18.7	41	31 26.8	49	0 37 35.0
2	1 32.0	10	7 40.2	18	13 48.4	26	19 56.5	34	26 4.7	42	32 12.8	50	0 38 22.0
3	2 18.1	11	8 26.2	19	14 34.4	27	20 42.5	35	26 50.7	43	32 58.9	55	0 42 11.1
4	3 4.1	12	9 12.2	20	15 20.4	28	21 28.6	36	27 36.7	44	33 44.9	60	0 46 1.2
5	3 50.1	13	9 58.3	21	16 6.4	29	22 14.6	37	28 22.7	45	34 30.9	70	0 53 41.4
6	4 36.1	14	10 44.3	22	16 52.4	30	23 0.6	38	29 8.8	46	35 16.9	80	1 1 21.6
7	5 22.1	15	11 30.3	23	17 38.5	31	23 46.6	39	29 54.8	47	36 2.9	90	1 9 1.8
8	6 8.2	16	12 16.3	24	18 24.5	32	24 32.6	40	30 40.8	48	36 49.0	100	1 16 42.0

TABLE XLVIII. To find Part Second of Preceſſion in Right Aſcenſion, and the Preceſſion in Declination.

Arg. Al.	Rt. Star.	COMPLETE YEARS. 1	2	3	4	5	6	7	8	9	10	20	50	Arg. Al.	Rt. Star.
+	−	″	″	″	″	″	″	″	″	″	″	″	″	+	−
0	180	0.0	0.0	0.0	0.0	0.0	0.0	0.0	0.0	0.0	0.0	0.0	0.0	180	360
1	181	0.3	0.7	1.0	1.4	1.7	2.1	2.4	2.8	3.1	3.5	7.0	17.5	179	359
2	182	0.7	1.4	2.1	2.8	3.5	4.2	4.9	5.7	6.3	7.0	14.0	35.0	178	358
3	183	1.0	2.1	3.1	4.2	5.2	6.3	7.3	8.4	9.4	10.5	21.0	52.5	177	357
4	184	1.4	2.8	4.2	5.6	7.0	8.4	9.8	11.2	12.6	14.0	28.0	69.9	176	356
5	185	1.7	3.5	5.2	7.0	8.7	10.5	12.2	13.9	15.7	17.5	34.9	87.4	175	355
6	186	2.1	4.2	6.3	8.4	10.5	12.6	14.7	16.8	18.9	21.0	41.9	104.8	174	354
7	187	2.4	4.9	7.3	9.8	12.2	14.6	17.0	19.5	21.9	24.4	48.9	122.2	173	353
8	188	2.8	5.6	8.4	11.2	13.9	16.7	19.5	22.3	25.1	27.9	55.8	139.5	172	352
9	189	3.1	6.3	9.4	12.5	15.7	18.8	21.9	25.1	28.2	31.4	62.7	156.8	171	351
10	190	3.5	7.0	10.4	13.9	17.4	20.9	24.4	27.9	31.4	34.8	69.6	174.1	170	350
11	191	3.8	7.7	11.5	15.3	19.1	23.0	26.8	30.6	34.5	38.3	76.5	191.3	169	349
12	192	4.2	8.3	12.5	16.7	20.8	25.0	29.2	33.3	37.5	41.7	83.4	208.4	168	348
13	193	4.5	9.0	13.5	18.0	22.6	27.1	31.6	36.1	40.6	45.1	90.2	225.5	167	347
14	194	4.8	9.7	14.5	19.4	24.3	29.1	34.0	38.8	43.7	48.5	97.0	242.5	166	346
15	195	5.2	10.4	15.6	20.7	25.9	31.1	36.3	41.5	46.7	51.9	103.8	259.5	165	345
16	196	5.5	11.1	16.6	22.1	27.7	33.2	38.7	44.2	49.8	55.3	110.5	276.3	164	344
17	197	5.9	11.7	17.6	23.5	29.3	35.2	41.1	46.9	52.8	58.6	117.2	293.1	163	343
18	198	6.2	12.4	18.6	24.8	31.0	37.2	43.4	49.6	55.8	62.0	123.9	309.8	162	342
19	199	6.5	13.1	19.6	26.1	32.6	39.2	45.7	52.2	58.8	65.3	130.6	326.4	161	341
20	200	6.9	13.7	20.6	27.4	34.3	41.2	48.0	54.9	61.7	68.6	137.2	342.9	160	340
21	201	7.2	14.4	21.6	28.7	35.9	43.1	50.3	57.5	64.7	71.9	143.7	359.3	159	339
22	202	7.5	15.0	22.5	30.0	37.6	45.1	52.6	60.1	67.6	75.1	150.2	375.5	158	338
23	203	7.8	15.7	23.5	31.3	39.2	47.0	54.8	62.7	70.5	78.3	156.7	391.7	157	337
24	204	8.1	16.3	24.4	32.6	40.8	48.9	57.1	65.2	73.4	81.5	163.1	407.7	156	336
25	205	8.5	16.9	25.4	33.9	42.3	50.8	59.3	67.8	76.3	84.7	169.5	423.7	155	335
26	206	8.8	17.6	26.4	35.1	43.9	52.7	61.5	70.3	79.1	87.9	175.8	439.5	154	334
27	207	9.1	18.2	27.3	36.4	45.5	54.6	63.7	72.8	81.9	91.0	182.1	455.1	153	333
28	208	9.4	18.8	28.2	37.6	47.1	56.5	65.9	75.3	84.7	94.1	188.3	470.6	152	332
29	209	9.7	19.4	29.2	38.9	48.6	58.3	68.0	77.7	87.5	97.2	194.4	486.0	151	331
30	210	10.0	20.0	30.1	40.1	50.1	60.1	70.1	80.1	90.2	100.2	200.5	501.2	150	330
31	211	10.3	20.6	31.0	41.3	51.6	61.9	72.2	82.6	92.9	103.2	206.5	516.3	149	329
32	212	10.6	21.2	31.9	42.5	53.1	63.7	74.3	85.0	95.6	106.2	212.5	531.2	148	328
33	213	10.9	21.8	32.8	43.7	54.6	65.5	76.4	87.3	98.3	109.2	218.4	546.0	147	327
34	214	11.2	22.4	33.6	44.8	56.1	67.3	78.5	89.7	100.9	112.1	224.2	560.6	146	326
35	215	11.5	23.0	34.5	46.0	57.5	69.0	80.5	92.0	103.5	115.0	230.0	575.0	145	325
36	216	11.8	23.6	35.4	47.1	58.9	70.7	82.5	94.3	106.1	117.9	235.7	589.3	144	324
37	217	12.1	24.1	36.2	48.3	60.3	72.4	84.5	96.5	108.6	120.7	241.3	603.3	143	323
38	218	12.3	24.7	37.0	49.4	61.7	74.0	86.4	98.7	111.1	123.4	246.9	617.2	142	322
39	219	12.6	25.2	37.9	50.5	63.1	75.7	88.3	100.9	113.6	126.2	252.4	630.9	141	321
40	220	12.9	25.8	38.7	51.5	64.4	77.3	90.2	103.1	116.0	128.9	257.8	644.4	140	320
41	221	13.1	26.3	39.5	52.6	65.8	78.9	92.1	105.2	118.4	131.5	263.1	657.7	139	319
42	222	13.4	26.8	40.3	53.7	67.1	80.5	93.9	107.3	120.8	134.2	268.3	670.8	138	318
43	223	13.7	27.3	41.0	54.7	68.4	82.0	95.7	109.4	123.1	136.7	273.5	683.7	137	317
44	224	13.9	27.9	41.8	55.7	69.6	83.5	97.5	111.4	125.4	139.3	278.6	696.4	136	316
45	225	14.2	28.4	42.5	56.7	70.9	85.0	99.2	113.4	127.6	141.7	283.4	708.9	135	315
46	226	14.4	28.8	43.3	57.7	72.1	86.5	100.9	115.3	129.8	144.2	288.5	721.2	134	314
47	227	14.7	29.3	44.0	58.7	73.3	88.0	102.7	117.3	132.0	146.6	293.3	733.2	133	313

TABLE XLVIII.—Continued.

Arg. Rt. Af. star. +	−	COMPLETE YEARS. 1	2	3	4	5	6	7	8	9	10	20	50	Arg. Rt. Af. star. +	−
		"	"	"	"	"	"	"	"	"	"	"	"		
48	228	14.9	29.8	44.7	59.6	74.5	89.4	104.3	119.2	134.1	149.0	298.0	745.0	132	312
49	229	15.1	30.3	45.4	60.5	75.7	90.8	105.9	121.1	136.2	151.3	302.6	756.6	131	311
50	230	15.4	30.7	46.1	61.4	76.8	92.2	107.5	122.9	138.2	153.6	307.2	768.0	130	310
51	231	15.6	31.2	46.8	62.3	77.9	93.5	109.1	124.7	140.3	155.8	311.6	779.1	129	309
52	232	15.8	31.6	47.4	63.2	79.0	94.8	110.6	126.4	142.2	158.0	316.0	790.0	128	308
53	233	16.0	32.0	48.0	64.0	80.1	96.1	112.1	128.1	144.1	160.1	320.2	800.6	127	307
54	234	16.2	32.4	48.7	64.9	81.1	97.3	113.5	129.8	146.0	162.2	324.4	811.0	126	306
55	235	16.4	32.8	49.3	65.7	82.1	98.5	115.0	131.4	147.8	164.2	328.5	821.2	125	305
56	236	16.6	33.2	49.9	66.5	83.1	99.7	116.3	133.0	149.6	166.2	332.4	831.1	124	304
57	237	16.8	33.6	50.4	67.3	84.1	100.9	117.7	134.5	151.3	168.2	336.3	840.8	123	303
58	238	17.0	34.0	51.0	68.0	85.0	102.0	119.0	136.0	153.0	170.0	340.1	850.2	122	302
59	239	17.2	34.4	51.6	68.7	85.9	103.1	120.3	137.5	154.7	171.9	343.7	859.3	121	301
60	240	17.4	34.7	52.1	69.5	86.8	104.2	121.5	138.9	156.3	173.6	347.3	868.2	120	300
61	241	17.5	35.1	52.6	70.1	87.7	105.2	122.7	140.3	157.8	175.3	350.7	876.8	119	299
62	242	17.7	35.4	53.1	70.8	88.5	106.2	123.9	141.6	159.3	177.0	354.1	885.1	118	298
63	243	17.9	35.7	53.6	71.5	89.3	107.2	125.0	142.9	160.8	178.6	357.3	893.2	117	297
64	244	18.0	36.0	54.1	72.1	90.1	108.1	126.1	144.2	162.2	180.2	360.4	901.0	116	296
65	245	18.2	36.3	54.5	72.7	90.9	109.0	127.2	145.4	163.5	181.7	363.4	908.6	115	295
66	246	18.3	36.6	54.9	73.3	91.6	109.9	128.2	146.5	164.8	183.2	366.3	915.8	114	294
67	247	18.5	36.9	55.4	73.8	92.3	110.7	129.2	147.6	166.1	184.6	369.1	922.8	113	293
68	248	18.6	37.2	55.8	74.4	92.9	111.5	130.1	148.7	167.3	185.9	371.8	929.5	112	292
69	249	18.7	37.4	56.2	74.9	93.6	112.3	131.0	149.7	168.5	187.2	374.4	935.9	111	291
70	250	18.8	37.7	56.5	75.4	94.2	113.0	131.9	150.7	169.6	188.4	376.8	942.0	110	290
71	251	19.0	37.9	56.9	75.8	94.8	113.8	132.7	151.7	170.6	189.6	379.2	947.9	109	289
72	252	19.1	38.1	57.2	76.2	95.3	114.4	133.5	152.6	171.6	190.7	381.4	953.4	108	288
73	253	19.2	38.3	57.5	76.7	95.9	115.0	134.2	153.4	172.5	191.7	383.5	958.7	107	287
74	254	19.3	38.5	57.8	77.1	96.4	115.6	134.9	154.2	173.4	192.7	385.5	963.7	106	286
75	255	19.4	38.7	58.1	77.5	96.8	116.2	135.5	154.9	174.3	193.7	387.3	968.3	105	285
76	256	19.4	38.9	58.3	77.8	97.3	116.7	136.2	155.6	175.1	194.5	389.1	972.7	104	284
77	257	19.5	39.1	58.6	78.1	97.7	117.2	136.7	156.3	175.8	195.3	390.7	976.8	103	283
78	258	19.6	39.2	58.8	78.4	98.1	117.7	137.3	156.9	176.5	196.1	392.2	980.6	102	282
79	259	19.7	39.4	59.0	78.7	98.4	118.1	137.8	157.4	177.1	196.8	393.6	984.1	101	281
80	260	19.7	39.5	59.2	79.0	98.7	118.5	138.2	158.0	177.7	197.4	394.9	987.3	100	280
81	261	19.8	39.6	59.4	79.2	99.0	118.8	138.6	158.4	178.2	198.0	396.1	990.2	99	279
82	262	19.8	39.7	59.5	79.4	99.3	119.1	139.0	158.8	178.7	198.5	397.1	992.7	98	278
83	263	19.9	39.8	59.7	79.6	99.5	119.4	139.3	159.2	179.2	199.0	398.0	995.0	97	277
84	264	19.9	39.9	59.8	79.8	99.7	119.6	139.6	159.5	179.5	199.4	398.8	997.0	96	276
86	266	20.0	40.0	60.0	80.1	100.0	120.0	140.0	160.0	180.0	200.0	400.0	1000.1	94	274
88	268	20.0	40.1	60.1	80.3	100.2	120.3	140.2	160.3	180.3	200.3	400.7	1001.9	92	272
90	270	20.0	40.1	60.1	80.4	100.2	120.3	140.3	160.4	180.4	200.5	401.0	1002.5	90	270

TABLE XLIX. Decimal Numbers for multiplying the annual Precession of a Star in Right Ascension.

M	D	dec.	M	D	dec.	M	D	dec.	M	D	dec.	M	D	dec.	M	D	dec.
January.	1	.01	March.	2	.18	May.	16	.35	July.	5	.52	September.	4	.69	Nov.	18	.86
	3	.02		7	.19		19	.36		8	.53		9	.70		21	.87
	6	.03		12	.20		22	.37		11	.54		14	.71		24	.88
	9	.04		17	.21		25	.38		14	.55		19	.72		27	.89
	12	.05		22	.22		28	.39		17	.56		24	.73		30	.90
	15	.06		27	.23		31	.40		20	.57		29	.74	December.	3	.91
	19	.07	April.	1	.24	June.	3	.41		23	.58	October.	4	.75		6	.92
	22	.08		6	.25		6	.42		27	.59		9	.76		9	.93
	25	.09		11	.26		9	.43		30	.60		14	.77		12	.94
	28	.10		15	.27		12	.44	August.	3	.61		18	.78		15	.95
February.	1	.11		20	.28		15	.45		6	.62		23	.79		18	.96
	4	.12		24	.29		18	.46		10	.63		27	.80		21	.97
	8	.13		28	.30		21	.47		14	.64		31	.81		23	.98
	12	.14	May.	2	.31		24	.48		18	.65	Nov.	4	.82		26	.99
	16	.15		5	.32		27	.49		22	.66		8	.83		29	1.00
	21	.16		9	.33		30	.50		26	.67		11	.84		31	1.01
	25	.17		12	.34	Jul.	2	.51		30	.68		15	.85			

TAB. L. Semian. Solar Eq. of North. Stars in Declin.

Arg. a lon. ☉ — rt. al. star.

D	0 VI − +	I VII − +	II VIII − +	
	"	"	"	
0	0.0	0.2	0.4	30
3	0.0	0.3	0.4	27
6	0.1	0.3	0.4	24
9	0.1	0.3	0.4	21
12	0.1	0.3	0.5	18
15	0.1	0.3	0.5	15
18	0.1	0.4	0.5	12
21	0.2	0.4	0.5	9
24	0.2	0.4	0.5	6
27	0.2	0.4	0.5	3
30	0.2	0.4	0.5	0
	+ − XI V	+ − X IV	+ − IX III	D

If declin. is S. change sign.

T A B L E S

For computing the Nutation of a Star in Right Ascension and Declination.

TABLE LI.
Or, Table I. of Nutation.

ARGUMENT.

For the Nutation in R. A. R. A. star — lon. moon's node

For the Nutation in Declin. R.A. ☆ + 3 signs — lon. ☽'s node.

D.	0 VI − +	I VII − +	II VIII − +	
	"	"	"	
0	8.33	7.21	4.16	30
1	8.33	7.14	4.04	29
2	8.33	7.06	3.91	28
3	8.32	6.99	3.78	27
4	8.31	6.91	3.65	26
5	8.30	6.82	3.52	25
6	8.28	6.74	3.39	24
7	8.27	6.65	3.26	23
8	8.25	6.56	3.12	22
9	8.23	6.47	2.98	21
10	8.20	6.38	2.85	20
11	8.18	6.29	2.71	19
12	8.15	6.19	2.57	18
13	8.12	6.09	2.43	17
14	8.08	5.99	2.29	16
15	8.05	5.89	2.15	15
16	8.01	5.78	2.01	14
17	7.97	5.68	1.87	13
18	7.93	5.57	1.73	12
19	7.88	5.46	1.59	11
20	7.83	5.35	1.45	10
21	7.78	5.24	1.30	9
22	7.72	5.13	1.16	8
23	7.67	5.01	1.01	7
24	7.61	4.90	0.87	6
25	7.55	4.78	0.73	5
26	7.49	4.66	0.58	4
27	7.42	4.53	0.44	3
28	7.35	4.41	0.29	2
29	7.28	4.29	0.15	1
30	7.21	4.16	0.00	0
	− + XI V	− + X IV	− + IX III	D.

TABLE LII.
Or, Table II. of Nutation.

ARGUMENT.

For the Nutation in R. A. R.A. star + lon. moon's node

For the Nutation in Declin. R.A. star + 3 signs + long. moon's node.

D.	0 VI − +	I VII − +	II VIII − +	
	"	"	"	
0	1.22	1.06	0.61	30
1	1.22	1.05	0.59	29
2	1.22	1.03	0.57	28
3	1.22	1.02	0.55	27
4	1.22	1.01	0.53	26
5	1.21	1.00	0.52	25
6	1.21	0.99	0.50	24
7	1.21	0.97	0.48	23
8	1.21	0.96	0.46	22
9	1.20	0.95	0.44	21
10	1.20	0.93	0.42	20
11	1.20	0.92	0.40	19
12	1.19	0.91	0.38	18
13	1.19	0.89	0.36	17
14	1.18	0.88	0.34	16
15	1.18	0.86	0.32	15
16	1.17	0.85	0.29	14
17	1.17	0.83	0.27	13
18	1.16	0.82	0.25	12
19	1.15	0.80	0.23	11
20	1.15	0.78	0.21	10
21	1.14	0.77	0.19	9
22	1.13	0.75	0.17	8
23	1.12	0.73	0.15	7
24	1.11	0.72	0.13	6
25	1.11	0.70	0.11	5
26	1.10	0.68	0.08	4
27	1.09	0.66	0.06	3
28	1.08	0.65	0.04	2
29	1.07	0.63	0.02	1
30	1.06	0.61	0.00	0
	− + XI V	− + X IV	− + IX III	D.

TABLE LIII.
Equa. of Equinoxes in R.A.

ARGUMENT.

Longitude of the moon's node.

D.	0 VI − +	I VII − +	II VIII − +	
	"	"	"	
0	0.0	8.2	14.2	30
1	0.3	8.4	14.3	29
2	0.6	8.7	14.5	28
3	0.9	8.9	14.6	27
4	1.1	9.2	14.7	26
5	1.4	9.4	14.8	25
6	1.7	9.6	15.0	24
7	2.0	9.8	15.1	23
8	2.3	10.1	15.2	22
9	2.6	10.3	15.3	21
10	2.8	10.5	15.4	20
11	3.1	10.7	15.5	19
12	3.4	11.0	15.6	18
13	3.7	11.2	15.7	17
14	4.0	11.4	15.7	16
15	4.2	11.6	15.8	15
16	4.5	11.8	15.9	14
17	4.8	12.0	16.0	13
18	5.1	12.2	16.0	12
19	5.3	12.4	16.1	11
20	5.6	12.5	16.1	10
21	5.9	12.7	16.2	9
22	6.1	12.9	16.2	8
23	6.4	13.1	16.2	7
24	6.7	13.2	16.3	6
25	6.9	13.4	16.3	5
26	7.2	13.6	16.3	4
27	7.4	13.7	16.3	3
28	7.7	13.9	16.4	2
29	7.9	14.0	16.4	1
30	8.2	14.2	16.4	0
	+ − XI V	+ − X IV	+ − IX III	D.

To find the Nutation of a Star in Right Ascension.

To the log. of the sum or difference of the equations from tables LI, LII, answering to their proper arguments, add the log. tangent of the star's declination; the sum will be the log. of part first of nutation, and if the declination is south, change the sign—to which apply the equation from table LIII, answering to the longitude of the moon's node, and the sum or difference will be the nutation in right ascension.

To find the Nutation of a Star in Declination.

Increase the arguments of tables LI. and LII. each by three signs, and the sum or difference of the corresponding equations will be the nutation in declination.

T A B L E S

For computing the Aberration of a Star in Right Afcenfion and Declination.

TABLE LIV.

Or, Table I. of Aberration

ARGUMENT.

For the Aberration in R.A. R. A. ✻ — long. ☉

For the aberration in decl. R. A. ftar + 3 figns — fun's longitude.

D.	0 VI − +	I VII − +	II VIII − +	
	"	"	"	
0	19.17	16.60	9.59	30
1	19.17	16.43	9.30	29
2	19.16	16.26	9.00	28
3	19.15	16.08	8.70	27
4	19.13	15.89	8.40	26
5	19.10	15.71	8.10	25
6	19.07	15.51	7.80	24
7	19.03	15.31	7.49	23
8	18.99	15.11	7.19	22
9	18.94	15.90	6.87	21
10	18.88	14.69	6.56	20
11	18.82	14.47	6.24	19
12	18.75	14.25	5.93	18
13	18.68	14.02	5.61	17
14	18.60	13.79	5.28	16
15	18.52	13.56	4.96	15
16	18.43	13.32	4.64	14
17	18.33	13.08	4.31	13
18	18.23	12.83	3.99	12
19	18.13	12.58	3.66	11
20	18.02	12.32	3.33	10
21	17.90	12.07	3.00	9
22	17.78	11.80	2.67	8
23	17.65	11.54	2.34	7
24	17.52	11.27	2.00	6
25	17.38	11.00	1.67	5
26	17.23	10.72	1.34	4
27	17.08	10.44	1.00	3
28	16.93	10.16	0.67	2
29	16.77	9.87	0.33	1
30	16.60	9.59	0.00	0
	− + XI V	− + X IV	− + IX III	D.

TABLE LV.

Or, Table II. of Aberration.

ARGUMENT.

For the Aberration in R. A. R. A. ftar + fun's long.

For the Aberration in Decl. R. A. ftar + 3 figns + fun's longitude.

D.	0 VI + −	I VII + −	II VIII + −	
	"	"	"	
0	0.83	0.72	0.41	30
1	0.83	0.71	0.40	29
2	0.82	0.70	0.39	28
3	0.82	0.69	0.38	27
4	0.82	0.68	0.37	26
5	0.82	0.67	0.35	25
6	0.82	0.67	0.33	24
7	0.82	0.66	0.32	23
8	0.82	0.65	0.30	22
9	0.82	0.64	0.29	21
10	0.82	0.63	0.28	20
11	0.82	0.62	0.27	19
12	0.82	0.61	0.25	18
13	0.81	0.60	0.24	17
14	0.81	0.59	0.23	16
15	0.80	0.58	0.22	15
16	0.80	0.57	0.20	14
17	0.80	0.56	0.19	13
18	0.79	0.55	0.17	12
19	0.78	0.54	0.15	11
20	0.78	0.53	0.14	10
21	0.77	0.52	0.12	9
22	0.76	0.51	0.11	8
23	0.76	0.50	0.10	7
24	0.75	0.49	0.09	6
25	0.75	0.47	0.07	5
26	0.75	0.46	0.06	4
27	0.74	0.45	0.05	3
28	0.73	0.44	0.03	2
29	0.72	0.43	0.02	1
30	0.72	0.41	0.00	0
	+ − XI V	+ − X IV	+ − IX III	D.

TABLE LVI.

Or, Tab. III. of Aberration.

ARGUMENT.

For part 2d Aber. in Dec. Sun's long. + ftar's dec.

For part 3d Aber. in Dec. Sun's long. — ftar's dec.

D.	0 VI − +	I VII − +	II VIII − +	
	"	"	"	
0	3.98	3.45	1.99	30
1	3.98	3.42	1.93	29
2	3.98	3.38	1.87	28
3	3.98	3.34	1.81	27
4	3.97	3.30	1.75	26
5	3.97	3.26	1.68	25
6	3.96	3.22	1.62	24
7	3.95	3.18	1.56	23
8	3.94	3.14	1.49	22
9	3.93	3.10	1.43	21
10	3.92	3.05	1.36	20
11	3.91	3.01	1.30	19
12	3.90	2.97	1.23	18
13	3.89	2.92	1.17	17
14	3.87	2.87	1.10	16
15	3.85	2.82	1.03	15
16	3.83	2.77	0.97	14
17	3.81	2.72	0.90	13
18	3.79	2.67	0.83	12
19	3.77	2.62	0.76	11
20	3.74	2.56	0.69	10
21	3.72	2.51	0.63	9
22	3.70	2.46	0.56	8
23	3.67	2.40	0.49	7
24	3.64	2.34	0.42	6
25	3.61	2.28	0.35	5
26	3.58	2.23	0.28	4
27	3.55	2.17	0.21	3
28	3.52	2.11	0.14	2
29	3.49	2.05	0.07	1
30	3.45	1.99	0.00	0
	− + XI V	− + X IV	− + IX III	D.

To find the Aberration of a Star in Right Afcenfion.

To the log. of the fum or difference of the equations from tables LIV, LV, anfwering to their proper arguments, add the log. fecant of the ftar's declination, the fum will be the log. of the aberration in right afcenfion.

To find the Aberration of a Star in Declination.

Find the fum or difference of the equations anfwering to the former arguments, encreafed by III figns—to the log. of which, add the log. fine of the ftar's declination; the fum will be the log. of part 1ft of the aberration. Take parts fecond and third of aberration from table LVI, which applied to the former will give the aberration in declination. If the ftar's decll. is fouth, change the fign of parts 2d. and 3d.

TABLE LVII.

The Longitudes and Latitudes of Stars, chiefly Zodiacal, adapted to the beginning of the Year 1793.

Names of Stars.	Mag.	Longitude. s ° ′ ″	ann. part of sec. var	Latitude. ° ′ ″	ann. part of sec. var
γ Pegasi	2	0 6 16 26	—.26	12 35 39 N.	+.12
δ Piscium	4	0 11 15 25	—.17	2 10 24 N.	+.16
α Andromedæ	2	0 11 25 37	—.37	25 41 3 N.	+.16
ε Piscium	4	0 14 38 52	—.16	1 5 46 N.	+.38
μ Piscium	5	0 20 13 43	—.13	3 3 55 S.	—.22
ν Piscium	4.3	0 23 35 18	—.12	5 21 56 N.	+.22
ξ Ceti	4	1 1 9 14	—.13	4 16 58 S.	—.30
α Arietis	2	1 4 46 4	—.22	9 57 36 N.	+.33
μ Ceti	4	1 9 2 1	—.12	5 34 39 S.	—.35
Menkar	2	1 11 25 41	—.08	12 35 54 S.	—.37
δ Arietis	4	1 17 57 9	—.16	1 48 25 N.	+.40
Electra	5	1 26 31 20	—.17	4 10 7 N.	+.43
Celeno	6	1 26 32 45	—.17	4 20 43 N.	+.43
Taygeta	5	1 26 40 30	—.17	4 27 54 N.	+.43
Maia	6	1 26 47 25	—.17	4 22 13 N.	+.43
Alcyone	3	1 27 6 8	—.17	4 1 52 N.	+.43
Atlas	5	1 27 27 57	—.17	3 53 49 N.	+.43
Pleione	6	1 27 28 21	—.17	3 58 41 N.	+.43
f Tauri	4	2 3 58 25	—.14	3 59 28 S.	—.45
ε Tauri	3.4	2 5 34 3	—.15	2 35 19 S.	—.46
Aldebaran	1	2 6 53 45	—.17	5 28 51 N.	+.46
τ Tauri	5	2 9 15 50	—.15	0 41 24 S.	—.47
ι Tauri	4	2 13 53 36	—.15	1 13 22 S.	—.47
Rigel	2	2 13 56 15	—.11	31 8 54 S.	—.47
Capella	1	2 18 57 57	—.16	22 51 57 N.	+.48
Orionis	2	2 19 28 23	—.14	23 34 45 S.	—.48
β Tauri	2	2 19 41 2	—.15	5 21 57 N.	+.48
ε Orionis	2	2 20 34 29	—.15	24 31 56 S.	—.48
ζ Orionis	2	2 21 47 33	—.15	25 19 9 S.	—.48
ζ Tauri	3	2 21 53 39	—.15	2 13 12 S.	—.48
Betelgeuse	1	2 25 51 46	—.16	16 3 3 S.	—.48
Propus	4	2 28 3 30	—.15	0 11 28 S.	—.48
η Geminorum	4	3 0 33 1	—.15	0 54 46 S.	—.48
μ Geminorum	3	3 2 24 31	—.15	0 50 16 S.	—.47
ν Geminorum	4	3 3 54 49	—.16	3 4 52 S.	—.47
γ Geminorum	2.3	3 6 12 43	—.17	6 45 54 S.	—.47
ε Geminorum	3	3 7 2 54	—.15	2 2 44 N.	+.46
Sirius	1	3 11 13 44	—.28	39 33 30 S.	—.45
ζ Geminorum	3	3 12 6 3	—.16	2 3 48 S.	—.45
x Geminorum	5	3 14 0 29	—.15	2 30 58 N.	+.45
δ Geminorum	3	3 15 37 51	—.15	0 12 5 S.	—.44
Castor	1.2	3 17 21 28	—.12	10 4 4 N.	+.43
Pollux	2	3 20 21 31	—.13	6 40 3 N.	+.42
κ Geminorum	4.5	3 20 46 36	—.14	3 3 32 N.	+.42
Procyon	1	3 22 55 42	—.03	15 58 14 N.	+.41
ψ 2 Cancri	4	3 26 21 8	—.13	5 19 18 N.	—.40
ζ Cancri, double	5	3 28 26 48	—.16	2 16 55 S.	—.39
	5	3 28 26 49	—.16	2 16 52 S.	—.39
γ Cancri	4	4 4 39 16	—.16	3 10 33 N.	+.35
δ Cancri	4	4 5 49 39	—.23	0 4 25 N.	+.33
α 1 Cancri	4.5	4 10 13 53	—.19	5 29 40 S.	—.32
α 2 Cancri	4.3	4 10 45 4	—.19	5 5 46 S.	—.32
κ Cancri	4.5	4 13 16 55	—.19	5 35 9 S.	—.30
ξ Leonis	4	4 18 45 45	—.18	3 9 46 S.	—.26
o Leonis	3.4	4 21 22 2	—.18	3 45 53 S.	—.25
Alphard	2	4 24 23 56	—.33	22 23 56 S.	—.22
ν Leonis	4.5	4 24 26 56	—.15	0 2 26 N.	+.22
η Leonis	3.4	4 25 0 48	—.12	4 51 15 N.	+.22
π Leonis	4	4 26 25 35	—.18	3 55 14 S.	—.21
Regulus	1	4 26 57 5	—.15	0 27 30 N.	+.21
ρ 2 Leonis	4	5 3 30 1	—.15	0 8 34 N.	+.15

TABLE LVII.

The Longitudes and Latitudes of Stars, chiefly Zodiacal, adapted to the beginning of the Year 1793.

Names of Stars.	Mag.	Longitude. s ° ′ ″	an. part. of sec. var.	Latitude. ° ′ ″	an. part. of sec. var.
λ Leonis	4	5 11 38 23	—.14	1 20 55 N.	+.09
σ Leonis	4.5	5 15 49 19	—.14	1 41 54 N.	+.05
τ Leonis	4	5 18 37 [illegible]	—.15	0 33 17 S.	—.03
β Leonis	1.2	5 18 44 39	—.05	12 17 0 N.	+.03
υ Leonis	4	5 22 9 2	—.15	3 2 49 S.	—.00
β Virginis	3	5 24 13 38	—.15	0 41 38 N.	—.02
η Virginis	4.3	6 1 56 47	—.14	1 22 24 N.	—.08
γ Virginis	3	6 7 17 11	—.13	2 48 49 N.	—.12
δ Virginis	4	6 15 20 59	—.14	1 45 30 N.	—.19
Spica Virginis	1	6 20 57 10	—.17	2 2 23 S.	+.23
Arcturus	1	6 21 20 41	+.10	30 52 43 N.	—.24
κ Virginis	4	7 1 36 21	—.13	2 55 19 N.	—.31
λ Virginis	4	7 4 3 50	—.15	0 30 30 N.	—.32
Alphacca	2.3	7 9 23 21	+.11	44 20 46 N.	—.35
α 2 Libræ	2.3	7 12 11 45	—.15	0 21 38 N.	—.37
ν Libræ	4	7 15 53 1	—.15	1 13 17 N.	—.39
ι Libræ	4	7 18 7 2	—.16	1 49 23 S.	+.40
α Serpentis	2.3	7 19 10 10	—.03	25 31 31 N.	—.40
ζ Libræ	4	7 21 0 32	—.14	2 7 21 N.	—.41
γ Libræ	3.4	7 22 14 6	—.13	4 24 31 N.	—.41
υ Libræ	4	7 24 27 52	—.14	4 1 44 N.	—.42

TABLE LVIII. Preceſſion of the Equinoctial Points in Longitude for complete Years.

Y.	Preceſ.	Y.	Preceſ.	Y.	Preceſ.	Y.	Preceſ.	Y.	Preceſ.	Y.	Preceſ.
1	0′ 50″3	15	12′ 35″2	29	24′ 20″1	43	36′ 5″0	57	47′ 49″9	71	0° 59′ 34″8
2	1 40.7	16	13 25.6	30	25 10.5	44	36 55.4	58	48 40.3	72	1 0 25.2
3	2 31.0	17	14 15.9	31	26 0.8	45	37 45.7	59	49 30.6	73	1 1 15.5
4	3 21.3	18	15 6.3	32	26 51.2	46	38 36.1	60	50 21.0	74	1 2 5.9
5	4 11.7	19	15 56.6	33	27 41.5	47	39 26.4	61	51 11.3	75	1 2 56.2
6	5 2.1	20	16 47.0	34	28 31.9	48	40 16.8	62	52 1.7	76	1 3 46.6
7	5 52.4	21	17 37.3	35	29 22.2	49	41 7.1	63	52 52.0	77	1 4 36.9
8	6 42.8	22	18 27.7	36	30 12.6	50	41 57.5	64	53 42.4	78	1 5 27.3
9	7 33.1	23	19 18.0	37	31 2.9	51	42 47.8	65	54 32.7	79	1 6 17.6
10	8 23.5	24	20 8.4	38	31 53.3	52	43 38.2	66	55 23.1	80	1 7 8.0
11	9 13.8	25	20 58.7	39	32 43.6	53	44 28.5	67	56 13.4	85	1 11 19.7
12	10 4.2	26	21 49.1	40	33 34.0	54	45 18.9	68	57 3.8	90	1 15 31.5
13	10 54.5	27	22 39.4	41	34 24.3	55	46 9.2	69	57 54.1	95	1 19 43.2
14	11 44.9	28	23 29.8	42	35 14.7	56	46 59.6	70	58 44.5	100	1 23 55.0

TABLE LIX. Preceſſion of the Equinoctial Points in Longitude for Months and Days.

Days.	Jan.	Feb.	Mar.	Apr.	May	June	July.	Aug.	Sept.	Oct.	Nov.	Dec.
1	0″6	5″6	9″0	12″1	15″5	20″2	25″5	30″5	34″4	37″4	40″9	45″4
3	1.0	5.8	9.2	12.3	15.8	20.6	25.8	30.8	34.6	37.6	41.2	45.8
5	1.3	6.1	9.4	12.5	16.1	20.9	26.2	31.1	34.8	37.8	41.4	46.1
7	1.6	6.4	9.6	12.7	16.3	21.3	26.5	31.3	35.0	38.0	41.7	46.4
9	2.0	6.7	9.8	12.9	16.6	21.6	26.8	31.6	35.3	38.3	42.0	46.8
11	2.3	6.9	10.0	13.1	16.9	22.0	27.2	31.9	35.5	38.5	42.3	47.1
13	2.6	7.2	10.2	13.4	17.2	22.3	27.5	32.2	35.7	38.7	42.6	47.5
15	3.0	7.4	10.4	13.6	17.5	22.7	27.8	32.4	35.9	38.9	42.9	47.8
17	3.3	7.6	10.6	13.8	17.8	23.0	28.2	32.7	36.1	39.1	43.2	48.2
19	3.6	7.9	10.8	14.0	18.1	23.4	28.5	32.9	36.3	39.3	43.5	48.6
21	3.9	8.1	11.0	14.3	18.4	23.7	28.8	33.2	36.5	39.6	43.8	48.9
23	4.2	8.3	11.2	14.5	18.8	24.1	29.1	33.4	36.7	39.8	44.1	49.3
25	4.5	8.6	11.4	14.8	19.1	24.4	29.4	33.6	36.9	40.0	44.4	49.6
27	4.8	8.8	11.6	15.0	19.4	24.8	29.7	33.9	37.1	40.3	44.8	50.0
29	5.1		11.8	15.3	19.7	25.1	30.0	34.1	37.2	40.5	45.1	50.4
31	5.4		12.0		20.1		30.3	34.3		40.8		50.7

In the Months of January and February, in Leap Year, take out the Preceſſion for the Day preceding the given Day.

TAB. LX. Secular Var. of Stars. Arg. Longitude of Star.

D	− 0 / + VI	− I / + VII	− II / + VIII	+ III / − IX	+ IV / − X	+ V / − XI
0	47″5	37″8	18″0	6″7	29″5	44″5
2	47.3	36.7	16.4	8.4	30.8	45.1
4	47.0	35.6	14.8	10.0	32.1	45.6
6	46.6	34.5	13.2	11.6	33.4	46.1
8	46.1	33.3	11.6	13.2	34.5	46.5
10	45.6	32.1	10.0	14.8	35.7	46.9
12	45.1	30.8	8.4	16.4	36.8	47.3
14	44.5	29.5	6.7	18.0	37.8	47.5
16	43.8	28.2	5.0	19.5	38.8	47.7
18	43.2	26.8	3.3	21.0	39.8	47.8
20	42.4	25.4	1.6	22.5	40.7	47.9
22	41.5	24.0	0.0	24.0	41.5	48.0
24	40.7	22.5	1.7	25.4	42.4	48.0
26	39.7	21.0	3.4	26.8	43.1	47.9
28	38.8	19.5	5.0	28.2	43.8	47.8
30	37.8	18.0	6.7	29.5	44.5	47.5

For var. in lon. mult. by tan. of lat.
- - In lat. add 3 ſigns to the arg.
In either caſe, if the lat. is S. change the ſign.

LXI. Eq. equin. in lon. Arg. Lon. ☽'s node.

D	− 0 / + VI	− I / + VII	− II / + VIII	D
0	0″0	8″9	15″5	30
2	0.6	9.5	15.8	28
4	1.2	10.0	16.1	26
6	1.9	10.5	16.3	24
8	2.5	11.0	16.6	22
10	3.1	11.5	16.8	20
12	3.7	12.0	17.0	18
14	4.3	12.4	17.2	16
16	4.9	12.8	17.3	14
18	5.5	13.2	17.5	12
20	6.1	13.7	17.6	10
22	6.7	14.1	17.7	8
24	7.3	14.5	17.8	6
26	7.8	14.8	17.8	4
28	8.4	15.1	17.9	2
30	8.9	15.5	17.9	0
	− V / + XI	− IV / + X	− III / + IX	D

LXII. Aberra. in Lon. Arg. Lon. ☉ — lon. ſtar.

D	− 0 / + VI	− I / + VII	− II / + VIII	D
0	20″0	17″3	10″0	30
2	20.0	17.0	9.4	28
4	20.0	16.6	8.8	26
6	19.9	16.2	8.1	24
8	19.8	15.8	7.5	22
10	19.7	15.3	6.8	20
12	19.6	14.9	6.2	18
14	19.4	14.4	5.5	16
16	19.2	13.9	4.8	14
18	19.0	13.4	4.2	12
20	18.8	12.9	3.5	10
22	18.5	12.3	2.8	8
24	18.3	11.8	2.1	6
26	18.0	11.2	1.4	4
28	17.7	10.6	0.7	2
30	17.3	10.0	0.0	0
	+ V / − XI	+ IV / − X	+ III / − IX	D

APPENDIX.

CONTAINING,

The Method of finding the Apparent Time of the Rising and Setting of the Sun, Moon, and fixed Stars.

INTRODUCTION.

AN obſervation of the riſing or ſetting of the Sun, affords a ready method of aſcertaining time at ſea.— This method is, however, affected by an error ariſing from the uncertainty of the horizontal refraction at that time:—But this error diminiſhes with the latitude and declination; and, in moſt caſes, will ſeldom exceed half a minute.— The riſing or ſetting of a fixed Star may, alſo, be applied to the ſame purpoſe.— The time of the Moon's riſing is neceſſary to be known, in order to determine if a certain obſervation of that object will be viſible, and to be prepared for ſuch an obſervation.— Upon theſe accounts, it is thought proper to ſubjoin this Appendix, in order to render this work ſtill more uſeful.

TABLE A.

Correction of the Approximate Time of the Rising or Setting of the Sun

The Latitude and Declination being

Of the same Name.

Latit.	Declination 0°	3°	6°	9°	12°	15°	18°	20°	22°	23°	23½°
°	″	″	″	″	″	″	″	″	″	″	″
0	0	0	0	0	0	0	0	0	0	0	0
2	1	1	1	1	1	1	1	0	0	0	0
4	2	2	2	2	2	1	1	1	1	0	0
6	2	2	2	2	2	2	2	2	1	1	0
8	3	3	3	3	3	3	2	2	1	1	0
10	4	4	4	4	4	4	3	3	2	1	0
12	5	5	5	5	5	4	4	3	2	1	0
14	6	6	6	6	5	5	4	4	3	2	0
16	7	7	7	7	6	6	5	4	3	2	0
18	8	8	8	7	7	7	6	5	4	2	0
20	9	9	9	8	8	8	7	6	4	2	0
22	10	10	10	9	9	8	7	6	4	3	0
24	11	11	11	10	10	9	8	7	5	3	0
26	12	12	12	11	11	10	9	8	5	3	0
28	13	13	13	12	12	11	10	8	6	4	0
30	14	14	14	14	13	13	11	9	7	4	0
32	15	15	15	15	14	14	12	10	7	4	0
34	16	16	16	16	16	15	13	11	8	5	0
36	17	17	18	17	17	16	15	12	9	5	0
38	18	18	19	19	18	17	16	13	10	6	0
40	20	20	21	21	20	19	17	15	11	6	0
41	21	21	22	21	22	20	18	15	11	6	0
42	21	22	22	22	22	21	19	16	12	7	0
43	22	22	23	23	23	22	20	17	12	7	0
44	23	23	24	24	24	22	21	18	13	7	0
45	24	24	25	25	25	23	22	19	13	8	0
46	24	25	26	26	26	24	23	20	14	9	0
47	25	26	27	27	27	26	24	21	15	9	0
48	26	27	28	28	28	27	25	22	16	10	0
49	27	28	29	29	29	28	26	23	16	10	0
50	28	29	30	31	30	30	27	24	17	10	0
51	29	30	31	32	32	31	29	26	19	11	0
52	30	31	32	33	33	32	30	27	20	11	0
53	31	32	33	34	34	34	32	28	21	12	0
54	32	34	35	36	36	36	34	30	22	13	0
55	34	35	36	37	38	38	36	32	24	14	0
56	35	37	38	39	40	40	38	34	25	15	0
57	36	38	40	41	42	42	40	36	27	17	0
58	38	40	42	43	44	44	43	39	30	19	0
59	39	42	44	45	46	47	46	42	32	20	0
60	41	43	46	47	49	50	49	45	35	22	0

Of contrary Names.

Latit.	Declination 0°	3°	6°	9°	12°	15°	18°	20°	22°	23°	23½°
°	″	″	″	″	″	″	″	″	″	″	″
0	0	0	0	0	0	0	0	0	0	0	0
2	1	1	1	1	1	1	1	0	0	0	0
4	2	2	2	2	2	2	1	1	1	0	0
6	2	2	2	2	2	2	2	2	1	1	0
8	3	3	3	3	3	3	2	2	1	1	0
10	4	4	4	4	4	3	3	2	2	1	0
12	5	5	5	5	4	4	3	3	2	1	0
14	6	6	6	6	5	5	4	3	2	1	0
16	7	7	6	6	6	5	5	4	3	2	0
18	8	8	7	7	6	6	5	4	3	2	0
20	9	8	8	8	7	6	6	5	3	2	0
22	10	9	9	9	8	7	6	5	4	2	0
24	11	10	11	10	9	8	7	6	4	2	0
26	12	11	11	10	10	9	7	6	4	2	0
28	13	12	12	11	10	9	8	6	5	3	0
30	14	13	13	12	11	10	9	7	5	3	0
32	15	14	14	13	12	11	9	7	5	3	0
34	16	15	15	14	13	11	10	8	6	3	0
36	17	16	16	15	14	12	11	9	6	4	0
38	18	18	17	16	15	13	11	9	7	4	0
40	20	19	18	17	16	14	12	10	7	4	0
41	21	20	18	17	16	14	12	10	7	4	0
42	21	20	19	18	17	15	13	11	7	4	0
43	22	21	20	19	17	15	13	11	8	4	0
44	23	22	21	[illegible]	18	16	14	11	8	4	0
45	24	23	22	[illegible]	18	16	14	11	8	4	0
46	24	24	22	21	19	17	15	12	8	5	0
47	25	24	23	21	19	17	15	12	9	5	0
48	26	25	24	22	20	18	15	13	9	5	0
49	27	26	25	23	21	18	16	13	9	5	0
50	28	27	25	24	22	19	16	13	9	5	0
51	29	28	26	24	22	20	17	14	9	6	0
52	30	29	27	25	23	21	17	14	10	6	0
53	31	30	28	26	24	21	18	15	10	6	0
54	32	31	29	27	24	22	18	15	10	6	0
55	34	32	30	28	25	23	19	16	10	6	0
56	35	33	31	29	26	23	19	16	11	6	0
57	36	35	32	30	27	24	20	17	11	6	0
58	38	36	33	31	28	25	20	17	11	6	0
59	39	37	35	32	29	26	21	18	11	7	0
60	41	38	36	33	30	27	22	18	11	7	0

TABLE B.

The Time answering to a Change of Altitude of One Degree at the Horizon.

Latit.	DECLINATION. 0°	3°	6°	9°	12°	15°	18°	21°	24°	27°	30°
°	′ ″	′ ″	′ ″	′ ″	′ ″	′ ″	′ ″	′ ″	′ ″	′ ″	′ ″
0	4 0	4 0	4 1	4 3	4 6	4 9	4 13	4 17	4 22	4 29	4 37
2	4 0	4 0	4 1	4 3	4 6	4 9	4 13	4 17	4 23	4 30	4 37
4	4 1	4 1	4 2	4 4	4 6	4 9	4 13	4 18	4 24	4 30	4 38
6	4 1	4 1	4 2	4 4	4 7	4 10	4 14	4 18	4 25	4 31	4 39
8	4 2	4 2	4 3	4 5	4 8	4 11	4 15	4 20	4 26	4 33	4 41
10	4 4	4 4	4 5	4 7	4 10	4 12	4 17	4 22	4 28	4 35	4 43
12	4 6	4 6	4 7	4 9	4 12	4 15	4 19	4 24	4 30	4 37	4 46
14	4 8	4 8	4 9	4 11	4 14	4 17	4 22	4 27	4 33	4 40	4 49
16	4 10	4 10	4 11	4 13	4 16	4 19	4 24	4 29	4 36	4 43	4 52
18	4 13	4 13	4 14	4 16	4 19	4 22	4 26	4 32	4 39	4 47	4 57
20	4 16	4 16	4 17	4 19	4 22	4 26	4 30	4 36	4 43	4 52	5 2
22	4 19	4 19	4 21	4 23	4 26	4 30	4 35	4 41	4 48	4 57	5 7
24	4 23	4 23	4 25	4 27	4 30	4 34	4 39	4 45	4 53	5 3	5 14
26	4 27	4 27	4 29	4 31	4 35	4 39	4 44	4 51	4 59	5 10	5 21
28	4 32	4 33	4 34	4 36	4 40	4 44	4 50	4 57	5 6	5 17	5 30
30	4 37	4 38	4 40	4 42	4 45	4 50	4 57	5 5	5 14	5 25	5 39
32	4 43	4 44	4 45	4 48	4 52	4 57	5 4	5 12	5 23	5 35	5 50
34	4 49	4 50	4 52	4 55	4 59	5 5	5 12	5 21	5 32	5 46	6 3
36	4 57	4 58	5 0	5 3	5 8	5 13	5 21	5 32	5 43	5 58	6 17
38	5 5	5 6	5 7	5 11	5 16	5 22	5 31	5 42	5 56	6 13	6 34
40	5 13	5 14	5 16	5 20	5 26	5 33	5 42	5 55	6 10	6 29	6 54
41	5 18	5 19	5 21	5 25	5 31	5 39	5 49	6 1	6 18	6 38	7 4
42	5 23	5 24	5 26	5 30	5 36	5 45	5 53	6 9	6 27	6 48	7 16
43	5 28	5 29	5 32	5 36	5 42	5 51	6 3	6 18	6 36	6 59	7 30
44	5 33	5 35	5 38	5 42	5 49	5 58	6 10	6 26	6 45	7 12	7 44
45	5 39	5 41	5 44	5 48	5 55	6 5	6 17	6 34	6 55	7 23	8 0
46	5 45	5 47	5 50	5 54	6 1	6 12	6 26	6 44	7 6	7 37	8 18
47	5 52	5 54	5 57	6 1	6 8	6 20	6 35	6 55	7 19	7 52	8 38
48	5 59	6 1	6 4	6 8	6 16	6 29	6 45	7 7	7 34	8 8	9 6
49	6 6	6 8	6 12	6 16	6 25	6 38	6 56	7 20	7 50	8 27	9 25
50	6 14	6 16	6 20	6 25	6 34	6 48	7 8	7 33	8 7	8 47	9 54
51	6 22	6 24	6 28	6 34	6 44	6 59	7 20	7 48	8 26	9 11	10 28
52	6 31	6 33	6 38	6 44	6 55	7 11	7 33	8 4	8 43	9 37	11 8
53	6 30	6 42	6 47	6 54	7 6	7 23	7 47	8 20	9 3	10 7	11 58
54	6 49	6 52	6 58	7 5	7 18	7 36	8 1	8 37	9 24	10 43	12 57
55	6 58	7 1	7 7	7 16	7 30	7 49	8 16	8 55	9 48	11 25	14 14
56	7 8	7 11	7 18	7 28	7 43	8 4	8 32	9 18	10 18	12 15	15 58
57	7 19	7 23	7 30	7 41	7 58	8 21	8 59	9 46	10 57	13 18	
58	7 31	7 35	7 43	7 55	8 14	8 37	9 20	10 19	11 47	14 38	
59	7 45	7 49	7 57	8 10	8 30	8 59	9 42	10 55	12 40		
60	8 0	8 5	8 14	8 27	8 48	9 11	10 6	11 36	11 45		

To find the Time answering to a given Change of Altitude at the Horizon.

RULE. To the Prop. log. of the time, answering to the latitude and declination, add the P. log. of the given change of altitude, and the constant log. 9.5229. the sum will be the P. log. of the time required.

EXAMPLE. Required the interval of time between the rising of the upper and lower limbs of the sun, August 11, 1794, in lat. 57° N.

Time answering to lat. and decl.	8′	21″	P. L.	1.3336
Sun's diameter, - - - -	31	41	P. L.	0.7544
Constant log. - - - - - - -				9.5229
Time required to rise, - - -	4	20½	P. L.	1.6109

Problem I.

To find the Apparent Time of the Rising or Setting of the Sun.

Rule.

Reduce the ſun's declination to the meridian of the given place, by problem v. page 74. to the log. tangent of which, add the log. tangent of the latitude, the ſum will be the log. coſine of an arch, which, reduced to time, will be the approximate time of the riſing of the ſun, and its complement to 12 hours, will be that of ſun ſetting, the latitude and declination being of the ſame name; but if of contrary names, the above arch reduced to time will be the approximate time of ſun ſetting, and its complement to 12 hours that of ſun riſing.

Take the number from Table A. anſwering to the latitude and declination, which is to be added to the approximate time of ſun riſing or ſetting, if the declination is increaſing; otherwiſe, it is to be ſubtracted therefrom.

Enter Table B. with the latitude and declination, and take out the correſponding time, which is that anſwering to a change of altitude at the horizon of 1°. Now, find the proportional part of this time, anſwering to the horizontal refraction, and dip of the horizon, which is to be ſubtracted from the approximate time of ſun riſing, and added to that of ſun ſetting. Hence, the apparent time of the riſing or ſetting of the ſun will be obtained.

Example.

Required the apparent time of ſun riſing, April 1, 1793, in latitude 33° 42′ N. longitude, 16° 20′ W. the height of the eye above the ſea being 28 feet?

Sun's declination - -		4°	30′ N.		
Eq. to long. table XIII. - -		+	1		
Reduced declination -		4	51 N.	- tangent	8.92866
Latitude - - -		33	42 N.	- tangent	9.82407
Arch - - -		86	45½	- - coſine	8.75273
			4		
Approx. time of riſing	5	47	1		
Equation table A. -	+		16		
P. part to hor. ref. & dip = 38′ table B.	−	3	4		
Apparent time of riſing	5	44	13		

Problem

PROBLEM II.

To find the Apparent Time of the Paſſage of a Star over the Meridian of a known Place.

RULE.

Reduce the right aſcenſion of the ſtar to the propoſed day, from which, increaſed by 24 hours, if neceſſary, ſubtract the ſun's right aſcenſion, the remainder will be the approximate time of the ſtar's paſſage over the meridian.

From the approximate time of tranſit, ſubtract the equation from table XVIII. anſwering to that time, and the ſun's right aſcenſion—and apply that from the ſame table, anſwering to the longitude, by addition or ſubtraction, according as it is eaſt or weſt, and the aggregate will be the apparent time of the paſſage of the ſtar over the meridian of the given place.

EXAMPLE.

Required the apparent time of tranſit of Algol, December 25, 1792, in longitude 50° eaſt ?

Right aſcenſion of Algol - - -		2h.	54′	43″
Sun's right aſcenſion - - - -		18	19	40
Approximate time of tranſit - -		8	35	3
Eq. from tab. XVIII. to approx. time and ☉'s R. A.	—		1	34
- - - to long. and ☉'s R. A. -	+			36
Apparent time of tranſit - - -		8	34	5

PROBLEM III.

To find the Apparent Time of the Riſing or Setting of a fixed Star.

RULE.

Find the apparent time of the ſtar's tranſit, by the laſt problem, and compute the ſtar's ſemidiurnal arc, by problem I. which is to be applied to the time of tranſit, by addition or ſubtraction, according as the time of ſetting or riſing is required. Take the equation from table XVIII. anſwering to the ſemidiurnal arc and ſun's right aſcenſion, which is to be added to the time of riſing, or ſubtracted from that of ſetting. Apply the proportional part, anſwering to the ſum of the horizontal refraction and dip, from table B. as in problem I. and the apparent time of the riſing or ſetting of the ſtar will be obtained.

Example.

Required the apparent time of the ſetting of Menkar, April 10, 1793, at Bridgetown, Barbadoes; height of the eye 40 feet?

Right aſcenſion Menkar			2h.	51′	29″
Sun's right aſcenſion			1	17	47
Approximate time of tranſit			1	33	42
Eq. to approx. time, table xviii.			—		15
to longitude			—		36
Apparent time of tranſit			1	32	51
Menkar's declination	3° 16½′ N.	tangent	8.75719		
Latitude	13 5 N.	tangent	9.36624		
	89 14½	coſine	8.12343		
	4				
	5 56 57				

Semidiurnal arc	6	3	3
App. time of tranſit	1	32	51
Approx. time of ſetting	7	35	54
Eq. ſemidi. arc, t. xviii.	—		56
Effect ref. and dip, t. B.	+	2	41
App. time ſett. Menkar	7	37	39

Problem IV.

To find the Apparent Time of the Riſing or Setting of the Moon.

Rule.

To the apparent time of the moon's paſſage over the meridian of Greenwich, apply the number from table xx. anſwering to the longitude and daily variation of paſſage, by addition or ſubtraction, according as the longitude is weſt or eaſt, and the reduced time of tranſit will be obtained.

Compute the ſemidiurnal arc of the moon, uſing the latitude and moon's declination at noon; which, ſubtracted from the reduced time of tranſit, will give the approximate time of the moon's riſing; or, added thereto, will give that of ſetting.

To this time, reduced to the meridian of Greenwich, find the moon's declination, with which and the latitude, recompute the moon's ſemidiurnal arc: to which apply the number from table xx. anſwering thereto, and the daily variation of tranſit, by addition or ſubtraction, according as time of the riſing or ſetting is required; from the ſame arc, ſubtract the proportional part from table B. anſwering

to

to the exceſs of the moon's horizontal parallax, above the ſum of the horizontal refraction and dip—hence, the apparent ſemidiurnal arc will be obtained.

Subtract the apparent ſemidiurnal arc from the reduced time of tranſit, the remainder will be the apparent time of the riſing of the moon, and their ſum will be that of the moon's ſetting.

Example.

Required the apparent time of the riſing of the moon, at Paris, December 20, 1793?

Apparent time of tranſit over mer. of Green. -	14h. 57′
Equation from table 22. anſwering to long. - -	— 0
Reduced time of tranſit - - - -	14 57

To latitude 48° 57′ N. and moon's declination at noon, the ſemidiurnal arc is about 7h. 22′, which, ſubtracted from 14h. 57′, the reduced time of tranſit, gives 7h. 35′, the approximate time of the riſing of the moon:—Now, the moon's declination at this time, reduced to the meridian of Greenwich, is 16° 31′ N.

Moon's declination - -	16° 31′	- tangent	9.47207
Latitude - - -	48 50	- tangent	0.05829
	70 11	- coſine -	9.53036
	4		
	4 40 44		

Semidiurnal arc - - -		7 19
Eq. table xx. to ſem. arc - -		+ 14
Moon's hor. par. -	54′ 51″	
Ref. and dip = 33′+0′=33.		
Difference - - - -	21. 51. the effect of which, by table B. is -	— 2
Apparent ſemidiurnal arc - -		7 31
Reduced time of tranſit - - -		14 57
App. time of the riſing of the moon -		7 26 Afternoon.

From hence, the method of computing the apparent time of the riſing or ſetting of a planet will be obvious.

General Remark.

If the polar diſtance of the object exceeds the latitude of the place, that object will not ſet, the latitude and declination being of the ſame name; but if of a contrary name, the object will not riſe above the horizon of that place.

THE END.

PROPOSALS

For PUBLISHING, *by* SUBSCRIPTION,

BY A. MACKAY, A.M. F.R.S.E.

I.

A TREATISE ON THE THEORY AND PRACTICE OF NAVIGATION;

In which will be contained, all the new Discoveries in that useful Art, together with a complete Set of Tables.

In One Volume, 8vo.

Price, to Subscribers, *Five Shillings* and *Sixpence* in Boards, payable on Delivery; to Non-subscribers the Price will be advanced to *Seven Shillings*.

A List of the Subscribers will be printed.

II.

A TREATISE ON THE THEORY AND PRACTICE OF ASTRONOMY;

Containing, all the modern Improvements in that sublime Science; the Description and Use of Astronomical Instruments; the various Calculations in the Practice of Astronomy; and a complete Set of Tables of the Sun, Moon, Planets, fixed Stars, &c.

In Three Volumes, 4to

Price, to Subscribers, *One Pound Thirteen Shillings*, in Boards, payable on Delivery; to Non-subscribers the Price will be advanced to *Two Guineas*.

Each Volume will be delivered as soon as it is printed; and a List of the Subscribers will be given with the last Volume.

It is requested, that those who incline to patronize either of the above Works, will be pleased to intimate their Intention, as early as possible, to the Author, at *Aberdeen*.

www.ingramcontent.com/pod-product-compliance
Lightning Source LLC
Chambersburg PA
CBHW030316170426
43202CB00009B/1017
* 9 7 8 3 7 4 1 1 8 0 2 8 6 *